The Marine Environment and United Nations Sustainable Development Goal 14

Center for Oceans Law and Policy

Series Editor

Myron H. Nordquist
John Norton Moore

VOLUME 22

The titles published in this series are listed at *brill.com/colp*

The Marine Environment and United Nations Sustainable Development Goal 14

Life below Water

Edited by

Myron H. Nordquist
John Norton Moore
Ronán Long

BRILL
NIJHOFF

LEIDEN | BOSTON

Library of Congress Cataloging-in-Publication Data

Names: Nordquist, Myron H., editor. | Moore, John Norton, 1937- editor. | Long, Ronán J., editor.
Title: The marine environment and United Nations sustainable development goal 14 : life below water / edited by Myron H. Nordquist (University of Virginia School of Law); John Norton Moore (University of Virginia School of Law); Ronán Long (World Maritime University - Sasakawa Global Institute).
Description: Leiden ; Boston : Brill Nijhoff, 2018. | Series: Center for oceans law and policy, ISSN 1872-7158 ; VOLUME 22 | Includes index.
Identifiers: LCCN 2018033944 (print) | LCCN 2018035197 (ebook) | ISBN 9789004366619 (E-book) | ISBN 9789004364202 (hardback : alk. paper)
Subjects: LCSH: Marine resources conservation--Law and legislation. | Sustainable development--Law and legislation. | Marine pollution--Law and legislation. | Marine biodiversity. | United Nations Convention on the Law of the Sea (1982 December 10)
Classification: LCC K3485 (ebook) | LCC K3485 .M39 2018 (print) | DDC 346.04/695616--dc23
LC record available at https://lccn.loc.gov/2018033944

Typeface for the Latin, Greek, and Cyrillic scripts: "Brill". See and download: brill.com/brill-typeface.

ISSN 1872-7158
ISBN 978-90-04-36420-2 (hardback)
ISBN 978-90-04-36661-9 (e-book)

This book is printed on acid-free paper and produced in a sustainable manner.

Printed by Printforce, the Netherlands

Contents

PART 2
Biodiversity beyond National Jurisdiction

PART 3
Status of Deep Seabed Minerals

Preface

This volume titled *The Marine Environment and United Nations Sustainable Development Goal 14: Life below Water* was suggested by Indonesian Ambassador Arif Havas Oegroseno as the theme for the 41st Annual Conference of the Center for Oceans Law and Policy, University of Virginia School of Law. The Conference was co-hosted with Indonesia's Coordinating Ministry for Maritime Affairs and was held May 17–18, 2017, in Yogyakarta, Indonesia. The book is almost entirely based on oral presentations given at the 41st Conference.

Fittingly, the first featured remarks on "The Marine Environment and UN Sustainable Goal 14" were made by Ambassador Arif Havas Oegroseno. He focused on concrete issues such as striking a balance between marine protection and sustainable uses, developing norms for governance gaps, accommodating the challenges facing small island States, and evading potential conflicts over resources, especially in disputed waters.

The next featured remarks were by Dr. Hasjim Djalal who for many years led an Indonesian team to Workshops on Managing Potential Conflicts in the South China Sea area. This highly influential "track two" process involving key participating authorities fostered many cooperative efforts and laid the foundation for a Code of Conduct in the region. The third featured remarks were by H.E. N. Hassan Wirajuda, former Foreign Minister of Indonesia, who earned an S.J.D. from the University of Virginia School of Law. His remarks addressed the future using a dialogue on sustainable development to advance regional and global order in the oceans. The final featured speaker was H.E. Satya N. Nandan who, among other prestigious posts, was Under-Secretary-General of the United Nations and Special Representative of the Secretary-General for the Law of the Sea. The President of the Conference, Tommy Koh of Singapore, and Satya Nandan of Fiji, contributed as much as any two persons to the success of the Third U.N. Conference on the Law of the Sea. This article contains priceless reflections on those efforts.

Part 1 of the program dealt with "Conservation and Sustainable Use of Oceans and Their Resources". The lead off speaker was Rena Lee, Senior State Counsel, Singapore Attorney-General's Chambers. She reviewed recent U.N. activities and their synergic processes to overcome fragmentation in ocean governance. The next contribution was by International Tribunal for Law of the Sea Judge Liesbeth Lijnzaad. Her topic posed the question "Conservation and Sustainable Use of Marine Biodiversity: Siamese Twins?", which she answers that further clarity is required before inclusion in a multilateral treaty. Indonesian Professor Melda Kamil Ariadno spoke on achieving sustainable fisheries development. Her basic message was that Indonesia must exercise stronger enforcement to achieve its goal of being a “maritime fulcrum”. The last paper was prepared by Professor Alistair McIlgorm and Brooke Campbell, Research Fellow, at the University of Wollongong’s Australian National Centre for Ocean Resources and Security (ANCORS). They point out that Indonesia has cracked down on IUU fishing and now needs to increase its domestic fishery management activities, including human capacity, to further its future economic development in offshore fisheries.

Part 2 of the book discusses “Biodiversity beyond National Jurisdiction” starting with the “Action Goals” under the 2030 agenda on sustainable development. Professor Ronán Long, who now directs the Sasakawa Global Ocean Institute at the World Maritime University, co-authored this paper with Mariamalia Rodriguez Chaves of the Pew Charitable Trusts. They offer comprehensive, wide-ranging comments on the 2030 Agenda’s political impetus and added context to the implementation of international law for the international community. The second presentation was by Professor Robin Warner (ANCORS) on the elements of an international legally binding instrument (ILBI) for conservation and sustainable use of marine diversity in areas beyond national jurisdiction (ABNJ). Her article examines key features of the existing global and regional ocean governance framework in the southern hemisphere in light of ILBI. The third paper in Part 2 was presented by Kristine Dalaker Kraabel from the K.G. Jebsen Centre for the Law of the Sea, The Arctic University of Norway, Tromsø. She covered the U.N. Preparatory Committee developing elements for the ILBI, particularly the proposed institutional arrangements pertaining to the “hybrid approach”. Two South Korean scholars, Dr. Su Jin Park of the Korea Maritime Institute, and Ki Hyeon Kim of the Ministry of Foreign Affairs, examined legal aspects of marine protected areas (MPAS) in the ABNJ. They discussed the definition, designation process, monitoring review process and effectiveness of MPAS and their relation to existing relevant international instruments and frameworks. The last speaker on the ABNJ was A. Gusman Siswandi, Faculty of Law, Bandung, Indonesia. His presentation was on Marine Genetic Resources in the ABNJ from the perspective of developing countries. He urged a more integrated approach between sustainable development goals and relevant international instruments.

Part 3 is titled “Status of Deep Seabed Minerals” with a panel moderated by Alfonso Ascencio-Herrera, Legal Counsel and Deputy to the Secretary-General of the International Seabed Authority (ISA). He contributed a paper that identified the work of the ISA and its relevance to Agenda 2030 Targets. The first major panel paper in this part was co-authored by Professor Seokwoo

Lee of Inha University Law School, Korea and Director Hee Cheol Yang, Ocean Policy Institute, Korea Institute of Ocean Science and Technology, Korea. They reviewed efforts to enhance deep seabed activities and domestic Korean law on exploration for exploitation of resources in the deep seabed area. Dr. Tomohiko Fukushima of the Japan Agency for Marine-Earth Science and Technology then spoke on the role of science for environmental impact evaluation resulting from ocean mining. The author elaborated on a more deliberate process of decision-making. The fourth presentation on deep sea minerals was by Steven Potter, Esq., who principally focused on the ISA's current process in finalizing the exploitation regime regulations and stakeholder concerns in the revised draft thereof.

Part 4 is titled "Marine Pollution and Coastal Ecosystems". The first paper is by Director Brahmantya Satyamurti Poerwadi and Abdul Muhari of Indonesia's Ministry of Marine Affairs and Fisheries. These co-authors described the impact of marine debris on marine habitats and the practical activities in the national action plan to deal with the problems. Professor Stephen A. Macko, Department of Environmental Sciences, University of Virginia, next provided a detailed review of a broad range of pollutants whose heightened levels will reduce coastal diversity and productivity and result in a reduction of food resources. He urged greater public awareness of these sources and the detrimental effects of these materials. The third paper was given by Professor Chie Kojima of Musahino University, Tokyo who offered a human rights perspective on marine pollution. She advocated for the potential of UNCLOS to safeguard human rights by protecting coastal ecosystems from land-based marine pollution.

Part 5 addresses "Climate Change and the Oceans". The first paper in this part by Karen N. Scott of the University of Canterbury, New Zealand is on ocean acidification and sustainable development Goal 14. She posed the question whether sustainable development was a goal but not a target? Her reply was no specific actions, including targets to limit ocean acidification had been established and she concluded with a number of proposals for reform. The second paper by J. Ashley Roach discussed the on-going work of two International Law Association (ILA) Committees, one on Baselines and another on Sea Level Rise. He discussed the legal issues raised at ILA meetings with respect to sea level rise on maritime zones, maritime boundaries and proposed solutions. Next, Dr. Elise Johansen from the K.G. Jebsen Centre for the Law of the Sea focused on climate change and the Arctic. She explained how opportunities for economic growth in the region must be balanced against environmental risks and examined the regulatory framework's capability to deal with climate change challenges in the Arctic. The final paper provided an overview of ocean acidification relationships by Professor Stephen A. Macko and his doctoral student at the University of Virginia, Christina M. Fantasia. The authors refer to rapid changes in the acidity of the oceans and robust future pH predictions chiefly because of the amount of carbon dioxide entering the atmosphere and being absorbed by the oceans.

Acknowledgments

The Marine Environment and United Nations Sustainable Development Goal 14: Life below Water is part of a series of publications on oceans law and policy sponsored by the Center for Oceans Law and Policy, University of Virginia School of Law, in connection with its annual conference. This volume is largely based on presentations made at the Center's 41st conference, which was held 17–18 May, 2017 in Yogyakarta, Indonesia. The principal organizers were the Virginia Center and the Coordinating Ministry of Maritime Affairs of the Republic of Indonesia. Principal associated sponsors were the Australian National Centre for Ocean Resources and Security, the Institute for China-America Studies, the Korea Maritime Institute, the K.G. Jebsen Centre for the Law of the Sea, the International Seabed Authority, and the Centre for International Law, Singapore.

Featured Remarks

The Marine Environment and UN Sustainable Development Goal 14

*Arif Havas Oegroseno**

Excellencies, Distinguished Speakers, Distinguished Participants, Ladies and Gentlemen:

> First of all, I would like to take this opportunity to express my warm welcome and gratitude for your attendance and active participation in this excellent event. The annual conference of Oceans Law and Policy of the University of Virginia is indeed respected and attracts distinguished congregations of oceans scholars. Today, in this 41st Annual Conference of the Centers for Oceans Law and Policy in Yogyakarta, joining us are 120 scholars and participants from around the world.

This conference, with the theme of Marine Environment and UN Sustainable Development Goal 14, takes place in a very timely situation. In recent years, international communities witnessed increasing attention toward the health and sustainable use of our ocean. The ocean is indeed high on the agenda of international communities. In 2017, for instance, Indonesia was the host for the World Ocean Summit and the Leaders' Summit of IORA member countries as well as the host of the 2nd IORA Blue Economy Ministerial Conference. In October 2017 various oceans' multi-stakeholders also convened in Malta to join the Our Ocean Conference hosted by the EU. But before that, international communities had a grand meeting and discussion on multiple oceans issues during the UN Ocean Conference which was held in June 2017 in New York.

Increasing interest and global attention toward our oceans is certainly a good sign that we should celebrate. For years, discussions of the health, management, and utilization of oceans were mainly the domain of scientists, academics, diplomats and lawyers. Private sectors outside the fishing or mining industry and common citizens of the world rarely joined the discussions. In

* H.E. Arif Havas Oegroseno is the Republic of Indonesia's Ambassador to Germany and former Deputy Coordinating Minister, Coordinating Ministry for Maritime Affairs.

 | DOI:10.1163/9789004366619_002

fact, even today among scholars and scientists, the attention towards oceans is significantly low in comparison to the considerable resources and attention given to outer space. It is, indeed, noteworthy that humans have only observed less than five percent of our oceans.

Indonesia is of the view that this moment should be better used to drive concrete action to preserve and sustainably use our ocean. This is why I praise the theme of our meeting which focuses on the issues of the marine environment in supporting the accomplishment of the sustainable development goals. Indeed, our discussion should be guided to focus more on concrete issues and challenges, which are confronted by countries and communities in the sustainable management of oceans.

Following the thrust to drive concrete ocean policy and action to better manage the oceans, Indonesia has just announced its National Ocean Policy. This document aims to guide overall nation policies of Indonesia. The Policy is founded upon seven pillars:

a. Management of Marine Resources and the Development of Human Resources;
b. Defense, Security, Law Enforcement and Safety at Sea;
c. Ocean Governance and Institution;
d. Economy and Infrastructure of the Marine Sector and the Prosperity Enhancement;
e. Management of Ocean Space and the Protection of Marine Environment;
f. Maritime Culture; and
g. Maritime Diplomacy

The roles of scholars, policy makers, and students are crucial in developing the solutions—solutions that are not only founded merely on the theoretical thinking and logical assumptions of researchers and policy makers alike, but also on solutions that evolve within a community to confront a challenge, adapt and successfully resolve problems in an original and authentic manner. That is, local solutions founded upon the wisdom of a community. These are solutions that can be observed in many regions of the world, including in many islands of Indonesia.

Having said the above, I also truly believe that the global community should also set a better framework of actions that can better accommodate and encapsulate various efforts of nations, private companies and civil society organizations into a better focus and coordinated actions.

Besides the above matters, I also, invite and encourage awareness of some crucial questions that should be better discussed among us such as:

1. How do we strike a balance between marine protection and preservation and sustainable use of the oceans to achieve sustainable development goals as a whole?

Indeed, in a short observation, these two ideas seem to have no conflicting thrust. They are linear and self-supporting notions that strengthen one another. Yet, if we look into more details in the implementation level, the dynamics can be very challenging. We should be aware that the marine environment thrust is founded upon the principle of preservation, whereas the sustainable development thrust may not necessarily be founded upon the similar benevolent idea. In some countries, the so called sustainable policy to use of our oceans could be founded upon the need to justify more utilization of our oceans due to depletion of land-based resources. This could result in more destruction of the ocean.

Nevertheless, the sustainable policy certainly cannot overlook the economic function of resources for a nation and community and the existence of multiple players surrounding ocean resources that can lead to competition over resources. Such conditions should certainly be better managed.

2. The second crucial question that needs to be addressed is: How do we develop measures and set norms that address gaps in current ocean governance with regard to the marine environment and SDG 14?

Since the creation of the UN Convention on the Law of the Sea, international communities have been equipped with cardinal principles and standards that set out rules for various forms of use of the oceans. Nevertheless, as we are moving to a new frontier of ocean exploration, especially in deep-sea activities, it is clear that some measures are increasingly critical to be developed to protect the marine environment. In this regard, we should cautiously and thoroughly note that our ocean is a public good, a common heritage that should not be treated in a first come, first served, manner that could lead to the tragedy of the common.

It is also noteworthy that areas require measures to fill the gaps at the implementation level is not limited only to the exploitation of Areas Beyond National Jurisdiction, but also exists in other issues, for instance, in resolving the issue of transboundary pollution originated from the offshore oil platforms. Indonesia, in this regard, has led the establishment of a guideline at the IMO to better manage such situations.

Regarding the question of developing norms, I personally am delighted that in this conference we will have an opportunity to directly listen to respected individuals which represent the generation who actively participated in the negotiation of UNCLOS. I believe their insights will provide an excellent pool of knowledge that we can learn from in assisting our own endeavor.

3. The third question is: How do we better accommodate the needs and challenges confronting the archipelagic and small island developing States with regard to marine protection and sustainable development?

It is clear that in the event of marine pollution and the effects of global warming and rising sea level, developing states and small island states are the most prone nations. At the same time, many of these small nation states are also the epicenter of marine biodiversity, which confront challenges between preserving and protecting their resources or exploiting them.

While marine ecotourism seems to be the immediate answer for resolving such an issue, the improvement of capacity and the level of the comprehensiveness of regulatory frameworks in developing marine tourism vis-à-vis preservation or protection of marine environment should be carefully designed and given required further assistance. Most important is to ensure that marine tourism activities really benefited the communities surrounding the resources that generate incentives for further marine protection and preservation.

4. The last question we should note is: How do we evade potential conflict arising from competition over resources, especially in fisheries, in adjacent and/or disputed waters?

One of the pressing issues of preserving our ocean and ensuring the realization of SDG goal 14 is to ensure that the fish stock is not depleted in our waters due to illegal fishing and criminal fisheries activities. Countries are inevitably moving toward more stringent measures and polices to tackle IUU Fishing to protects its fishermen's livelihoods as well as its sovereignty. These legal actions, however, could also lead into tension between countries, especially if the state of the adjacent waters is not yet legally defined.

In my view, this is not merely a management issue that can be solely resolved through the established RFMOs system. More cooperation in areas such as enhancement of maritime domain awareness and mutual legal assistance should be instituted to comprehensively complete the existed arrangement.

Indonesia is proposing a new regional cooperation agreement to tackle this issue in a more sensible and applicable manner. We are certainly open to the valuable input of distinguished scholars and policy makers in order to better protect the marine environment and achieve goal 14 of the SDGs.

I am truly glad that some elements of these four questions have been selected as the theme of our two-day conference. I believe this conference will meet its objective as an avenue to exchange views and ideas as well as to put some crucial questions that we should carefully consider in order to better protect marine environment and use it sustainably.

In following the spirit, I also propose this conference come up with a joint conclusion of academia to support an action-driven dialogue to the accomplishment of sustainable development goal 14 and its interrelated goals.

Finally, I thank Center of the Oceans Law and Policy at the University of Virginia that allowed Indonesia to co-host this distinguished conference in Yogyakarta. I also gratefully thank other associate sponsors, the Centre for the International Law of The National University of Singapore, the Korean Maritime Institute, Australian National Centre for Ocean Resources and Security (ANCORS), Institute for China America Studies (ICAS), International Seabed Authority (ISA), and the KG Jebsen Center for the Law of the Sea (JCLOS) for your generous support in realizing this conference.

I truly hope this conference and all of us will greatly contribute to preservation and protection of the marine environment and the realization of SDG goal 14.

Thank you very much. Wassalamu 'Alaikum Warrahmatullahi Wabarakatuh.

Biodiversity Studies

*Hasjim Djalal**

Speaking on Biodiversity, I remember the Anambas Expedition carried out around the Indonesian islands of Anambas in March 2002 aboard the Indonesian research vessel *Baruna Jaya VII*. The Expedition was the result of the agreement reached in the Workshop on Managing Potential Conflicts in the South China Sea (SCS). The Anambas Biodiversity Expedition was participated in by 29 experts and researchers from Malaysia, The Philippines, China, Chinese-Taipei, Thailand, Vietnam, Singapore, and Indonesia. Before the expedition there were a lot of debates among the participants that the expedition would involve territorial disputes in the SCS. In order to overcome this problem, I suggested that it could be held in and around Anambas Island, where there are no territorial disputes. I was of the opinion that the conduct of the scientific expedition should not be upset by territorial disputes.

The Expedition was the first of its kind organized voluntarily by the participating authorities in the Workshop on Managing Potential Conflicts in the South China Sea. The expenses for the Anambas Expedition were contributed voluntarily by the participating authorities.

The Expedition was intended to identify the biodiversity resources in the South China Sea with the hope and expectation that the knowledge and the findings of the Expedition would contribute significantly to the knowledge of biodiversity and living resources in the South China Sea. It was also hoped that the joint expedition would promote cooperation in the area, regardless of disputes that may exist in or regarding the specific area of the South China Sea. The Expedition collected some 300 kilograms of biological speciments consisting of about 1,000 species of various marine organisms from 60 sites. The Expedition also uncovered numerous new records and even some species.

The South China Sea Workshop Process on Managing Potential Conflicts in the area started in 1990. So far, it has been held yearly in Indonesia, participated by all the littoral South China Sea authorities, namely Brunei Darussalam, Indonesia, Malaysia, The Philippines, China, Chinese-Taipei, Thailand, Vietnam, Singapore, Cambodia and Laos.

In addition, numerous meetings of the various TWGs and GEMs (Group of Experts Meetings) have also been organized and held in various locations and cities around the South China Sea. Depending on the topics being discussed,

* Director of the Center for Southeast Asian Studies.

 | DOI:10.1163/9789004366619_003

some experts from outside of the region had also been invited from time to time as resource persons. From its inception until 2001, the process had been supported by Canadian International Development Agency (CIDA) through the South China Sea Informal Working Group at the University of British Columbia in Vancouver. After that, the process had been carried out with the support of the authorities around the South China Sea.

The aims of the workshops process are to promote dialog among the participating authorities, to develop confidence building process, and to devise cooperative efforts among the participants. It was understood that to avoid confrontation the authorities in the area must learn how to cooperate. Devising cooperative efforts in which every authority could participate in an inclusive approach was regarded as one of the ways of converting potential conflicts from confrontation to cooperation that would benefit everyone.

The Workshop Process is informal in nature and attended by the participants in their personal capacities but appointed by their authorities. The Workshop has established five Technical Working Groups and several Experts Groups Meetings, thus expanding the networks of cooperative efforts in the South China Sea area in various fields. The five Technical Working Groups (TWGs) are: the TWG on Marine Scientific Research; on Resources Assessment; on Marine Environmental Protection; on Safety of Navigation, Shipping and Communication, and on Legal Matters. Each of the TWG has devised a number of cooperative efforts in their respective fields. The Anambas Biodiversity Expedition was one of the cooperative efforts designed by the TWG on Marine Scientific Research.

The Anambas Expedition was expected to be followed by the Palawan Expedition in 2004. It was hoped that the expedition would augment scientific knowledge of the biodiversity in the area as well as foster a cooperative spirit in the region. It was also hoped that cooperative efforts that have been or are being designed by other TWGs would also be implemented in due course. Unfortunately, the Palawan Expedition, which was originally agreed to follow the informal model of the Anambas Expedition, later was changed unilaterally by the Philippines to a formal track I activity that made it a problem to many participants to attend. However, the workshop process has also agreed to conduct similar biodiversity expedition in the other parts of the South China Sea, and it is now waiting for the possibility to do it in and around the Pratash Island in the northeast of the South China Sea.

Finally, I hope that the experience of collaborative efforts in the SCS, particularly on biodiversity studies, could also be emulated in the other maritime areas worldwide, particularly on the high seas beyond national jurisdiction in the area where there are no territorial disputes, such as in the Indian and Pacific Oceans.

UNCLOS 35 Years Later: We Are Still at Sea

*N. Hassan Wirajuda**

The law of the sea is one of the most dynamic areas of the law. We can see the development of this area of law since the beginning of the twentieth century. Even in the last five to ten years, many things have changed nationally, regionally, as well as internationally.

Sixty years ago, Indonesia began its campaign to achieve international recognition of its archipelagic State concept.[1] It took 25 years for the archipelagic State concept to be recognized and finally enshrined by the United Nations Convention on the Law of the Sea (UNCLOS).[2] But only now, under the administration of President Joko Widodo, has Indonesia launched the World Maritime Fulcrum initiative focusing on, among others, its pillar of maritime culture and maritime infrastructure development, and for that purpose, has established the Coordinating Ministry for Maritime Affairs.[3]

We also see that the South China Sea conflicts have entered a new phase with the construction activities, the Arbitration,[4] and various subsequent state

* Based on the remarks given by H.E. Dr. N. Hassan Wirajuda at the 41st Annual Conference of the Center for Oceans Law and Policy, 17 May 2017, in Yogyakarta, Indonesia. H.E. Dr. N. Hassan Wirajuda was the Foreign Minister of Indonesia from 2001 to 2009, and member of the Council of Presidential Advisor from 2010 to 2014. He is a patron to the Institute of Peace and Democracy, Bali Democracy Forum. H.E. Dr. N. Hassan has held several important posts including Director-General of Political Affairs of the Ministry of Foreign Affairs (2000–2001), Ambassador and Permanent Representative to the United Nations and other international organizations in Geneva (1998–2000), Ambassador Extraordinary and Plenipotentiary to Egypt (1997–1998), and Director of International Organizations of the Department of Foreign Affairs (1993–1997).

1 Indonesia proclaimed itself as an archipelagic State on 13 December 1957. This is also known as the Djuanda Declaration, named after Indonesian Prime Minister Djuanda Kartawidjaja. The Declaration was later affirmed through Law No. 4/Prp of 1960 on Indonesian Waters.

2 The United Nations Convention on the Law of the Sea (UNCLOS), 10 December 1982, entered into force 16 November 1994.

3 The "*Poros Maritime Dunia*" or "World Maritime Fulcrum", sometimes informally translated as "world maritime axis", was introduced by President Joko "Jokowi" Widodo during his speech at the 9th East Asia Summit, 13 November 2014 in Nay Pyi Taw, Myanmar. The concept was further elaborated through Presidential Regulation No. 16 of 2017 on Indonesian Ocean Policy, 20 February 2017. The Coordinating Ministry for Maritime Affairs was created during President Jokowi's administration, *see* Presidential Decree No. 121/P of 2014 on Establishment of Ministries and Inauguration of Ministers of the Working Cabinet for 2014–2019 Period.

4 An Arbitral Tribunal Constituted under Annex VII to the 1982 United Nations Convention on the Law of the Sea in the Matter of the South China Sea, between the Republic of the

 | DOI:10.1163/9789004366619_004

responses. On a positive note, President Xi Jinping, during his state visit to Jakarta in October 2013, announced a new initiative called the Maritime Silk Road, an attempt to revive the old Maritime Silk Road[5]—known in Indonesia as the Old Maritime Spice Road—that connected China, the Sriwijaya Maritime Kingdom based in South Sumatra, Chola in South Eastern India and the Middle East (Egypt and Syria).[6]

Internationally, we can see various new oceans-related initiatives were generated. One of the notable initiatives is the United Nations Sustainable Development Goal No. 14,[7] which is the central theme of the 41st Annual Conference of the Center for Oceans Law and Policy held in May 2017.

Despite this progress in reaching legal order in the world's oceans, the process has not been linear, as we have also experienced setbacks. Therefore, after almost 35 years since the adoption of UNCLOS, we are still at sea.

1 Introduction to UNCLOS, Lack of Universality

The 35th Anniversary of UNCLOS will marked in 2017. Despite the fact that UNCLOS was praised as the Constitution for the Oceans,[8] up until today, UNCLOS has not achieved universal acceptance.[9] Among others, Turkey, Peru, Cambodia, and most notably, the United States, the largest maritime power, have yet to become party to UNCLOS. The goal of UNCLOS to create international order in the world's oceans relies heavily on universal acceptance.

While it is possible for non-parties to declare they are bound by principles of customary international law in UNCLOS, it does not prevent their "pick and

---

Philippines and the People's Republic of China, PCA Case No. 2013–19, award on merits rendered on 12 July 2016.

5 *See* Speech by Chinese President Xi Jinping to Indonesian Parliament, Jakarta, 2 October 2013, transcript available at http://www.asean-china-center.org/english/2013-10/03/c_133062675.htm.

6 For brief information about the Spice Road or Spice Routes, *see The Trade in Spices,* also available at http://en.unesco.org/silkroad/sites/silkroad/files/knowledge-bank-article/the%20trade%20in%20spices.pdf.

7 The Sustainable Development Goal, adopted by the UN General Assembly by Resolution A/Res/70/1 "Transforming Our World: the 2030 Agenda for Sustainable Development", 25 September 2015. There are 17 goals in total, see http://www.un.org/sustainabledevelopment/sustainable-development-goals/.

8 The term "Constitution for the Oceans" was first used by Ambassador Tommy Koh of Singapore, President of the Third United Nations Conference on the Law of the Sea, as the title of his remarks following the adoption of the final text of the UNCLOS in 1982.

9 As of 17 July 2017, UNCLOS has 168 parties including the European Union. *See* www.un.org/Depts/los/index.htm.

choose" attitude. UNCLOS in its entirety is a package, preserving the delicate balance of various interests of states.

Therefore, from a norm-shaping and sharing perspective, it is important for all states to become parties to UNCLOS in order to create stability of expectations, a key to international order. In this regard, I am happy that relevant stakeholders in non-party states, particularly the academics, continue their endeavor so that one day their country will become a party.

2 The Problem of Observance by Parties of UNCLOS, Paramount of Negotiations, and Lack of Dispute Settlement Culture

The next layer of challenge faced by UNCLOS is its observance by the state parties. Although they have become parties to UNCLOS, we can see that not all parties abide by the rules explicitly stated by UNCLOS. Problems between parties may also arise with regard to their interpretation of the UNCLOS provisions.

These problems of non-compliance were anticipated. Hence, UNCLOS has provided a comprehensive dispute settlement mechanism in Part XV. UNCLOS has provided a wide variety of mechanisms, starting from non-binding to binding mechanisms.

Negotiations remain a paramount starting point for dispute settlement, for states to settle their differences directly and come up with an acceptable solution. For example, on the Exclusive Economic Zone and the Continental Shelf, UNCLOS mandated that disputes shall be settled through agreement.[10] But, inhabiting a country which shares its maritime boundaries with 10 neighboring countries,[11] in Indonesia we knew very well that even persuading our counterparts to come to the negotiation table is not easy. And when they do come to the negotiation table, it is possible that negotiations will hit a deadlock or that negotiations will take a very long time.

Indonesia-Vietnam negotiations on their continental shelf boundary delimitation took 32 years to conclude.[12] As a junior officer, I joined the negotiation at its 10th session in 1979 and the agreement was signed by me in my

10 *See* UNCLOS Articles 74 and 83.

11 Indonesia shares maritime borders with India, Thailand, Malaysia, Singapore, Vietnam, the Philippines, Palau, Papua New Guinea, Australia and Timor-Leste. Subject to the recommendation of the Commission on the Limits of the Continental Shelf, Indonesia and the Federated States of Micronesia may share continental shelf borders.

12 Agreement between the Government of the Republic of Indonesia and the Government of the Socialist Republic of Vietnam concerning the delimitation of the continental shelf boundary, 26 June 2003 (entered into force 29 May 2007).

capacity as Foreign Minister. I echo the advice from Ambassador Hasjim Djalal: "patience, patience and more patience".[13]

UNCLOS has prepared for this, as it also created a third-party binding dispute settlement mechanism. UNCLOS even provides options of forum that parties can choose from, namely the International Court of Justice (ICJ), the newly created International Tribunal for the Law of the Sea (ITLOS), and an *ad-hoc* arbitration.[14] Almost all of East Asian states are parties to UNCLOS but only a handful of states utilize the dispute settlement mechanisms.

Out of *25 cases* submitted to the International Tribunal for the Law of the Sea (ITLOS), *only four have involved East Asian States.*[15] Out of *13 cases* submitted to an *ad hoc* arbitration under UNCLOS, only *five East Asian States* were involved.[16] Moreover, only *six East Asian States* became party to proceedings in the ICJ.[17] That shows that the culture for peaceful dispute settlement through binding mechanisms is not rooted in East Asian States.

If we look at the statistics closely, the Southeast Asian States are more inclined to settle their disputes through adjudication compared to their Northeast Asian brothers. I once stated that although these figures show good statistics on adjudication trends, there seems to be a setback. After the judgment of the tribunal was rendered, the domestic backlash from the so-called "losing party" indicated reluctance to return to the adjudication procedure in the foreseeable future. I am referring to my own country, Indonesia.[18]

3 East Asia: Microcosm of the World's Oceans and Its Relation Problems

If we look at the map of East Asia, both in the Northeast as well as in Southeast Asia, dispute settlement culture has not developed as fast as the region's economic development. Ironically, East Asia is one of the most diverse regions in the world in terms of geographic maritime configuration (and interests). It is

---

13 During the 41st COLP Annual Conference, Ambassador Hasjim Djalal delivered his remarks before H.E. Dr. N. Hassan Wirajuda.

14 *See* UNCLOS Article 287 (1).

15 These states are Japan, Malaysia, Singapore, and Myanmar, the list of ITLOS cases can be found on https://www.itlos.org/cases/list-of-cases/.

16 These states are Japan, Malaysia, Singapore, and the Philippines. The fifth state is the People's Republic of China, who refused to participate in the arbitration against the Philippines.

17 These states are Indonesia, Malaysia, Singapore, Cambodia, Thailand, and Japan. The list of ICJ cases can be found on http://www.icj-cij.org/en/list-of-all-cases.

18 *See* Case concerning Sovereignty over Pulau Ligitan and Pulau Sipadan (Indonesia/Malaysia), ICJ Reps. 2002 p. 625.

comprised of archipelagic States, land-locked States and in between, geographically disadvantaged States, and states with enclosed and semi-enclosed seas. China is a continental state with its huge landmass but has limited sea space.

Geography dictates competing marine activities and conflicting maritime interests that lead to tensions and potential conflicts in the maritime domain. Due to its geographical configuration and its diversity, East Asia is often referred to as the "Balkans of Asia".[19]

Against this backdrop, and confronted by the absence of a culture of peaceful conflict resolution on ocean affairs, the East Asia region is faced with a situation of instability and growing tensions. It is ironic that this is happening in a region that is increasingly considered the region of the 21st century, and this is mainly seen from its economic perspective. In the last two decades, the process of community building in the region has been mainly focused on economic integration, and much less on political and security cooperation. This imbalance is often called "the Asian Paradox".

The recent South China Sea Arbitration, where China decided not to participate in the proceeding and ultimately rejected the award, appears to be an alarming signal.[20] Despite the fact that there are disagreements on the jurisdiction or even the substance, non-participation clearly shows the absence of a dispute settlement culture that may lead to disorder. Even worse, referring a case to adjudication proceedings has been considered as a provocative action and the bigger party readily "flexes its muscles", making this region comparable to Europe of the Middle Ages. This situation makes the path of dialog become more crucial to prevent open conflict.

4 Conclusion—The Need to Promote Political-Legal Order

Legal order is an important part and parcel of any order, whether world or regional. Without a minimum political order, it is difficult to create legal order and eventually promote regional order. Hence, there is a need for creativity to construct a mixed approach of both political and legal order.

---

19 *See* e.g. C. Fitzgerald, *China and Southeast Asia Since 1945*, 1973 at 97.

20 For elaboration of China's position in rejecting the arbitration, *see* "Position Paper of the Government of the People's Republic of China on the Matter of Jurisdiction in the South China Sea Arbitration Initiated by the Republic of the Philippines", 7 December 2014, available at http://www.fmprc.gov.cn/mfa_eng/zxxx_662805/t1217147.shtml. *See also* "Statement of the Ministry of Foreign Affairs of the People's Republic of China on the Award of 12 July 2016 of the Arbitral Tribunal in the South China Sea Arbitration Established at the Request of the Republic of the Philippines", 12 July 2016, available at http://www.fmprc.gov.cn/mfa_eng/zxxx_662805/t1379492.shtml.

Despite the setback in recent years—which has caused the region to be "still at sea"—the commitment encapsulated by the parties to the 2002 ASEAN-China Declaration on the Conduct of Parties in the South China Sea,[21] demonstrated both political will of self-restraint, as well as assurances of their legal commitments, in order to maintain peace in the region. A stronger commitment in a form of a binding code of conduct is yet to be seen, and it may only be achieved with stronger political and legal commitment.

In the same vein, the 1976 Treaty of Amity and Cooperation[22] is another example of a legal-political commitment rooted from the Southeast Asia region. Understanding the volatility of relations among state parties, there is an urgent need to promote its implementing rules and regulations, in particular on peaceful conflict resolution.

5 Moving Forward—The Environment as a Platform for Regional Order

As a way forward, it is time for states to embark on a parallel approach. States need to use the marine environment as well as sustainable development as a basis to develop a good habit of dialogue, while at the same time developing a comprehensive regional order of the ocean.

Sustainable Development Goal No. 14 (SDG-14) may become a new platform, a way out to reduce tension in the region and to create regional order. As there is a universal acceptance of SDG-14, its issues are considered to be less sensitive. To achieve SDG-14, states are required to closely cooperate with one another, which can only be achieved by putting aside sensitive political issues relating to sovereignty, sovereign rights, and jurisdiction.

The East Asia region has done tremendously well in terms of its economic development. This should be balanced with development in the political and legal sphere to create a stable regional order. In the absence of a shared world order, and for that matter, an effective global order, there is an urgent need for this region to develop its own regional order including regional order of the ocean. In order to do so, let us start with encouraging a habit of dialogue concerning the environment, then expand to other sensitive areas.

With this spirit in mind, I am confident that we will no longer be at sea in the near future.

21 The full text of the document can be found at http://www.asean.org/13163.htm.

22 Treaty of Amity and Cooperation in Southeast Asia, 24 February 1976, entered into force 21 June 1976.

Revisiting the Law of the Sea Negotiations

*Satya N. Nandan**

Excellencies, Fellow Participants, Distinguished Guests, Ladies and Gentlemen:

> I am very pleased to be invited by Professor Myron H. Nordquist, Associate Director of the Center for Oceans Law & Policy, and Ambassador A. Havas Oegroseno, Deputy Coordinating Minister of the Coordinating Ministry for Maritime Affairs of the Republic of Indonesia to address this luncheon gathering. I am very grateful for this honor and also for the opportunity to meet so many old and new friends.

The 1982 United Nations Convention on the Law of the Sea has reached an important milestone in that it has received 168 ratifications. We have come a long way since the 10th of December 1982 when the Convention was signed in Montego Bay, Jamaica. It is not necessary for me to recount the achievements of the Convention and its global importance, since these are well known.

It has been very encouraging to hear the latest developments that are taking place. The Convention is working well, but there is still work to be done. First, there are those matters that require further action and implementation under the Convention, including, among others (i) measures that could be taken by States Parties under the Convention to come together with the global community to address the great harm to our oceans caused by land-based pollution, and (ii) actions that could be taken by coastal States to better protect and preserve the marine environment, and in the case for developing States in particular, better implementation of the capacity building and technology transfer provisions under the Convention that would allow them to conserve and sustainably use the marine resources within their jurisdictions.

---

* Ambassador Satya N. Nandan, C.F., C.B.E., was the Secretary-General of the International Seabed Authority 1996–2009. He served earlier as Secretary for Foreign Affairs of Fiji and as Under-Secretary-General of the United Nations and Special Representative of the Secretary-General for the Law of the Sea. He is currently Distinguished Senior Fellow of the Center for Oceans Law and Policy, University of Virginia School of Law.

 | DOI:10.1163/9789004366619_005

Second, there are issues of which we were not aware during the Conference, such as those relating to the conservation and sustainable use of marine biodiversity in areas beyond national jurisdiction. We need to work collectively to find solutions to both sets of issues. We are, therefore, at a very important crossroad.

I have been listening closely to the panels and the various discussions. For me, what is interesting is how far we have come since 1982. For example, the extension of jurisdiction has allowed for progress to be made in ocean management, in particular to the conservation and management of fisheries. Yet, there is still a propensity to overfish.

Given the recent focus on marine biodiversity in areas beyond national jurisdiction, I am especially pleased to hear that we are discussing marine biodiversity within coastal ecosystems as well. These areas include organisms from which samples can be collected and useful applications derived. Given that the bulk of marine biodiversity is on the continental shelf and within 200 nautical miles, I think much can be gained by looking more closely at the Convention on Biological Diversity, which in part encourages coastal States to develop their own rules and standards. There is more work that can be done in these areas that would have significant impact—by not only using the Convention, but also existing conventions and legal regimes.

Recently, I have been thinking about these current oceans issues and developments within the context of the making of the Convention as part of a book project I have been working on with the Centre for International Law (CIL) at the National University of Singapore. It began with discussions between Ambassador Tommy T.B. Koh, the President of the Third Conference, who now serves as the Chairman of the Governing Board of CIL, and Associate Professor Robert C. Beckman, former CIL Director and Head of CIL's Ocean Law and Policy programme. In our preliminary discussions, the primary aim of the book project was to provide an anecdotal history of the making of the Convention.

In the course of the development of the project, we realised that it would need to focus on the how and why behind the making of the Convention, rather than just retelling of events. This project was further inspired by the time I spent at the Institute of Peace in Washington DC from 1992 to 1993, where my research project was on the use of consensus and various other negotiating methods I had used during the Third Conference. But fate intervened in 1993 and I was called back to the United Nations to serve as Chairman of the Conference to negotiate the UN Fish Stocks Agreement. This book has become the book I never had the chance to write. Furthermore, as Under Secretary-General of the United Nations responsible for the Law of the Sea, I also had the responsibility on behalf of the Secretary General to initiate consultations to resolve the outstanding issues in Part XI of the 1982 Convention.

Our primary aim with this book project is to provide a resource that will complement and enrich the existing literature on the law of the sea by defining my approach—or what we have called the "Nandan approach"—to consensus-building diplomacy, as well as providing a personal recollection and narrative of the events surrounding the making of the modern law of the sea, all of which constitute a story that has not been told before. Through the definition and illustration of the "Nandan approach", our secondary aim is that these working methods and devices may inform future lawyers and diplomats on how to approach multilateral negotiations for the resolution of commons problems.

As we continue to discuss these issues of marine biodiversity as part of this Conference and try to find solutions, I thought I may add to the discussion by sharing some of my reflections on my approach to the making of the Convention with the hope that studying the past may illuminate a way forward.

The Third United Nations Conference on the Law of the Sea was notable for the delegates' willingness to break from traditional treaty making and codification exercises and create innovative procedures and methods for the work of the Conference. At the final session of the Third Conference at the signing of the United Nations Convention on the Law of the Sea, in Montego Bay, Jamaica on 10 December 1982, I noted the "unique nature of the negotiating process" that allowed for these innovations.[1] While the consensus-based "procedures" set the ground rules for decision making in the Conference proceedings, it was through the creative adaptation and evolution of the Third Conference's "working methods" that those who participated in the Conference furthered the progressive development of international diplomacy and treaty making.

As rapporteur of the Second Committee, I was too busy in the midst of the Conference proceedings to reflect on my negotiating style and methods, but as I have looked back upon my career through this project—I would say that my approach first began to develop with the preparation and submission of the proposal on innocent passage in the 1973 Summer Session in Geneva of the Seabed Committee. At that juncture, I realized the power that a well-conceived and crafted text could have on the negotiating process. This power was reaffirmed by my insistence on the development of the "Main Trends" paper. With over 4000 proposals having been submitted to the Second Committee alone, "Main Trends" consolidated those proposals by presenting identical proposals in clusters.

1 187th Plenary Meeting, 7 December 1982, *Third United Nations Conference on the Law of the Sea, Official Records*, Vol. xvii, p. 44, para. 84.

Further examination of my approach revealed a pattern whereby I used the following three devices or methods, either singly or in combination, to bring about consensus-based solutions: one, the use of a draft text; two, the application of an equitable approach; and three, the exercise of creativity. Although we have elaborated more fully on my approach in the course of the CIL project, I offer a brief description and example or anecdote of each of these three "working methods".

First and foremost, I am a strong proponent of the use of a text. My underlying method is having a piece of paper prepared. By leading with a text, it gives people something to focus on. The use of a text as one of my working methods was first evident in the paper I submitted to the Seabed Committee on innocent passage. It was further developed in my preparation of "Main Trends"—and was fully realised in my writing of the Second Committee's Single Negotiating Text. In these instances, I used a text in large Conference proceedings to consolidate support around certain ideas and concepts in an effort to find a compromise position. I used the same working method in the many small, informal negotiating groups that I chaired in the course of the Third Conference.

Regardless of whether I used the text at a meeting of the Second Committee or in the corridors with a small group, my purpose in using this working method was to create momentum in the negotiations. As Chair, I would move the participants through the agenda or the text, as the case may be, in an attempt to find a solution. I led the same way on the Part XI Implementation Agreement and the conference proceedings of the UN Fish Stocks Agreement that I chaired.

The best example of my use of a text involves archipelagoes—for in all things regarding the law of the sea for Fiji—delivering the archipelagic State concept was of paramount importance. Given that we are meeting in this beautiful Indonesian archipelago, this example is especially appropriate and relevant.

At the beginning of the Third Session of the Third United Nations Conference on the Law of the Sea in Geneva in May 1975, there was expectation that agreement would be reached on many of the important issues. However, these expectations soon faded because the Conference seemed to lack a sense of direction, in particular, in the Second Committee, dealing with national jurisdictions. The lack of a sense of direction was mainly due to a lack of draft treaty texts on the basis of which negotiations could be undertaken. The result was that delegations kept referring to their own national proposals and repeating their well-known positions.

The hope that the Third Session of the Third Conference in Geneva would proceed expeditiously to reduce the innumerable alternate and sometimes irreconcilable proposals into a single text soon vanished in a series of indecisions about the means by which this text could be prepared. As the session proceeded, it became clear that the process of consultation and negotiation could not achieve this, because no delegation was prepared to compromise on important interests except in the final stages. As I reported to my Government at the time:

> It became apparent by early April [of 1975] that unless the Second Committee in particular had before it a basic negotiating text indicating a total package on all matters relating to national jurisdiction, the Conference was in jeopardy. Concerned that if this happens then all the good work we have done in promoting the archipelagic concept would never be reflected in a consolidated text of the Conference, and with a desire to save the Conference, I, as rapporteur of the Second Committee, began to propagate the idea that the Chairmen of the Committees should be asked to prepare a single negotiating text.[2]

It was not until 18 April 1975 that the Plenary agreed to authorise the Chairmen of the three Main Committees to prepare single negotiating texts for their respective committees. As (mis)fortune would have it, I was walking from one conference room to another in the UN building in Geneva with the Second Committee Chairman Galindo Pohl of El Salvador, when the Chairman—deep in conversation—walked headfirst into a glass door. He was seriously injured and unable to prepare the text. The whole of the task, therefore, fell on me to draft the Second Committee's Single Negotiating Text on behalf of the Chairman.

From Fiji's point of view and that of other archipelagic States—Indonesia, the Philippines and the Bahamas—this was tremendous progress because I had included in the Second Committee's Single Negotiating Text a comprehensive set of draft articles on archipelagic States, which I knew from my consultations to be in an acceptable form to a large number of States, including the major maritime States. I considered this to be an indication of an irreversible trend in favour of the acceptance of the archipelagic State concept in international law. I reported to my Government at the time, "Although a negotiating

2 S.N. Nandan, "Report of the Third Session of the Conference on the Law of the Sea" (Geneva, 17 March–10 May 1975) (*Papers of the United Nations Conferences on the Law of the Sea at the University of Virginia*, Arthur J. Morris Law Library Special Collections, Box 69 of 204), pp. 1–2.

text, the document nevertheless reflects an important development in favor of the [archipelagic State] concept in international law of the sea".[3]

The second working method that I have relied upon in my career is the application of an equitable approach, which has been central to my style of consensus-building diplomacy. In working on this project, my use of a sense of equity became readily apparent. While I was in law school lectures at the Council of Legal Education in London organised by the Inns of Court, I discovered that the idea of finding equitable solutions to seemingly intractable issues in international law was a useful and generally acceptable way to resolve outstanding issues.

Lastly, creativity—when exercised pragmatically—was a key component of my approach to multilateral negotiations, as demonstrated in my frequent use of language to reach consensus. I have a passion for language and I often consulted three different dictionaries during the Third Conference, searching for synonyms to create agreement. I found that the right choice of words could be used to bridge differences between opposing views. For instance, consider the wording of Article 38 on transit passage: "right of passage, which shall not be impeded … for the purpose of continuous and expeditious transit of the strait" and compare it with the wording of Article 53 on archipelagic sea-lanes passage: "rights of navigation and overflight … solely for the purpose of continuous, expeditious and unobstructed transit".[4] This similar, but different, wording—which describes essentially the same passage regime—serves as a very apt example of my use of synonymous terms and phrases to forge agreement.

Putting aside my approach to the law of sea negotiations, what has become increasingly clear in working on this CIL project is that—more often than not—what saved the Third Conference from near collapse on several occasions during its nine-year negotiations were the friendships among the delegates. It is therefore no coincidence that I made mention of it in my statement at the final session of the signing of the Convention in Montego Bay, Jamaica: "The personal interrelationships which developed during the course of our work resulted in an atmosphere of mutual trust that allowed individuals and groups of individuals to make their contributions".[5]

Many States retained their representative—for the most part—throughout the Third Conference. While some States did so for reasons of necessity because

---

3 *Ibid.* at p. 2.

4 *United Nations Convention on the Law of the Sea*, 10 December 1982, 1833 UNTS 3 (entered into force 16 November 1994), Arts. 38 and 53.

5 187th Plenary Meeting, *supra* note 1 at p. 44, para. 85.

they did not have a deep bench of international law-making expertise, others recognised the importance of maintaining stable delegate membership, despite the long duration of the Conference. However, others—a few of the major maritime powers in particular—kept to their usual staff rotation schedules and consequently lost significant time and momentum in the negotiations, in addition to personal ties and information with every rotation.

I, on the other hand, came to the United Nations in New York on a one-year contract in 1970 and I participated in the Seabed Committee meetings—and then stayed for the duration of the Third Conference—and beyond. On the day of the signing of the Convention in Montego Bay, Jamaica, I had a flight home to Fiji in the early afternoon. After depositing Fiji's documents at the morning session to be the first country to ratify the Convention, Ambassador Tommy Koh, the President of the Conference, and I went for a swim in the ocean. We went back to the hotel to work on Tommy's statement, "A Constitution for the Oceans", before I headed to the airport. I said goodbye to Tommy and all of my friends from the Conference. In the course of the eleven sessions, the friendships that developed for me in the Conference cut across oceans, cultures and boundaries and have endured these many years. As I boarded the plane, I thought I would return to Fiji and be done with the law of the sea for good. How wrong I was...

In closing, I would ask the next generation of lawyers, diplomats and scientists to look past experience and ideology and work together to find common ground to solve the problems that imperil the health of the world's oceans. In doing so, my hope is that the Convention, together with the various methods and devices we used to arrive at the Convention, will serve as a framework that will guide and inform future negotiations and decision making.

PART 1

Conservation and Sustainable Use of Oceans and Their Resources

CHAPTER 1

Promoting the Conservation and Sustainable Use of the Oceans through Cooperative Decision Making

*Rena Lee**

Abstract

The recent years have seen an uptick in activities within the United Nations related to oceans, including the adoption of Sustainable Development Goal 14, the Ocean Conference, and the development of a legally binding instrument under UNCLOS on the conservation and sustainable use of marine biological diversity of areas beyond national jurisdiction. These activities have served to enhance awareness of the need to conserve and sustainably use our oceans and the resources within. The importance of global cooperation to enhance coherent management of the oceans and their resources cannot be overestimated. The paper takes a brief look at the synergies process, which is one model for more coherent global governance. It then looks at what lessons may be learnt to enhance greater cooperative decision making.

1 Introduction

In 2012, the Joint Inspection Unit of the United Nations (UN) reported on its evaluation of UN-Oceans,[1] which is a coordinating mechanism for oceans and coastal issues. One of the conclusions of the report was that oceans affairs had, to date, received "low visibility and priority in the United Nations system".[2] One wonders if the inspectors would draw the same conclusion today, given the many activities within the UN in relation to oceans today. In 2015, the UN adopted the Sustainable Development Goals for 2030,[3] including Sustainable Development Goal 14, which is to conserve and sustainably use the oceans, seas

* Rena Lee, Senior State Counsel, International Affairs Division, Attorney-General's Chambers, Republic of Singapore. The views expressed here are the views of the author and do not reflect the views of the Government of Singapore.

1 JIU/REP/2013/3.

2 *Supra* note 1, page iv, Executive Summary: main findings and conclusions.

3 A/RES/70/1.

 | DOI:10.1163/9789004366619_006

and marine resources for sustainable development. In June 2017, the UN held a UN Conference to Support the Implementation of Sustainable Development Goal 14, named the Ocean Conference.[4] The Preparatory Committee to develop an international legally binding instrument under the United Nations Convention on Law of the Sea (UNCLOS) on the conservation and sustainable use of marine biological diversity of areas beyond national jurisdiction (BBNJ) also recently concluded its fourth session and forwarded its report to the UN General Assembly.[5] And in September 2017, the UN Secretary-General appointed a Special Envoy for the Ocean, Mr Peter Thomson of Fiji, to galvanize efforts to follow up on the Ocean Conference.[6]

This paper looks at how we can enhance conservation and sustainable use of our oceans and their resources through greater cooperation. It looks at the importance of cooperation on a global basis, and examines a model for coherent global governance. The paper then considers what lessons may be learnt in trying to enhance greater cooperative decision-making.

2 Conservation and Sustainable Use

Sustainable Development Goal 14 calls for us to conserve and sustainably use the oceans, seas and marine resources for sustainable development. In particular, goal 14.c recognises that conservation and sustainable use can be enhanced by implementing international law, as reflected by UNCLOS.[7]

Article 197 of UNCLOS requires states to cooperate on a global basis and, as appropriate, on a regional basis. In other words, states are in the driver's seat, so to speak, when it comes to cooperation in relation to the marine environment.

Article 197 recognises that cooperation undertaken by states may be effected through "competent international organisations". This was recognised in the outcome document of the Ocean Conference, "Our ocean, our future: call for action", which calls on all stakeholders to "strengthen cooperation,

4 See <https://oceanconference.un.org>, the official website of the Ocean Conference.

5 <www.un.org/depts/los/biodiversity/prepcom.htm>. The report may be accessed in the list of documents under the Fourth Session of the Prep Com.

6 <https://www.un.org/sg/en/content/sg/personnel-appointments/2017-09-12/mr-peter-thomson-fiji-special-envoy-ocean>.

7 *Supra* note 3. Goal 14.c reads as follows: "Enhance the conservation and sustainable use of oceans and their resources by implementing international law as reflected in the United Nations Convention on the Law of the Sea, which provides the legal framework for the conservation and sustainable use of the oceans and their resources, as recalled in paragraph 158 of 'The future we want'".

policy coherence and coordination among institutions at all levels, including between and among international organisations, regional and sub-regional organisations and institutions, arrangements and programmes".[8]

2.1 *Importance of Cooperation on a Global Basis*

Looking at the world's oceans and their resources today, one thing is evident; there are many players involved. These include international organisations, regional organisations, states, civil society, researchers, corporations, and individuals. In some fashion or another, these players undertake activities in the oceans and/or utilise the resources of the oceans. However, this is not intended to imply that these players have done so in an unregulated fashion. Far from it; there are many regulations applicable to the oceans and their resources. But the current governance framework employs a sectoral and/or regional approach. This has led to concerns being expressed about the fragmented nature of governance or oversight over the oceans and their resources. Such fragmentation, which may be a result of limited mandates in terms of geographic or subject matter coverage, has resulted in gaps in governance. This is an issue that the ongoing BBNJ negotiations is trying to grapple with—how to close the gaps so as to enhance conservation and sustainable use in a more coherent fashion. In considering how the gaps may be closed, negotiators must take into account the mandate given to them pursuant to UN General Assembly Resolution 69/292. In particular, operative paragraph 3 requires that negotiators should not "undermine existing relevant legal instruments and frameworks and relevant global, regional and sectoral bodies".[9]

The issue is how governance and oversight gaps may be closed while still respecting "existing relevant legal instruments and frameworks and relevant global, regional and sectoral bodies". This is a very complex issue with multiple facets, particularly given that the current governance framework means that the bodies involved in a particular aspect of conservation and sustainable use may not be the same bodies involved in other aspects. For example, the International Maritime Organisation, which regulates shipping, may be involved in the consideration of issues relating to the utilisation of area based management tools but are not likely to be involved in questions surrounding the utilisation of marine genetic resources. A number of proposals have been put forward by delegations in the context of the BBNJ negotiations.[10]

---

8 A/RES/71/312, Annex, paragraph 13(b).

9 A/RES/69/292, OP 3.

10 <www.un.org/depts/los/biodiversity/prepcom.htm>. See the submissions received from delegations under the heading of "Additional information".

It is not the intention to discuss those proposals but to look at a model outside the context of oceans which may have lessons to offer in promoting cooperation among states via an institutional framework. This model is the synergies process.

2.2 *The Synergies Process*

The synergies process is an institutional framework designed to apply to what are known as the "chemicals conventions". These three conventions are the Basel Convention on the Control of Transboundary Movements of Hazardous Wastes and their Disposal, the Rotterdam Convention on the Prior Informed Consent Procedure for Certain Hazardous Chemicals and Pesticides in International Trade, and the Stockholm Convention on Persistent Organic Pollutants.

The synergies process came about as a means to improve international environmental governance after calls were made by the international community to reduce the fragmentation of governance in environmental issues. The "synergies decision" undertaken by the three conventions noted, in PP6, a series of earlier decisions "calling for reduced fragmentation on environmental issues".[11] The synergies process focuses on enhancing cooperation and coordination among the three conventions, with measures "aimed at strengthening implementation of the three conventions at the national, regional and global levels, promoting coherent policy guidance, enhancing efficiency in the provision of support to Parties with a view to reducing their administrative burden and maximising the effective and efficient use of resources at all levels".[12]

In essence, the process seeks to explore and make full use of the synergies that exist among the three conventions. This includes arranging for the Conference of the Parties (COPS) of each of the three conventions to take place back-to-back with each other. Joint sessions are conducted during these COPS for issues to be jointly considered. There are also simultaneous extraordinary COPS or EX-COPs, which allows for collective consideration of issues that affect all three conventions. At such EX-COPs, decisions are taken simultaneously by all three conventions. However, this is not to say that the three conventions have "merged" into a single decision-making entity. On the contrary, parties of the three conventions are "mindful of the legal autonomy" of each convention and this statement is reflected in the preamble of all decisions.[13]

---

11 "Cooperation and coordination among the Basel, Rotterdam and Stockholm conventions", BC IX/10, RC-4/11 and SC-4/34. See PP 6 of the decisions.

12 *Supra* note 6. See PP 7 of the decisions.

13 See for example, PP 1 of the "Omnibus decision adopted by the Conference of the parties to the Basel Convention", BC.Ex-1/1, a decision taken at the first EX-COPs of the three

A number of decisions involve administrative and financial issues, including the establishment of a joint secretariat that administers all three conventions. These administrative and financial issues are specific to the context of the three conventions and to that extent, may be of less relevance in the oceans context. However, beyond administrative and financial issues, a number of technical, scientific and other substantive issues are considered jointly as well. These include undertaking cross-cutting and joint activities, such as enhanced cooperation and coordination between the technical bodies of the three conventions, technical assistance, such as providing capacity-building programmes at the regional level, joint outreach and the establishment of a joint clearing-house mechanism for the exchange of information.[14]

2.3 *Synergies in Oceans?*

The above is only a very brief account of the synergies process and there is, obviously, a lot more to the synergies process. It should be stressed that it is not the intention of this paper to suggest that the synergies process should be replicated in the oceans context. The synergies process is limited currently to three conventions which have a high degree of what is termed as "congruence" in this paper.[15] Such congruence is reflected both in terms of subject matter and in terms of number of parties. All three conventions share a common objective of protecting human health and the environment.[16] There are 183 parties to the Basel Convention, 181 parties to the Stockholm Convention,

---

conventions, which was mirrored in the decisions taken by the parties to the Rotterdam and Stockholm conventions, RC Ex-1/1 and SC Ex-1/1.

14 See for example, the note by the Secretariat on the "Implementation of joint activities contained in the programme budgets of the Basel, Rotterdam and Stockholm conventions for 2012", UNEP/FAO/CHW/RC/POP/EXCOPS.2/INF/10.

15 In 2013, the omnibus decision on enhancing cooperation and coordination among the Basel, Rotterdam and Stockholm conventions signalled the readiness of the process to "cooperate and coordinate with the Minamata Convention on Mercury". See OP 27 of BC. Ex-2/1, RC.Ex-2/1 and SC.Ex-2/1. Even if the Minamata Convention is included in the synergies process, it is suggested that there is a high congruence in terms of subject matter with the three conventions and over time, a high congruence in number of parties may also be achieved.

16 See the last preambular paragraph of the Basel Convention, which refers to the parties' determination "to protect, by strict control, human health and the environment". See also Article 1 of the Rotterdam Convention and the Stockholm Convention, which sets as the objective of the two conventions the protection of human health and the environment. Article 1 of the Minamata Convention on Mercury also provides that its objective is to "protect the human health and the environment" from mercury emissions and releases.

and 154 parties to the Rotterdam Convention.[17] This high degree of congruence means that there are several issues that can be jointly discussed and joint action undertaken. In addition, the congruence in the number of parties to the three conventions means that effectively, there are only 29 states which are not involved in decisions of all three conventions.

In the oceans context, the picture is a lot more complex and the degree of congruence is not as high as for the synergies process. The BBNJ webpage maintained by the UN Division for Ocean Affairs and the Law of the Sea (DOALOS) contains an indicative list of global treaties and an indicative list of regional treaties for reference.[18] These lists do not purport to be comprehensive but even without being comprehensive, there are approximately eighty instruments set out in the two lists that are related in some fashion or another with conservation and sustainable use of the oceans and its resources. Even if one were to conduct a study to weed out some of the "less relevant" instruments, the number of instruments remaining would still be relatively high. The subject matter is varied; ranging from the protection of biodiversity, to climate change, to the prevention of pollution from ships.

Beyond the range in subject matter, the range in the number of parties in the listed instruments is also very large. At one end of the spectrum, there are 197 parties to the United Nations Framework Convention on Climate Change. At the other end of the spectrum, there are only 13 parties to the Agreement on the Conservation of Albatrosses and Petrels.

The above demonstrates the relatively lower degree of congruence, and the correspondingly higher degree of complexity of the web of instruments and frameworks that are relevant when considering issues of oceans governance. Moreover, the comparisons above on the subject matter and number of parties are based only on the indicative list of global treaties. Once the instruments in the list of regional agreements are included in the picture, the situation becomes even more complex, involving different regions and different subject matter.

2.4 *Lessons in Synergies*

As noted above, this paper does not suggest replication given the complexities in oceans governance, but neither is it suggested that there are no synergies to be pursued despite the known complexities. Even though the degree of congruence is not as high, there is scope for considering whether there are lessons

17 There are currently 84 states that have ratified the Minamata Convention on Mercury, which entered into force on 16 August 2017.

18 *Supra* note 5. The lists can be accessed under the heading of "Additional Information".

that one can take away from the synergies process. This paper suggests that there are a number of potential lessons.

The first potential lesson is the focus on national coordination. Although the synergies process was designed to enhance environmental governance at the international level, the process includes a focus on national coordination that centers around combining efforts at the country level. At the national level, parties are invited to establish or strengthen national processes or mechanisms for coordinating and to ensure close coordination among relevant sectors, ministries or programmes, including through preparation of national positions.[19] At present, the process has a special programme aimed at institutional strengthening at the national level for implementation of the three conventions, as well as the Minamata Convention.[20] Enhanced national coordination not only avoids duplication and utilises resources more efficiently, but also it could also lead to a greater degree of cooperative decision making in that delegations could attend the COPs and EX-COPs with better coordinated national positions. This could enhance the ability of the EX-COPs to take joint decisions.

Within the oceans context, if one were to examine the participants representing states parties in various meetings of relevant bodies such as the IMO, FAO or CBD, one would likely find that the participants represent different bodies within the country such as the Maritime Department or the Environment Ministry or the Fisheries Department. It is possible that in some countries, there is little or no coordination amongst the different agencies participating in the meetings of relevant bodies. This could result in diffused or fragmented positions taken by the country in different meetings. An emphasis on greater national coordination could be the driver for greater international cooperation, in that more coherent and coordinated national positions could be adopted across the various forums. By doing so, it opens the possibility for different bodies to work towards similar goals or undertake similar measures, as states are the primary decision makers in many international organisations.

A second potential lesson concerns the undertaking of joint activities. There are several joint activities occurring under the auspices of the synergies process. These include capacity building and training activities covering the three conventions,[21] as well as a strategy for developing and operating a joint clearing-house mechanism.[22] Within the oceans context, it could be worth

19 BC-IX/10, RC-4/11, SC-4/34, Section I(A): Coordination at the national level.

20 UNEP/CHW.13/INF/41.

21 See for example, UNEP/CHW.13/INF/34.

22 UNEP/CHW.13/INF/47.

considering whether there is scope for relevant bodies to cooperate in undertaking joint capacity building and training activities. This could be undertaken initially on an *ad-hoc* basis, which would allow for more informal arrangements, before developing more formal arrangements for cooperation. Such joint activities could reduce duplication of efforts as well as provide recipients with a more holistic understanding of the oceans and its conservation and sustainable use. In addition, in the BBNJ negotiations, there have been a number of proposals for a clearing-house mechanism. Some thought might be given as to how relevant institutions could participate in such a mechanism, which could enhance the sharing of information and increase knowledge about the oceans.

A third potential lesson involves joint meetings. While it might not be possible to replicate the EX-COPs process, consideration might be given to the possibility of a forum where states, as well as the relevant institutions could exchange views on issues of joint concern, share information on the kinds of activities being undertaken within the respective institutions and explore the possibilities for collaboration amongst the institutions.

3 Conclusion

The issue of decision making in relation to conservation and sustainable use of oceans and their resources is one in which there may not be a one-size-fits-all solution. The quest to overcome fragmentation is not new. The synergies process has its roots in the desire to overcome fragmentation in environmental governance. In the quest to overcome fragmentation in oceans governance, one might consider what food for thought the synergies process may have to offer as we try to close the gaps in oceans governance.

CHAPTER 2

The Conservation and Sustainable Use of Marine Biodiversity: Siamese Twins?

*Liesbeth Lijnzaad**

Abstract

This contribution focusses on the meaning of the formula "conservation and sustainable use" as it appears in the debate about the conservation of marine biodiversity in areas beyond national jurisdiction (BBNJ). As the UN General Assembly has decided to start a negotiation process for the establishment of an implementing agreement to the UN Law of the Sea convention on marine biodiversity beyond national jurisdiction, clarity about the relationship between conservation and sustainable use is important.

The paper reflects on a discussion that has so far not taken place. What do these requirements entail? This reflection is necessary given the likelihood that "conservation and sustainable use" will be understood as part of the implementing agreement's object and purpose (art. 31 VCLT).

1 Introduction

"Conservation and sustainable use" of marine biodiversity in areas beyond national jurisdiction plays a central role in the current discussions about an Implementing Agreement under the Law of the Sea Convention on Marine Biodiversity beyond National Jurisdiction (BBNJ).[1] It is at the heart of the debate about a new instrument on marine biodiversity. Though this language has been use in the debate since its inception, it has not been questioned so far. This formulation of conservation together with sustainable use seems to be

* Dr. Liesbeth Lijnzaad, member of the International Tribunal for the Law of the Sea (Hamburg), Professor Practice of International Law at Maastricht University. This contribution has benefitted from discussions at the 41st Annual Conference of the Center for Oceans Law and Policy, The Marine Environment and UN Sustainable Development Goal 14 (Life below Water), in Yogyakarta (May 2017); and further discussion at the Oceans and Climate Change Conference, in Malmö (August 2017). Any errors remain mine. The author's PowerPoint presentation may be found at http://www.virginia.edu/colp/pdf/yogya-lijnzaad.pdf.

1 Henceforth referred to as BBNJ.

 | DOI:10.1163/9789004366619_007

used as one concept and thus *one goal* for the new instrument. It is important to understand this concept in greater detail, particularly so as the decision to move to a formal diplomatic conference on this matter in the near future has recently been taken by the UN General Assembly.[2] The "conservation and sustainable use" language may be understood as the object and purpose of a future Implementing Agreement and will thus presumably play a role in its interpretation.

Consequently, I am curious as to the meaning of this formulation. And I assume this question must be of importance to the drafters of the new instrument as well. The aim of the current contribution is to investigate whether the notion of "conservation and sustainable use" of marine biodiversity consists of one goal or two, or what the relationship between both notions would be. Even if linguistically distinct concepts, the drafters may consider that both elements can be addressed jointly for the benefit of marine biodiversity. If "conservation and sustainable use" are inextricably linked, the future Implementing Agreement will have one goal only, and the notions of conservation and sustainable use will go together like Siamese twins.

An alternative reading would be that the ordinary meaning of the text "conservation" and "sustainable use" are distinct and separate goals that may require different actions to be formulated in the Implementing Agreement. Conservation and sustainable use are not necessarily the same type of activities (even if the intention may be similar) and may not serve the same purpose. Pursuant to this alternative approach, an Implementing Agreement will have two distinct goals, or indeed two distinct objects and purposes.

As we reflect on these notions, there may indeed be more ways of understanding this formula and the relationship between conservation and sustainable use. This debate is not about semantics, but rather about the aim of the future instrument and understanding how these notions of conservation and sustainable use go together.

2 Origins of the "Conservation and Sustainable Use" Language

An early discussion on the subject of conservation and sustainable use of marine biodiversity in areas beyond national jurisdiction took place within the General Assembly during the UNICPOLOS[3] session in 2004 where a panel

2 A/RES/72/249 of 24 December 2017.

3 United Nations Informal Consultative Process on the Oceans and Law of the Sea, initially known as UNICPOLOS, currently as ICP.

discussed "New Sustainable Uses of the Oceans, including the Conservation and Management of Biological Diversity of the Seabed in Areas beyond National Jurisdiction." The discussion covered issues such as the ecosystems of the deep seabed, hydrothermal vents and gas hydrates as well as the effects of high seas bottom fisheries on marine biodiversity and the uses of marine genetic resources.[4] This was followed-up in the annual Law of the Sea omnibus resolution of the same year in which the General Assembly in paragraph 73:

> Decides to establish an Ad Hoc Open-ended Informal Working Group to study issues relating to the conservation and sustainable use of marine biological diversity beyond areas of national jurisdiction:
> (a) To survey the past and present activities of the United Nations and other relevant international organizations with regard to the conservation and sustainable use of marine biological diversity beyond areas of national jurisdiction;
> (b) To examine the scientific, technical, economic, legal, environmental, socio-economic and other aspects of these issues;
> (c) To identify key issues and questions where more detailed background studies would facilitate consideration by States of these issues;
> (d) To indicate, where appropriate, possible options and approaches to promote international cooperation and coordination for the conservation and sustainable use of marine biological diversity beyond areas of national jurisdiction;[5]

This paragraph created the so-called BBNJ Working Group[6] that was to meet from 2006 to 2015, initially with two year intervals and twice a year towards the end of its existence.

It should be noted that between the UNICPOLOS meeting (in June 2004) and the omnibus resolution of that year (adopted in November 2004) the name of the subject matter had changed from "conservation and management"[7] of marine biodiversity to "conservation and sustainable use". Management is a neutral, and often commercially used term that presupposes a rational approach

4 See for this discussion A/59/122, para. 56–94.

5 UNGA Resolution 59/24, 17 November 2004.

6 It had a lengthy name. Ad Hoc Open-ended Informal Working Group to Study Issues Relating to the Conservation and Sustainable Use of Marine Biological Diversity beyond Areas of National Jurisdiction, or otherwise: AHOEIWG BBNJ.

7 "Conservation and management" is language used in the Law of the Sea Convention, see e.g. articles 120, 266(2), 277(a) and 297(3).

(to marine biodiversity), but sustainable use (of marine biodiversity) adds a specific goal to the management activities on the high seas. It is difficult to retrieve how or why this change in language happened, but it is likely that either delegates or Secretariat staff considered it useful to make the language tie in with language use in other international environmental discussions.

The Working Group had a relatively slow start, but towards the end of 2011 it shifted gears and the work became more focused. The General Assembly, at the recommendation of the Working Group, decided in the annual Law of the Sea resolution to initiate a process (in the Working Group) aimed at ensuring the legal framework addressed the conservation and sustainable use of marine biodiversity by identifying gaps in the implementation of existing instruments as well as the possible development of an agreement under UNCLOS.[8]

The Working Group ended its work in January 2015,[9] and later that year the General Assembly decided to formally move towards an Implementing Agreement and to establish a Preparatory Committee to this end.[10] This also implicitly marked the end of one of the central debates in the Working Group—whether the threats to marine biodiversity in areas beyond national jurisdiction were the consequence of a lack of implementation of existing rules or the absence of a regulatory framework. In heading towards an Implementing Agreement, this matter has been settled: the aim of the efforts is most of all to remedy a so-called regulatory gap.

The following phase of work in the Preparatory Committee was a further step in focusing the debate, even if it proved impossible to formulate a draft negotiating text.[11] The discussions did clarify that there was great interest in

---

8 See A/RES/66/231, para.166 and its Annex with the so-called Package of inter-related issues of relevance to the conservation and sustainable use of marine biodiversity in areas beyond national jurisdiction.

9 See the report in A/69/780 of 13 February 2015.

10 See A/RES/69/292 of 19 June 2015.

11 See the report of the Preparatory Committee established by General Assembly resolution 69/292: Development of an international legally binding instrument under the United Nations Convention on the Law of the Sea on the conservation and sustainable use of marine biological diversity of areas beyond national jurisdiction (doc. A/AC.287/2017/PC.4/2 of 31 July 2017) which in paragraph 38 reads:

The Preparatory Committee, ... , recommends to the General Assembly:

(a) That the elements contained in sections A and B below be considered with a view to the development of a draft text of an international legally binding instrument under the United Nations Convention on the Law of the Sea on the conservation and sustainable use of marine biological diversity of areas beyond national jurisdiction. *Sections A and B do not reflect consensus.* Section A includes non-exclusive elements that generated convergence among most delegations. *Section B highlights some of the main issues on which there is divergence of views.* Sections A and B are for reference purposes because

the possibility of access and benefit sharing with respect to marine genetic resources from areas beyond national jurisdiction. Section 3 of the report of the Preparatory Committee deals with marine genetic resources including questions on the sharing of benefits. It lists as objectives of the sharing of benefits both the contribution to the conservation and sustainable use of marine biodiversity; and capacity-building for developing countries with respect to access and use of marine genetic resources. Under the heading of principles and approaches guiding benefit-sharing are to benefit current and future generations and to promote marine scientific research.

However, as to benefits, the modalities of benefit-sharing or intellectual property rights in the text are rather vague. In part, that is understandable because not only are intellectual property rights a field very different from the law of the sea, but also intellectual property rights in areas beyond national jurisdiction are somewhat of a *terra incognita* to the majority of delegations. This is also reflected in the latter part of the report where Section B states: "With regard to marine genetic resources, including the question of the sharing of benefits, further discussions are required on whether the instrument should regulate access to marine genetic resources; the nature of these resources; what benefits should be shared; whether to address intellectual property rights; and whether to provide for the monitoring of the utilization of marine genetic resources of areas beyond national jurisdiction".[12] In sum, the direction as to access and benefit sharing is quite open and probably on purpose. No consensus could be reached as was stated in the report, leaving this aspect to be followed as negotiations continue.

2.1 *The Meaning of "Conservation and Sustainable Use" Language*

As the purpose of this contribution is to discuss the relationship between *conservation* on the one side and *sustainable use* of marine biodiversity on the other side, it is useful to understand how these notions have been used before. If clarity about these two notions exists elsewhere this may guide us towards an understanding of the relationship between the two. The marine biodiversity debate is not an isolated discussion, and both delegates and NGOs frequently refer to existing regulation as well as to the need to ensure consistency within the international legal system. The report of the Preparatory Committee agrees: "[T]he text would provide definitions of key terms, bearing in mind the need for consistency with those contained in the Convention and other

they do not reflect all options discussed. *Both sections are without prejudice to the positions of States during the negotiations;* [emphasis added].

12 A/AC.287/2017/PC.4/2 at 17/19.

relevant legal instruments and frameworks".[13] That seems like a good starting point, but it may be that things are not quite as simple as not all existing instruments use the same language.

The UN Convention on the Law of the Sea speaks about conservation mostly in relation to fisheries and protection of the marine environment referring to such notions as the "conservation of the living resources", the "conservation of stocks" or their "conservation and management". There are but a few places where both conservation and sustainability are mentioned in the same provision. Sustainability features only in provisions on marine living resources and refers to maximum sustainable yield (MSY) in articles 61(3) and 119(1)(a). MSY presupposes a rational and science-based management approach.

The Agreement on the Implementation of UNCLOS Part XI of 1994[14] does not refer in any way to either conservation or sustainable use. In today's perspective this may be surprising, as developments in the deep sea are the main concern driving the BBNJ discussions. The unknown consequences of anthropogenic interference with unknown ecosystems of the oceans and the deep sea are at the heart of the debate. This will likely also include the impact of deep seabed mining on the conservation of the marine ecosystems of the deep. However, this absence is understandable given the nature of the instrument that sought to urgently change the legal structure of deep seabed mining at the time of the entry into force of the Law of the Sea Convention. Little was known at the time about the wealth of biodiversity in distant oceans and at the deep seabed, but most of all that was not the perspective of the effort to formulate the Agreement.

The 1995 UN Fish Stocks Agreement[15] talks about conservation and management measures throughout, but does not refer to sustainability often. Article 2 though states that the objective of the Agreement is "to ensure the long-term conservation and sustainable use of straddling fish stocks and highly migratory fish stocks through effective implementation of the relevant provisions of the Convention". Article 5 on General Principles then speaks about MSY in paragraph (b) as does Annex II under point Two. Sustainable use of fisheries resources is dealt with in article 5(h) in the context of measures to prevent or eliminate overfishing and excess fishing capacity.

13 A/AC.287/2017/PC.4/2 at 8/19.

14 Agreement relating to the Implementation of Part XI of the United Nations Convention on the Law of the Sea of 10 December 1982; New York, 17 August 1994.

15 Agreement for the Implementation of the Provisions of the United Nations Convention on the Law of the Sea of 10 December 1982 relating to the Conservation and Management of Straddling Fish Stocks and Highly Migratory Fish Stocks; New York, 4 August 1995.

The other major treaty relevant to the marine biodiversity debate, the Convention on Biological Diversity[16] frequently uses the formulation "conservation and sustainable use". Leaving aside the preamble, article 1 distinguishes between the *conservation* of biological diversity in a general sense, and the *sustainable use* of its components in a more specific manner. At an early stage it is clear that a broad concept of biodiversity is what is aimed at with respect to conservation efforts, whereas sustainable use is a requirement specific to dealing with components of biodiversity. In this approach, both notions appear to have a distinct character as they aim at different aspects of biodiversity. While providing a definition of biological diversity and of sustainable use in article 2, the Convention does not define the notion "conservation" itself. Rather specific definitions are provided distinguishing between ex-situ and in-situ conservation. The definition of in-situ conservation is the more relevant one as a description of the aim for an Implementing Agreement on marine biodiversity: "the conservation of ecosystems and natural habitats and the maintenance and recovery of viable populations of species in their natural surroundings".[17] It does mean that biodiversity is protected by implication: protecting the ecosystems and the natural habitats, and the maintenance and recovery of viable populations in their natural surroundings implies conservation.

In article 5 CBD the obligation to cooperate is formulated in relation to the purpose of conservation and sustainable use of biological diversity. In dealing with in-situ conservation, article 8 (c) discusses the regulation and management of biological resources important for the conservation of biological diversity, with the aim of ensuring conservation and sustainable use.[18] It would seem that "conservation" is necessary for ensuring "conservation and sustainable use". This sounds somewhat circular. Understandably the approach to ex-situ conservation in article 9 appears to be about conservation only. Article 10 then discusses the sustainable use of components of biological diversity, stressing the need to integrate.

There are many more provisions in which the notions of conservation and sustainable use appear. However, they do not always appear together. Sometimes it is either the combination of "conservation and sustainable use", or

16 Convention on Biological Diversity; Rio de Janeiro, 5 June 1992.

17 The definition continues with "… and, in the case of domesticated or cultivated species, in the surroundings where they have developed their distinctive properties". This later part does not seem to be necessarily relevant to marine biodiversity.

18 CBD art.8(c) reads: "Regulate or manage biological resources important for the *conservation* of biological diversity whether within or outside protected areas, with a view to ensuring their *conservation and sustainable use*" [emphasis added].

"conservation" on its own. Sustainable use only appears on its own twice in the text of the Convention, in articles 7 (b) and 10(e).

The Cartagena Protocol on Biosafety[19] does refer to conservation and the sustainable use of biological diversity, and does so in tandem all the time—the two notions go together as one. Perhaps this is a shorthand version of the general biodiversity concern in the CBD, while the Cartagena Protocol is quite clearly a specialist instrument dealing with biosafety, that is to ensure an adequate level of protection in the field of safe transfer, handling and use of living modified organisms resulting from modern biotechnology. As such, this is a type of activity that—while related to CBD—is somewhat beyond the frame of the conservation of biodiversity.

Similar to its parent instrument, the Nagoya Protocol[20] also speaks of the conservation of biological diversity, and of the sustainable use of its components.[21] It is an instrument specifically aimed at access and benefit sharing. Hence, this implies a focus on the (sustainable) use of biodiversity. The instrument aims to provide guarantees for sustainability within the protective framework of CBD. It may be particularly relevant for the particular focus in the BBNJ debate on access- and benefit-sharing. These activities are by their nature not aimed at conservation only, but imply a commercial interest.

As to the notion of *sustainable* use, it is should be noted that before emphasis was put on "sustainability", other qualifiers had appeared in international instruments pointing in the same direction. The 1971 Ramsar Convention speaks about conservation, management and "wise" use of migratory water fowl, as well as the conservation, management and "wise" use of wetlands and their flora and fauna.[22] The Convention on the Conservation of Antarctic Marine Living Resources (CCAMLR) of 1980[23] states in Article II in the first paragraph that the objective of the Convention is the conservation of Antarctic marine living resources; and adds to the definition in paragraph 2 that "conservation" includes rational use. Consequently, the overarching aim is conservation,

19 Cartagena Protocol on Biosafety to the Convention on Biological Diversity; Montreal, 29 January 2000.

20 Nagoya Protocol on Access to Genetic Resources and the Fair and Equitable Sharing of Benefits Arising from their Utilization to the Convention on Biological Diversity; Nagoya, 29 October 2010.

21 See articles 1, 9, 10 and 22 (5) (h) Nagoya Protocol. Article 8 is the only mention of conservation and sustainable use of biological diversity that omits a reference to the components.

22 Convention on Wetlands of International Importance especially as Waterfowl Habitat; Ramsar, 2 February 1971; see articles 2 (6), 3(1), 6 (2)(d) and 6 (3).

23 Convention on the Conservation of Antarctic Marine Living Resources; Canberra, 19 May 1980.

which is defined as including rational (presumably sustainable) use. Interestingly, the word "use" only appears once in the treaty text, in this definition in article II (2). Paragraph 3 of article II then follows with general principles of conservation, based on notions such as the maximum sustainable yield, the ecosystems approach and the importance of sustained conservation. Rational use in the CCAMLR area, as part of the conservation requirement, is directly put in the context of contemporary (fisheries) management. The approach in this instrument is hierarchical; use is limited to what is rationally acceptable—beyond that conservation takes precedence, and use would no longer be legitimate. Clearly such a formulation requires further detail in secondary regulations, but the notion of "rational" use sets the direction for the interpretation of what is "conservation" within the meaning of CCAMLR. It is fair to assume that sustainability is both a wise and a rational choice, and that *use* of biodiversity may be possible, provided that care is taken of the maintenance of condition of reproduction of the species, as well as the protection of habitats and ecosystems.

While conservation as a goal is perhaps more straightforward, sustainable use has more dimensions to it. It is necessary to investigate what type of activities will need to be regulated in order to "sustainably" use marine biodiversity. To regulate marine biodiversity beyond national jurisdiction one needs to understand current and future types of activities to which the instrument will be applicable, and have an idea as to how invasive or detrimental to marine biodiversity such activities will be. Do we have a complete picture of future activities yet, or is that even possible at this stage? With respect to sustainable use the discussion so far appears to be focused on sustainable fisheries on the high seas, and the more fundamental (and highly political) question as to whether fisheries should or should not be covered by the new instrument. Will it be necessary to engage in a discussion on sustainable bio-prospecting, or can it be assumed that bio-prospecting will in general be an activity that will be sustainable anyway and will not *per se* endanger pristine ecosystems? Presumably, marine scientific research or bio-prospecting on the high seas are not, or need not be, intrusive and can be carried out in a sustainable manner. Will it perhaps be necessary to address futuristic ideas such as deep seabed tourism? Legislating for the high seas and the deep ocean implies envisaging future developments. Particularly when legislating for activities with an important technological component, such as human activities on the seabed or within the high seas, there is a need to understand the limitations of the technological knowledge of today and the speed with which technological change may have an impact on the international legal regulation of particular activities.

3 Conservation *and* (*or*) Sustainable Use of Marine Biodiversity?

Language is the main tool lawyers have to achieve what they set out to do, and the choice of language has implications. In multilateral negotiations lawyers spend long hours in drafting groups to find exactly the right words that express their intentions in the future instrument, language that will hopefully tie in with existing treaties and be as unambiguous as possible. While the instrument itself will in all likelihood be the product of a compromise between the different views around the table, the drafting group will try to formulate that text with such precision as to faithfully reflect the precisely formulated compromise and the intentions of the parties.

3.1 *Conservation* of Marine Biodiversity

There is a sense of instant understanding with regard to the need for conservation of marine biodiversity in areas beyond national jurisdiction. For anyone who knows about the Blob fish or the Dumbo octopus, it is obvious these animals and their habitats need protection, even if at this time knowledge about such species and their ecosystems may be limited. Conservation as the overarching goal is the key message of the NGOs participating in the debate.

Areas beyond national jurisdiction are far away and few people have been to the deep seabed. There is an increased awareness about the limited knowledge available about species and ecosystems, or the relationships between them. In fact, there is a growing understanding about how little we understand about the pristine and largely unknown world of the oceans, particularly at great depths.

From the overwhelming evidence of humankind's limited knowledge of the oceans, the precautionary approach appears to be the only rational standpoint to take when having the interests of marine biodiversity at heart as a lack of knowledge implies that conservation is a well-considered and rational strategy. When the potential consequences of human interference with the oceans and the deep sea are insufficiently understood and cannot clearly be predicted, a conservationist approach to marine biodiversity is well advised. Use must be conditional. It will only be acceptable when it can be determined beforehand that such use will be sustainable.

Thus, in the absence of sufficient knowledge non-interference is the better if not the obligatory approach. This "better safe than sorry" perspective is conceptually the simpler way to address the world's responsibility for marine biodiversity. Yet it hardly responds to concerns with respect to the difficulties of enforcement of international rules on the high seas.

3.2 *Sustainable Use* of Marine Biodiversity

The other element is sustainable use. Here again it is necessary to start with what the concept of "sustainable use" implies. Understanding "sustainable use" presumably requires making a distinction between consumption or one-time-only use (e.g., while harvesting, the object vanishes as a consequence of its use), and repetitive use. An example of the first would be fisheries where the sale and consumption of a single fish is one-time-only: the harvested fish disappears when used. However, if the stock as a whole is sustainably managed many fish may be harvested on a regular basis—this ambition is at the heart of contemporary fisheries management. Use may thus also have the meaning of *use and re-use*, utilization of a resource on the presumption of renewal. This is what the Convention on Biological Diversity speaks about in its article 2: "the use of components of biological diversity in a way and at a rate that does not lead to the long-term decline of biological diversity…" In this perspective, use in the sense of (commercial) utilization is possibly subject to the sustainable requirement.

It may well be possible to engage in bio prospecting in a relatively non-intrusive manner. Collecting a limited number of individuals of a species for analysis or genetic synthesizing may not be harmful to the marine ecosystem in itself and thus be sustainable. The intrusion in the habitat in the ocean may be of a limited scope, and further analysis and synthesizing of the components will take place in laboratories on land, where work develops disconnected from its marine origins. Marine scientific research, and indeed bio prospecting may be understood as having a limited effect in the water column and the seabed, but this will clearly depend on how collection and research take place. This may be a situation in which the goals of conservation and sustainable use of marine biodiversity coincide.

Sustainability is understood as a guarantee against long-term decline. It does not prevent or prohibit interference with marine biodiversity *per se*, but sets limits to scope and scale. It requires rational and wise management that includes monitoring whether conditions maintaining sustainability are adhered to. Sustainability is a relative goal—whether certain activities will be sustainable will depend on a number of factors. There is no absolute yardstick for sustainability, it is only specific with respect to species and requires amongst others scientific insights into the condition of species, locations and eco-systems. Particularly with insufficient knowledge, this is a challenging evaluation. It also highlights the question of whether "sustainable use" is a goal in itself or is rather a method of conservation. Other methods of conservation would be the use of closed areas and no-take zones, and whether they are on a par with sustainable use.

3.3 *Object and Purpose of an Implementing Agreement*

Lawyers and participating diplomats understand that the formulations chosen in the new instrument will determine whether the terms will be binding in the future. The Implementing Agreement will require specific and precise drafting not just to address concerns about marine biodiversity, but also precision about what is to be achieved, which activities are to be addressed and which concerns are to be remedied.

Once the treaty has entered into force, the interpretation rules of the Vienna Convention on the Law of Treaties (VCLT) will guide discussions about what has been intended by the drafters of the treaty. Lawyers will evaluate the end text in the light of the rules on the interpretation of treaties. Articles 31 and 32 VCLT are the yardstick used by courts and tribunals, and more importantly before that by States when considering ratification with a view not only to the domestic ratification process but also to future litigation.

Article 31 VCLT, entitled General Rule of Interpretation, reads as follows:

> 1. A treaty shall be interpreted in good faith in accordance with the ordinary meaning to be given to the terms of the treaty in their context and in the light of its object and purpose. ....
> 4. A special meaning shall be given to a term if it is established that the parties so intended.

Paragraphs (1) and (4) are of importance to the present discussion. Article 31 (1) VCLT is a standard rule of treaty interpretation and consists of three elements: good faith, ordinary meaning of the terms of the treaty in context, and object and purpose of the treaty. Assuming an interpretation to be in good faith, difficulties may arise with respect to the second and third element. If conservation and sustainable use are two distinct goals of the new instrument, they may in certain cases be contradictory. What may be required for conservation may restrict sustainable use, or any use at all. An example would be the establishment of no-take zones where extractive activities are prohibited for the sake of conservation and recovery. The context would have to identify whether or not "conservation and sustainable use" are to be read as one or two separate goals. If this is unclear, the next step in article 31 (1) is to refer to the object and purpose of the treaty. This again puts the relation between the notions of conservation and sustainable use in the spotlight. How are these to be understood and how are they related? Apart from being curious, there is a need to understand (or indeed pre-empt) the future interpretation of a treaty at the time of drafting.

Paragraph (4) of article 31 leaves the possibility open that expressions in a treaty may have a special meaning if parties so intended, and if this can be established. Clearly, the debate has not yet come to a formulation of definitions in the future instrument and at this stage it would appear that no special meanings have been intended by the delegations participating in the preparatory committee. However, it may be necessary to consider this at some point.

4 Possible Relationships between Conservation and Sustainable Use of Marine Biodiversity beyond National Jurisdiction

Thus, the question arises whether sustainable use is subordinate to the aim of ensuring conservation of marine biodiversity. Is conservation of marine biodiversity beyond national jurisdiction the main goal of the Implementing Agreement, and is sustainable use a way of achieving this, perhaps also as a reflection of the impossibility to prevent use in general? In that case, the idea would be that conservation of marine biodiversity may be attained by either sustainable use (the methods of which will be described in the future agreement), or no-use (as in no-take zones and similar instruments). Environmental Impact Assessments (EIAs) might provide an organized and science-based evaluation of marine biodiversity in certain locations. This would serve the goal of conservation in that such analysis would indicate whether sustainable use were possible. Marine Protected Areas could also serve the aim of conservation by either regulating sustainable use, or at times prescribing no-use or other measures specific to that area.

Let us go through a number of alternative relationships between these two important notions to better understand possible relationships.

"Conservation and sustainable use" as elements of a single goal

When "conservation and sustainable use" go together as one single goal, the presumption is that it is possible to attain both the conservation and the sustainable use of marine biodiversity during activities in areas beyond national jurisdiction. This reading of the formula would require a specification or definition of the notion of conservation in order for it to be useable. Presumably, the interpretation of conservation would not be leaving everything untouched and in a pristine condition, as that would by definition exclude sustainable use. If "conservation and sustainable use" go together as though they are Siamese twins this might create a high level of protection at the expense of possibilities to (commercially) use the marine biodiversity in areas beyond national jurisdiction.

"Conservation" or "sustainable use" as alternative goals served by the new instrument (not necessarily at the same time)

In this reading "conservation" and "sustainable use" are two distinct, and parallel goals. Perhaps sustainable use implies conservation, but conservation may not always imply that use would be possible, even if it were sustainable. It is quite the reverse. The conservation requirement may lead to the conclusion that use would be unsustainable *per se*, or unsustainable during a certain period given the vulnerability of a particular area or species and its deterioration or more in general because insufficient information is available to understand the consequences of anthropogenic activity. Even if use would be required to be sustainable, conservation would preclude an absolute guarantee that sustainable use would always be possible. It may be concluded that this reading of the correlation between both notions leads to bizarre consequences and must be considered as unlikely.

Only when "conservation" is guaranteed will "sustainable use" be possible, sustainable use is conditional

Another way to read the conservation and sustainable use notion is that these are consecutive, sequential requirements. First, it will need to be established that a particular resource or area is sufficiently protected and that conservation efforts have been successful, before any thought about sustainable use can be entertained. Sustainable use is then conditional upon the prior fulfilment of conservation requirements. At this time, such a sequential reading is not based on a specific text, but rather on interpretation of the debates. This would not necessarily follow from the ordinary meaning of the text, but it may be argued that this can be based on context and object and purpose unless of course the new instrument would spell this out.

Conservation may imply situations in which use is acceptable, or is acceptable under certain conditions. Environmental Impact Assessments play an important role in determining the possibility of sustainable use, in the sense of evaluating the carrying capacity of particular areas in relation to different types of activities. Tools such as (a system of) Marine Protected Areas may lead to some MPAs containing closed, or no-take areas necessary for the recuperation of certain ecosystems and habitats. But equally, MPAs may contain sets of more or less restrictive conditions under which use is possible. Use will not always be possible, whether sustainable or not.

Seen in this light, conservation is the superior goal of the marine biodiversity effort, and sustainable use is one of the methods through which the aim of conservation is achieved (but not just the only one). Both elements are not equal, conservation is at the heart of the effort, with sustainable use being possible depending on the specific situation in a specific area. Conservation is an obligation, but sustainable use is not a given. This begs the question whether in the long run such sequential criteria are not in fact hierarchical. If conservation is the decisive element, sustainable use may very often not be possible. This may be what some seek to achieve, but certainly not all delegations.

Conservation ≈ sustainable use, the relationship is unclear as it is a compromise formula

Yet, another way of reading the "conservation and sustainable use" formula, perhaps more cynically, would be to see this as a compromise underlying the cooperation between the Group of 77 and China with the European Union and some other Western States in building their coalition in support of a new instrument on marine biodiversity. States in this alliance agree that a regulatory gap exists with respect to the conservation of marine biodiversity, as opposed to a smaller group of States arguing the existence of an implementation gap only. While conservation of marine biodiversity was key to the EUs concerns, many in the G77 may have been increasingly concerned about sustainable use, and in particular the importance of marine genetic resources and the consequent access and benefit sharing. During debates in New York this led to reiterations about marine biodiversity beyond national jurisdiction belonging to the Common Heritage of Mankind,[24] referring to the regulatory system created for the Area by the Law of the Sea Convention. In doing so, reference is made to access and benefit sharing mechanisms created through the International Seabed Authority with respect to the non-living resources in the Area, and under the Nagoya Protocol. The prospect of (commercial) benefits from marine biodiversity in areas beyond national jurisdiction is a fundamentally different driver in the debate than the conservation perspective. While it is easy to see that discussions about marine genetic resources, their potential financial importance and related issues concerning intellectual property rights will be

24 See articles 136 and 1 (1) UNCLOS. See also A/AC.287/2017/PC.4/2 at 17/19: "With regard to the common heritage of mankind and the freedom of the high seas, further discussions are required".

hugely complicated, it is necessary to understand this as one of the messages behind the emphasis on the (sustainable) use.

5 Conclusions—Why Does This Question Matter?

The current debate before the start of a diplomatic conference on an Implementing Agreement under UNCLOS on marine biodiversity beyond national jurisdiction is generic and does not seem to distinguish between conservation and sustainable use. The importance of the notion(s) of conservation and sustainable use is taken for granted, without too much concern about the differences between the two concepts or their relationship.

Delegates have more or less been going through a series of headings of issues since the start of the BBNJ working group in 2006. During the 2016–2107 Preparatory Committee, discussions became more specific, and a lot of detail was added to the different headings of marine genetic resources, area-based management tools, environmental impact assessment, marine scientific research and bio prospecting as well as capacity-building and technology transfer. However, none of this is focused on whether specific activities would fall into either the conservation or the sustainable use category or indeed whether such a distinction is necessary. Also, no consensus has so far been reached on the content of the future instrument, and the Preparatory Committee did not present a draft of a future instrument for lack of consensus.

It is fair to say that—as is often the case during negotiations—delegations participating in the marine biodiversity discussions are driven by different considerations as to what the aim of the effort is, and indeed what they stand to gain from a new implementing agreement. For some, the overarching consideration is the protection of the marine biodiversity of the high seas, and indeed the flora and fauna of the deep seabed. That perspective focuses on the conservation of marine biodiversity. For others, the possibility of benefits from the genetic resources of the high seas is a looming prospect. These would be positions more on the side of sustainable use. Synthesizing genetic resources only requires a few of the species and this need not be a very intrusive activity.

The concept of "conservation and sustainable use" appears to have been taken as a shorthand version for "what this instrument will be about" without questioning its precise meaning. There are good reasons to discuss the relationship between the two fundamental notions at the heart of the marine biodiversity debate. These are core concepts in the discussions in the past in the BBNJ Working Group as well as recently in the Preparatory Committee. Conservation and sustainable use are likely to be considered the object and purpose of the new instrument. In the law of treaties, the object and purpose of a treaty increasingly takes prominence when interpreting a treaty. Consequently, it stands to reason to take a step back and consider what really is at the core aim of the treaty—already during negotiations. Art. 31(1) VCLT will be the tool used with respect to the interpretation of rights and obligations created in the Implementing Agreement, and it is prudent to obtain clarity as to how the drafters are seeing these core concepts.

To be more precise: is the sustainable use of marine biodiversity in areas beyond national jurisdiction a goal in itself or a rather a method towards furthering conservation? Views on that question are likely to differ. As the drafting of an Implementing Agreement is moving from the preparatory committee to a diplomatic conference, the importance of clarity of formulations should not be underestimated. And so the question remains: what do we talk about when we talk about conservation *and* sustainable use of marine biodiversity?

CHAPTER 3

Toward Seafood Resilience: How to Achieve Sustainable Fisheries Development

*Melda Kamil Ariadno**

Abstract

Indonesia has the goal of being a maritime fulcrum and to establish itself as a true maritime country. One of the biggest challenges is how to build national resilience of seafood and in doing that there must be a firm policy in developing sustainable fisheries in the country. When we talk about sustainable fisheries, it needs more than just a good government with good fisheries policies, but also a strong government with strong law enforcement. When Indonesia starts talking about how seafood is the main diet of the Indonesian people, how our welfare is tied to our coastal communities, how significant Indonesian contributions are to the world fish market, how our marine environment is clean and healthy, how well-guarded is Indonesian marine biodiversity, and how Indonesia combats IUU fishing successfully, then we will achieve the objective of being a maritime fulcrum.

1 Introduction

On 13 November 2014, the President of Indonesia, Mr. Joko Widodo, gave an interesting speech during the 9th East Asia Summit in Myanmar. The message conveyed to the participants of the Summit was clear that Indonesia aims to reach its ideal goal as the largest archipelagic State in the world to be a "maritime fulcrum", i.e., a country which is developed based on its maritime activities in various aspects of life. This intention is not easily accomplished because of various existing conditions in Indonesia that are far away from even a maritime country.

---

* Professor of International Law at the Faculty of Law Universitas Indonesia. She holds an LL.B. from Universitas Indonesia, and an LL.M. and Ph.D. from School of Law University of Washington, Seattle. Her expertise includes the law of the sea, the law of treaties and international environmental law. This paper is a revised version of a Professorship Lecture. Special thanks go to my dedicated students, Aldi and Diena. The author's PowerPoint presentation is available at http://www.virginia.edu/colp/pdf/yogya-ariado.pdf.

 | DOI:10.1163/9789004366619_008

Major issues being stressed to achieve Indonesia's aim are Indonesia's willpower to rebuild maritime culture, build seafood security, build maritime infrastructure and connectivity, build maritime cooperation as well as to form supremacy in maritime defense. These aims are also known as the "five grand designs" of the maritime fulcrum.

Such willpower has a clear consequence, not least to mention that the majority of Indonesian government regimes in the past did not put major attention on the sea for a long time. Accordingly, the "grand design" on marine policy was just launched by the Presidential Regulation No 16 of 2017, 60 years after we declared Indonesia an archipelagic State. If we try to look at the primary plan of Indonesia to be a "maritime fulcrum" then the fact that Indonesia is the largest archipelagic State in the world unfortunately does not guarantee the success of Indonesia to be a "maritime fulcrum". This paper endeavors to identify the hindrances for Indonesia to build up itself as a maritime country, especially in building seafood security. The paper also explores the main challenges to be overcome and, finally, discusses some strategic recommendations to enable Indonesia to achieve seafood security.

2 Current Situation and Condition

Indonesia is the largest archipelagic State in the world, consisting of more than 17,000 islands, and has the second largest coastline after Canada. Indonesia has an extensive ocean space of almost six million km2, which includes its archipelagic waters, territorial sea, and Exclusive Economic Zone.

Indonesia's archipelagic status started from the creation of the Djuanda Declaration 1957 (13 December 1957). This was a unilateral action of Indonesia claiming its surrounding waters, in between and waters connecting islands within (without looking at the width of the water area). Indonesia created a new regime that did not get any recognition. Indonesia had to fight to get recognition during the First United Nations Conference on the Law of the Sea in Geneva, which established four Conventions of the Law of the Sea. However, Indonesia failed to obtain recognition on its claim. Indonesia did not stop there, for in the year 1960, Indonesia formulated Law No. 4 Year 1960 confirming unilateral action against claims on Indonesian internal waters where Indonesian baselines are drawn from the outermost point of the outermost islands. The waters in between are considered internal waters. Indonesia's claim was not recognized or acquired by the international law of the sea: baselines could only be drawn from each island's low water line. Internal waters are landward from the baselines and the territory towards the sea is called the territorial sea.

Diplomatic struggles continued during the negotiations of the Third United Nations Conference on the Law of the Sea. The negotiations lasted more than ten years until they produced a very comprehensive convention on law of the sea, known as the 1982 United Nations Convention on the Law of the Sea (1982 UNCLOS). The Convention was ratified by Indonesia through the enactment of Law No. 17 Year 1985, which obtained *de jure* international recognition of the concept of archipelagic State that had been created and claimed by Indonesia unilaterally for 25 years.

A large body of water should actually enable Indonesia to become a powerful maritime country with such a large "territory" of sea, almost triple the size of its land. The water area has extraordinary living and non-living resources. International shipping also places Indonesia at the heart of an economic intersection of the world since the sea is navigable by a fleet engaged in world trade since hundreds of years ago. These combined factors should make the aspiration of President Joko Widodo not difficult to achieve. Unfortunately, we have to acknowledge the fact that Indonesia is not yet a maritime state and becoming a maritime fulcrum is a big challenge for Indonesia to achieve.

The above five grand designs contain a large agenda and careful readiness of Indonesia is required in order to make it happen. Some weaknesses of Indonesia relating to those five issues especially those related to fisheries are explained below:

1. Rebuilding maritime culture is not an easy task such as turning the palm of one's hand. The people of Indonesia have long been exposed to a mainland perspective, the most important of which is to build on the land. That is, all interests must start from the interest of the land, particularly the mainland. The sea is deemed as the uniting factor of the land but it is nonetheless only seen as the object not as the subject. Everything has departed from the interest of the protection of the land and the people there and the sea have to support all of those purposes. The sea is never deemed as the strategic power that Indonesia should build up to every aspect of its potentiality. Indonesian people are not accustomed to see themselves as a coastal community except by those who live by the coast. Thus, the sea is never deemed as part of people's lives that needs to be guarded and developed sustainably. The young generation are not even taught about the concept of archipelagic State. They heard about "wawasan nusantara" but they do not really understand what the rights and obligations that result from the recognition as an archipelagic State. The Law of Sea that is supposed to be understood by the key officials of government bodies, law enforcers, sovereignty guardians, as well practitioners in the marine fields. But the understanding is not properly

addressed or comprehended. In many law schools, for instance, the law of the sea is not a mandatory subject, but merely the subject for those taking international law specialization. The sea has been treated as the backyard where garbage is thrown out. Coral reefs are not well protected, sea water is polluted as the biggest trash can, and marine tourism is rarely selected as a favorite tour destination.

2. Building seafood security is an ideal condition. This condition needs to be embedded by establishing seafood as a staple food for the people of Indonesia. This means building on the habits of the public to eat fish as a main course in their everyday life, or as a "daily diet". Such a habit might not be a problem for the coastal people, but many Indonesian people live in non-coastal areas and do not have the habit of eating fish. This situation contrasts with the strong habit of consuming fish at other States consisting of islands, among which are Japan and Maldives. In 2011, the data reported by the FAO describes the level of fish consumption per capita in Indonesia as 12.8 kg consumption per capita, with differences for each region, from southern Maluku consuming as much as 24.6 kg per capita to Yogyakarta consuming 4 kg per capita. More than 70 percent of the fish consumed is derived from marine fish (the fish obtained from the mainland reached 25 percent).[1] The unavoidable reality is that even though Indonesia is one of the world's largest tuna exporters, it is unfortunate that Indonesia does not consume tuna as its daily fare. As a social unit, the fishing communities live, grow, and thrive in coastal areas.[2] People in coastal areas of Indonesia have the properties of a particular characteristic or unique characteristic, related to the nature of business in the field of fisheries itself.[3] This is due to the fact that the people who live in coastal areas generally suffer from dependency on environmental conditions, season, and the market.[4] Local wisdom has been growing, especially in *Adat* law community, but has not yet gotten a place as a source of law even though their existence has been recognized by the Constitution and other laws related the sea.

1 Food and Agriculture Organization of the United Nations, *The State of World Fisheries: Opportunities and Challenges* (Rome, 2014), p. 153.

2 Kusnadi, "Kebudayaan Masyarakat Nelayan" (disampaikan dalam kegiatan JELAJAH BUDAYA TAHUN 2010, dengan tema "Ekspresi Buadaya Masyarakat Nelayan di Pantai Utara Jawa", Balai Pelestarian Sejarah dan NIlai Tradisional, Kementrian Kebudayaan dan Pariwisata, Yogyakarta, 12–15 Juli 2010).

3 Yudi Wahyudin, "Sistem Sosial Ekonomi Dan Budaya Masyarakat Pesisir" (disampaikan pada Pelatihan Pengelolaan Kawasan Konservasi Perairan, di Kampus Pusat Diklat Kehutana, Institut Pertanian Bogor, Bogor, Jawa Barat, 5 Desember 2003), 2.

4 *Ibid.*, p. 2–4.

3. The development of maritime infrastructure and connectivity has actually been long proposed, including the enactment of a national shipping "cabotage principle". This provides that every point in shipping lines in Indonesia should only be served by the national fleet. As has long been known the "supply and demand rule" does not apply between islands in Indonesia. In the field of fisheries, most of the biggest fishing ports are located in Java; they are not in the outer parts of Indonesia. It is a big challenge to export fresh tuna from Indonesia to Japan, the European Union and the United States since exports mainly are done from the inner part of Indonesia. This means that the Indonesian export of our tuna fishery is not competitive. Similarly, the fish processing companies are mainly located in Java while most of the tuna fishing resources are caught in the Indonesian EEZ. Fishing ports are barely equipped with necessary facilities such as ice storage, quality control facilities, packing and handling facilities as well as nearby airports to accelerate the exporting of fresh tuna that is the most high value commercial fishing commodity.
4. Cooperation in the marine sector can only be done effectively if there is mutual respect and common interest within the State. In modern international relations, countries always give priority to diplomacy in solving problems together within a framework of cooperation and mutual complementarity. The potential of maritime diplomacy can no longer be avoided to find a solution in helping Indonesia. Currently, maritime diplomacy is a significant part of President Joko Widodo's agenda as a tool to solve major problems on the sea, namely illegal fishing, violation of the state border, border disputes, piracy and marine pollution. One of Indonesia's efforts in maritime diplomacy relating to fishery is to participate actively in the Indian Ocean RIM Association (IORA). IORA itself is an entity that has a focus related to defense and maritime security, trade and investment facilities, fisheries management, disaster control management, academic cooperation, technology and science, and cultural exchanges as well as tourist access.[5]

Indonesia has ratified the 1995 UN Fish Stocks Agreement (UNFSA). Indonesia has joined three Regional Fisheries Management Organizations (RFMOs), these are the Indian Ocean Tuna Commission (IOTC), Commission for the Conservation of Southern Bluefin Tuna (CCSBT), and the Western Central Pacific Fisheries Commission (WCPFC). The RFMOs have set up global and regional measures related to the use of tuna as a highly migratory species. Indonesia

5 Mervyn Piesse, *op. cit.*, p. 4.

has adopted the International plan of Action on Illegal, Unreported and Unregulated Fishing (IPOA-IUU) and also is very active in the Regional Plan of Action (RPOA) to Promote Responsible Fishing Practices Including Combating IUU fishing in the Region (RPOA-IUU).

Indonesia also has promoted the protection of coral reefs through the Coral Triangle Initiative (CTI). It was developed to establish mechanisms for cooperation among the countries which have the purpose and the same view regarding the management of the environment and maintaining the sustainability of natural marine resources in the coral triangle. The triangle includes six countries: Indonesia, Philippines, Malaysia, Timor Leste, Papua New Guinea, and Solomon Islands.[6]

In regards to the protection of the crew, Indonesia has acceded to the 1978 International Convention on Standards of Training, Certification and Watchkeeping for Seafarers in 1987 (1987 STCW) aimed to protect crew working in a commercial ship. Indonesia, however, still has not ratified the 1995 International Convention on Standards of Training, Certification and Watchkeeping for Fishing vessels personnel (1995 STCW-F), which is the same legal instrument but specifically for the crews of fishing vessels.

5. Forming supremacy in maritime defense is the most coveted element by most Indonesian law of the sea observers. As the largest archipelago, building supremacy in maritime defense and security is necessary. Most of our borders have been agreed upon. Those still under negotiation, however, include the border of the EEZ of Indonesia and Malaysia in the straits of Malacca, and the border between Indonesia and Malaysia in East Kalimantan. The latter has resulted in the arrest of one another's fishermen of both countries in the conflict over maritime areas. The sea is supposed to be secured from any illegal acts and safe to navigate. The designated law enforcers in the field of fisheries (Navy, Water Police, Civil Servant Investigators) are now equipped by the Maritime Security Agency (Badan Keamanan Laut/BAKAMLA), as established under Law No. 32 Year 2014 on Marine (Marine Act).

One of the means used by the Indonesian Government to settle disputes on maritime boundaries and territory is by conducting Border Diplomacy. Border Diplomacy is an implementation of the foreign policy of dealing with bound aries issues including delimitation of land-water territory as well as management issues of international dimension.[7] Indonesia is located adjacent to

6 *Ibid.*

7 Arif Havas Oegrosono, *Kebijakan Dasar Indonesia Dalam Penetapan Perbatasan Maritim* (Jakarta: Departemen Luar Negeri Republik Indonesia, 2006), p. 13.

ten neighboring States: Australia, Malaysia, Philippines, Singapore, Thailand, Timor-Leste, Papua New Guinea, Vietnam, India, and Palau.[8] Some of the issues on maritime boundaries that are still being disputed and unresolved (not yet settled), among others are:[9]

a. Malaysia, Singapore, and Timor-Leste for Territorial Sea;
b. India, Thailand, Malaysia, Vietnam, Palau, Timor-Leste and Australia for EEZ;
c. Malaysia, Philippines, Palau, and Timor-Leste for Continental Shelf.

There are many conflicts relating to fisheries due to the disputed areas within boundaries.

3 Main Challenges

A few challenges that Indonesia needs to answer regarding the proposed "maritime fulcrum grand design" particularly relating to fisheries are as follows:

3.1 *In Re-establishing a Maritime Culture*

Maritime perspective needs to be upheld by all parts of the nation and should be enshrined in state documents to provide directives for all state units to achieve its aims. Although the focus of maritime infrastructure development has been incorporated in the development agenda of Jokowi-Jusuf Kalla, it still needs to be proven how and when the targets will be achieved. Maritime development requires a lot of cost, availability of sufficient technology, and long-term commitment. However, "quick wins" that will accelerate and trigger a snowball effect in the re-establishment of a maritime culture in Indonesia needs to be searched for.

What needs to be understood is that culture is not changed as an instant overnight result. It will be formed when cultivated since early on and practiced continuously in the society. Then the young generation consisting of students will understand Indonesia's position as the world's largest archipelagic nation and the responsibility as well as the challenges following that status.

One of President Jokowi's agenda is to prepare human resources regarding the maritime world by mainstreaming maritime knowledge in the education process. Indonesia also needs to prepare expertise in many maritime sectors,

8 http://www.bphn.go.id/news/42/SEMINAR-TENTANG-MASALAH-HUKUM-BATAS-LAUT-INDONESIA.

9 Identifikasi Segmen-Segmen Batas Maritim antara Republik Indonesia dan 10 Negara Tetangga, Kementerian Luar Negeri RI, as of June 2015.

starting from technical, technological, to strategists and international maritime law experts. Government also needs to emphasize that Indonesia needs to strengthen Maritime Domain Awareness (MDA). This be achieved when studies regarding the seas and the laws that regulate it are not integrated in the national curriculum, from primary, secondary, till tertiary education. The seas cannot be taken for granted as a resource from God to be utilized. The sea requires attention and proper management to ensure a sustainable pool of its resources. The sea cannot be considered only as a large garbage can. The assumption that the seabed and the grounds beneath it can withstand so much pollution must be eliminated. The reason is that the sea is the habitat of fishes, which humans will consume. The Minimata tragedy in Japan must not be forgotten, nor the Buyat case in our country.

3.2 *In Building a Seafood Security*

Indonesia has the largest fish resources in Southeast Asia but it has not made Indonesia the largest exporter of seafood in Southeast Asia. Thailand is the "king of seafood" and "the biggest producers of canned tuna in the world" along with General Santos in the Philippines. But, is it true that their fish raw materials come from Indonesia? If it is done through legitimate trade channels, that would not be a problem. But what about the fact that Indonesian water territory and EEZ have been drained by "Indonesian" flag vessels which do not have a "genuine link" with Indonesian and an Indonesian legal entity? Of course this cannot be allowed so Indonesia should adopt the policy of a "licensing moratorium". Such a moratorium is certainly a good step but it should be followed by a bolder and more resolute step in which Indonesia does not rely on ex-foreign ships with fraudulent intent. The effort in combatting IUU fishing by sinking particularly foreign fishing vessels is necessary to create a deterrent against the theft of Indonesian fishery resources. This policy however needs to be equipped with an integrated standard of procedure that will be the guidelines for any law enforcers at sea as well as for the enhancement of proof of law infringement at sea that needs to be singled out from that on land.

If Bagan Siapi-api was once the center for the construction of fishing vessels, if Bugis can make a ship that sailed the oceans to Madagascar, then why isn't Indonesia currently developing itself at the center of fishing vessels construction? An increase in capacity could lead to over exploitation, and the challenge is how to develop fishing vessels equipped with environmentally friendly technologies.

High quality fish should also be enjoyed freely in the domestic market and not be only for export purposes. The Government of Japan has provided a subsidy to fishermen through output subsidy that guarantees the price of fish is not impacted by market fluctuations. The Indonesian government can certainly do the same without being justly accused of violating international trade laws.

This plan is to support fishermen in improving the quality of yield by intervening in the post-capture process to produce a variety of products including fish that can be favored by the people who live far from the sea.

3.3 *In Developing Maritime Infrastructure and Connectivity*

Indonesia must succeed in creating a development plan in all areas, not just in the west but also in the east. This will create sea traffic ensuring mutual trading so that the prices of basic commodities will be evenly distributed throughout Indonesia. Not only that, but because many islands are on the border and susceptible to influence from neighboring countries, international ports should be built and developed in accordance with the available resources. An example is the Port of General Santos developed by the Philippine Government for tuna fish and pineapple products.

If we reflect on the United States, it is seen that they develop all corners of the region by establishing large international ports in the north, south, west and east (Seattle, New York, Los Angeles and Florida). Fishing ports need to be built and developed in resource areas such as in Sabang, Natuna, Bitung, Biak, Pelabuhan Ratu, Bali, etc. Thereafter, the products generated by Indonesia, both goods and fishery products, can be immediately served and sent to various parts of the world without having to go through Singapore as now it is currently done. Indonesia is only the feeder of Singapore.

The most important thing, however, is for Indonesia to secure its fisheries interests from the negative impact of the international navigation done by foreign merchant and fishing vessels. Indonesia must take any precautionary steps to prevent the negative impact of international navigation damaging the Indonesian fishery sector. Sea transportation should not damage the marine environment particularly in fishing reserves area such as spawning grounds. Both foreign and local vessels should be threatened by severe sanctions if they litter at sea and cause marine pollution that eventually will harm the fishery resources.

3.4 *Cooperation in the Maritime Sector*

Indonesia should be a "leading actor" in the field of fisheries, because Indonesia has one with the biggest fish resources, at least in the Asia Pacific region. When "management measures" have been established in the global and regional order it is necessary for Indonesia to implement them at the national level, both in legislation and in practice in the field. This includes "combatting illegal, unreported and unregulated (IUU) fishing" and "transnational organized crime related to Fishing Vessels". This is to carry out Indonesia's obligations as a coastal State, flag State, and port State. In preventing IUU fishing practices, Indonesia should ratify the 1993 FAO Compliance Agreement, just as Indonesia has ratified the 2009 Port State Measures Agreement.

While Indonesia strictly regulates ships, Indonesia must also be prepared to protect and regulate Indonesian citizens who work as crews on fishing vessels. Indonesia should consider its participation in the STCW-F so that Indonesian crews scattered in various foreign fishing vessels can be well-protected and secured of their skills. The vessels should at least meet the standards of "seaworthy", "catch worthy" and "store worthy".

The fact that Indonesia has ratified UNFSA and is a member of the IOTC, CCSBT, and WCPFC does not by itself make Indonesia as a responsible state for sustainable fisheries, particularly for tuna. It is necessary to ensure implementation in the field and the creation of a common understanding among actors of fishery management, business sector and law enforcement in the field of fisheries. IUU fishing has become a global enemy and fishing vessels are often used to carry out transnational organized crime such as "people smuggling", "drug trafficking" and "human trafficking" including "forced labor". These practices are known as crimes related to fisheries.

3.5 *In Building a Maritime Defense Power*

The Government of Indonesia should control its maritime territory either by the Navy, the relevant agencies and BAKAMLA that is established under Law No. 32 of 2014. Every inch of the sea must be secured both as the sovereign territory of Indonesia (internal waters, archipelagic waters and territorial sea) or under the Indonesian jurisdiction (contiguous zone, EEZ and continental shelf).

Indonesia must keep entrances not only on the agreed borders but also to show a "peaceful display of sovereignty" in areas that are still in dispute with neighboring countries. Indonesia should be able to secure the sea, especially from the IUU fishing threat done by foreign fishing vessels in disputed areas. Indonesia needs to undertake surveillance in every part of Indonesian ocean spaces including those of the Indonesian EEZ and particularly within the international routes of navigation in Indonesian waters.

4 Recommendation

Finally, this paper tries to recommend steps that Indonesia must take to become a maritime fulcrum particularly related to fisheries:

1. **To rebuild the maritime culture**

In rebuilding the maritime culture, Indonesia must be able to perform strategic actions such as, among others:

a. Fostering awareness regarding the concept of an archipelagic State as early as possible, starting from primary education using methods that fit its level of absorption;
b. Making the understanding of maritime law mandatory at least on the university level and for government officials, especially those working in sectors related to the sea;
c. Protecting and developing the traditional knowledge of the coastal society, especially for those who are part of the *adat* law community, regarding responsible and sustainable fishing and adapting with climate change;
d. Institutionalizing the *adat* law that has been used for generations by codifying it into an authentic legal basis, and making it as "best practices" that can be brought forth to regional, national, and international forums;
e. Implanting communal culture related to marine utilization as a main culture; and
f. Fostering societal awareness to protect the sea and to prevent marine pollution or marine environmental destruction.

2. **To establish seafood security**

The necessary steps to establish seafood security are, among others:

a. Fixing the fisheries data gathering system, so it becomes reliable and accountable;
b. Fixing the licensing system, so it becomes effective and guarantees legal certainty by putting forward responsible and sustainable domestic fisheries;
c. Fixing the ship registration and flagging system so it will weigh not only formal truth such as formal requirements, but also material truth. In efforts to prevent "reflagging" practices, Indonesia must consider participation in the 1993 FAO Compliance Agreement;
d. Developing a national fishing ships manufacturing industry that is internationally qualified by updating fishing ships facilities with advanced technologies and avoiding the use of ex-foreign ships unless due to a very strong reason or due to a "genuine link" that is materially proven to exist between the ship and its owner, who must be an Indonesian citizen or an Indonesian legal entity;
e. Not opening chances for foreign industries to invest capital in the fish-gathering sector, and only allow them to move in the processing and marketing sector of the industry if: the national industry is insufficient, it

obliges "transfer of knowledge and technology", and if it is of significant benefit for the national income;

f. Expanding Indonesia from being the world's biggest producer of fresh, frozen, and processed fish to also become the world's biggest producer of canned fish;
g. Implementing responsible and sustainable fishing methods, including regulating fishing grounds, open/close season, marine protected areas, fishing gear, by-catch, on vessel storage and handling;
h. Developing functional fishing ports to expedite the landing of fish catches and fish processors equipped with cold storages and effective fish packaging and marketing facilities;
i. Guaranteeing the availability of infrastructures for fish distribution, which includes making fish catch landing points and fish catch delivery points closer to each other;
j. Increasing attractiveness for the domestic fisheries market;
k. Guaranteeing the domestic supply of the best quality fresh and processed fish;
l. Creating innovation in fish-based foods;
m. Continuously campaigning fish-consuming culture, which includes developing fish-based cuisine in the archipelago;
n. Protecting the interest of small scale fishermen and small scale fish workers by increasing their capacity in order to increase their quality of life, which includes expediting the trade chain so they avoid middlemen ("*tengkulak*") that monopolize fish catches and marketing;
o. Integrating the principle of sustainable utilization of fish resources in every national or local level policy as well as between related ministries;
p. Preventing and abolishing "Illegal, Unreported, and Unregulated (IUU) Fishing" by increasing the role of ports to prevent the entry of catches that fail to produce certificate of origin ("traceability") and results of "trans-shipment". To do that, Indonesia must ratify the *2009 Port State Measures Agreement*;
q. Creating national provisions regarding a more effective Monitoring, Controlling, and Surveillance (MCS), which includes synergizing our surveillance capacity that is currently being spread apart in multiple institutions (PPNS, TNI AL, Police, BAKAMLA);
r. Putting forward security coordination to cover larger areas in favor of sectarian ego such as "single agency multi-tasks" or "multi agencies single-task".
s. Making an integrated database supported with the latest technology regarding the threats to the fisheries resources ("illegal fishing" means) that could help create more immediate and effective countermeasures;

t. Creating criminal sanctions that induce a deterrent effect, which includes legitimizing the act of "burning and sinking both local or foreign vessels within the archipelagic waters, territorial waters, and EEZ", as last resort punishment ("Ultimatum remedium") against "Illegal Fishing", as long as there is a clear SOP to regulate its implementation to avoid violation of human rights; and

u. Developing a "prompt release upon bond posting" system that suits the characteristic of fish catching in Indonesia by drafting a clear SOP as mandated by the law.

3. **To develop maritime infrastructure and connectivity**

Several steps that need to be taken are:

a. Strengthening Indonesian fishing vessels by developing the national commercial vessel manufacturing industry that is highly competitive in quality so that Indonesia can be independent in manufacturing fishing vessels while qualifying for the international standard of seaworthiness;

b. Securing the sea from any form of navigational safety and security threats, which includes improving the navigational support and safety patrol facilities at sea; and

c. Increasing the capacity of fishing ports to become an international port that fulfills the criteria for the international fishery trade.

4. **To establish cooperation in the maritime sector**

Several steps that need to be taken are:

a. Cooperating to increase the safety of navigation, protection of the marine environment, prevention and countermeasures for marine pollution and maritime security;

b. Cooperating in both regional and international levels to protect Indonesia ship crews working in both Indonesian and foreign vessels by applying international standards, some of which are stipulated in the *1978 STCW* and *1995 STCW-F*;

c. Initiating the birth of new regional and international norms and "best practices" that can be followed by other states in the field of fisheries and sea voyage; and

d. Actively participating in fisheries MCS in high seas through RFMOs.

5. **To build the strength of maritime defense**

Several steps that need to be taken are:

a. Making sure that the definitive state borders are not breached by neighboring states;

b. Conducting active border diplomacy even before problems arise especially those correlated with fishermen's activities near the borders; and

c. Completing all pending border disputes while putting forward national interests.

5 Conclusion

A lot of homework must be done and it is impossible to specify all one by one in this paper. The big steps have been laid out, however, by the Indonesian government which is expected not to end as mere slogans without any realization in future government programs. The government along with the House of Representatives must be able to realize this vision and mission to become a world maritime fulcrum starting by fixing problems internally and expanding its influence externally.

Indonesia must be optimistic that with the huge modality of being the biggest archipelagic State in the world, but which is not obtained easily, Indonesia must become a strong and advanced maritime state. A strong and consistent political will should be generated that will not be changed between presidencies. This is not only the dream of President Joko Widodo, but also the dream of Indonesia, that everyone sitting in the government should fight for. The motto of the Indonesian Navy is just perfect, Indonesia has to reign over the sea because the sea shapes Indonesia, JALESVEVA JAYAMAHE!

References

Books

Adisanjaya, Nyoman Ngurah. "Potensi, Produksi Sumberdaya Ikan di Perairan Laut Indonesia dan Permasalahannya". Dibuat untuk Pendekatan Ekosistem dalam Pengelolaan Perikanan Indonesia, 2009.

Agoes, Etty R. *Batas Wilayah Laut Ditinjau dari Segi Hukum laut dan Kelembagaan.* 2003.

Dimyati, Muh. *Kontribusi Strategis Iptek Untuk Mewujdukan Poros Maritim Dunia.* Jakarta: Kementristek dan Dikti, 2004.

Kusnadi, "Kebudayaan Masyarakat Nelayan". Disampaikan dalam kegiatan JELAJAH BUDAYA TAHUN 2010, dengan tema "Ekspresi Buadaya Masyarakay Nelayan di Pantai Utara Jawa", Balai Pelestarian Sejarah dan Nilai Tradisional, Kementrian Kebudayaan dan Pariwisata, Yogyakarta, 12–15 Juli 2010.

Mellbye, Christian Svane, and Erik W. Jakobsen. *Norwegian Maritime Equipment Suppliers* 2014. Norsk Industri, 2014.

Nijdam, Michiel H., and dan Peter W. De Langen. "Leader Firm in the Dutch Maritime Cluster". Paper presented at the ERSA 2003 Congress. *Available at* www-sre.wu.ac.at/ersa/ersaconfs/ersa03/cdrom/papers/395.pdf.

Oegroseno, Arif Havas. *Kebijakan Dasar Indonesia Dalam Penetapan Perbatasan Maritim*. Jakarta: Departemen Luar Negeri Republik Indonesia, 2006.

Piesse, Mervyn. "The Indonesian Maritime Doctrine: Realising the Potential of the Ocean". *Headmark*, no.154, June 2015, pp. 10–15.

Purnomo, Kartika. *Kebijaksanaan Pengelolaan Pulau-Pulau Kecil Terluar dan Permasalahannya*. Jakarta: Departemen Dalam Negeri Republik Indonesia, 2003.

Starke, J.G. *Pengantar Hukum Internasional*. Bandung: Sinar Grafika, 1972.

Trace, Keith, Barend Frielink dan Denis Hew. *Maritime Connectivity in Archipelagic Southeast Asia: An Overview*. Manila: Asian Development Bank, 2009.

Articles

Anwar, Syaiful. *Mengenal Jasa Transportasi Laut dan Udara* (Agustus 2011). http://www.bppk.depkeu.go.id/webbc/images/stories/file/2011/artikel/upload%205%20agustus%202011/SYAIFUL%20ANWAR_Mengenal%20Jasa%20Transportasi%20Laut%20dan%20Udara.pdf. (1 September 2015).

Aprilianto, Rizky. Abdul H., Ainul H. "Implementasi Asas Cabotage dalam Kebijakan Pelayaran di Indonesia". *Jurnal Administrasi Publik* (*JAP*). (Vol. 2, No. 4), p. 758–764.

Fajar, Ario. *Pemberlakuan Asas Cabotage Dorong Bisnis Pengapalan Lokal* (13 Januari 2015). http://swa.co.id/business-strategy/management/pemberlakuan-asas-cabotage-dorong-bisnis-pengapalan-lokal. (1 September 2015). http://aim-services.co.id/jumlah-kapal-niaga-nasional-capai-13-326-unit-2/ (1 September 2015).

Lasabuda, Ridwan. "Pembangunan Wilayah Pesisir dan Lautan dalam Perspektif Negara Kepulauan Republik Indonesia". *Jurnal Ilmiah Platax* (Vol. 1-2, Januari 2013), p. 92–101.

Mardjoeki, H. "Pemberdayaan Masyarakat Pesisir Pantai Utara Daerah Kabupaten Cirebon". *Jurnal Ekonomi 1*, no 1 (2012): 52–57.

Maulana, Ridha. *Strategi Peningkatan dan Pengembangan Kinerja Pelabuhan di Indonesia*. (Fakultas Teknik Supil Unaya). http://www.academia.edu/5896016/Analisa_Stategei_Pengembangan_Pelabuhan (1 September 2015).

Muhsidin. "Pelabuhan Sorong dibangun Juli 2015". *Antara News* (*11 Mei 2015*). http://www.antaranews.com/berita/495558/pelabuhan-sorong-dibangun-juli-2015 (2 September 2015).

Nainggolan, Hermina, Krisna Fery Rahmantya, Anggie Destiti Asianto, Dadang Wibowo, Tri Wahyuni, and Walim Abdul Somad. *Kelautan Dan Perikanan Dalam Angka Tahun 2014*. Jakarta: Pusat Data, Statistik dan Informasi, 2014.

Nikijuluw, Victor P.H. "Identification of Indigenous Coastal Fisheries Management (ICFM) System in Sulawesi, Maluku, and Irian Jaya" *Jurnal Pesisir & Lautan* 1, no. 2 (1998): 40–52.

Pro, Supriyadu Pro. *Hubungan Perdagangan Antar Pulau di Indonesia*, 2013, http://www.sejarah-negara.com/2013/02/hubungan-perdagangan-antar-pulau-di.html. (pada 31 September 2015).

Sudirman. "Potensi Sumberdaya Laut Perairan Indonesia Timur dan Tingkat Pemanfaatannya Ke Depan Oleh Masyarakat Pantai dan Nelayan Setempat". Disajikan pada Semiloka Perikanan FK8PT di Universitas Nusa Cendana Kupang, 14 Oktober 2006.

Wahyono, A., dan F.I. Thufail Sudiyono, 1993. "Aspek-aspek Sosial Budaya Masyarakat Maritim Indonesia Bagian Timur. Hak Ulayat Laut Desa Para, Kecamatan Manganitu, Sangihe Talaud". *Seri Penelitian PMB-LIPI No. 4.*

Wahyudin, Yudi. "Sistem Sosial Ekonomi Dan Budaya Masyarakat Pesisir". Disampaikan pada Pelatihan Pengelolaan Kawasan Konservasi Perairan, di Kampus Pusat Diklat Kehutana, Institut Pertanian Bogor, Bogor, Jawa Barat, 5 Desember 2003.

Reports

A Report by the Navy League of United States, *America's Maritime Industry—The Foundation of American Seapower. Available at* https://navyleague.org/files/legislativeaffairs/americas-maritime-industry.pdf.

ASEAN-India Centre at RIS, *ASEAN-India Maritime Connectivity Report* (New Delhi: India Habitat Centre, 2014).

D.G. Mare, *Study on the Application and Implementation of the IUU Regulation—Final Report* (*2014*). *Available at* https://ec.europa.eu/fisheries/sites/fisheries/files/iuu_regulation_final-report_en.pdf.

Maritime Administration, *A Vision for the 21st Century* (Washington DC, 2007).

SmartComp Research Report No. 3, *Maritime Sector Development in The Global Market* (October 2013). *Available at* https://www.utu.fi/en/units/tse/units/PEI/research/Documents/SmartComp%20Research%20Report%20October%202013%20final.pdf.

CHAPTER 4

Future Opportunities and Challenges in Developing Sustainable Offshore Indonesian Fisheries

*Alistair McIlgorm and Brooke Campbell**

Abstract

In recent years, Indonesia has asserted its national sovereignty over the resources in its Exclusive Economic Zone (EEZ). The government has dealt severely with foreign Illegal, Unreported and Unregulated (IUU) fishing, sinking over 315 vessels since 2012. Indonesia has also banned foreign-crewed trawlers and foreign-owned transport ships involved in transhipment of catch, with over 700 vessels impounded. These measures have sent a strong message to those involved in IUU fishing, many of whom were based in countries neighbouring Indonesia. Going forward, how should Indonesia take up the opportunity to sustainably develop and maintain its fisheries that were previously harvested by IUU fishing?

This article discusses issues in domestic fishery development and describes the opportunity and the challenges facing Indonesia to realise the potential of their offshore fisheries. Deriving domestic economic benefits from fishing will require substantial capital investment in fishing vessels by Indonesians to access offshore fisheries, as well as in on-shore processing infrastructure. These investments may enable Indonesians to better capture the domestic economic benefits available to them from replacing the previous IUU fishing activity. However, securing these economic benefits beyond short-term horizons relies strongly on the development of more sustainable and restrictive fisheries management strategies that effectively support capital investment, while controlling the levels of fishing effort in Indonesian waters. The replacement of the IUU fishing offshore with Indonesian-driven initiatives can enable new fishery management approaches to be applied in a more context-appropriate corporate framework with full domestic industry involvement.

To achieve these future economic benefits, Indonesia needs to further develop its human capacity. This will involve training for the fishing industry, seafood processing

* Professor Alistair McIlgorm and Research Fellow Ms. Brooke Campbell are at the Australian National Centre for Ocean Resources and Security (ANCORS), University of Wollongong. PowerPoint presentation available at http://www.virginia.edu/colp/pdf/yogya-mcilgorm.pdf.

 | DOI:10.1163/9789004366619_009

sector, export marketing and also in sustainable fisheries management. Indonesians led fishing industry expansion is just beginning, and will require important national investment and policy decisions. However, several international domestic fishery development experiences overseas illustrate the need to preclude over-investment and overfishing by Indonesian nationals. This should not dissuade Indonesia's domestic fishery development; however, a staged approach to introducing vessels to the offshore area should be taken in order to avoid capital investment exceeding the sustainable capacity of the fish resources in the longer term. There is a need to implement a more restrictive fishery management framework that controls catch and excess effort and promotes sustainable investment. If this can be achieved, then there can be greater sustainable economic and social benefits from fish resources for the people of Indonesia in the coming decades.

1 Introduction

Since 2014, the government of Indonesia has taken a tough stance on national resource sovereignty, targeting primarily foreign Illegal Unreported and Unregulated (IUU) fishing taking place in Indonesian waters. This has led to 315 IUU fishing vessels being sunk and over 700 fishing vessels being impounded by the Indonesian authorities (Soeriaatmadja, 2016). The ban has impacted foreign-crewed trawlers and foreign-owned transport ships used for transhipping catch at sea (Alford 2015). There have been many newspaper and internet headlines on these developments (Alford 2015; Soeriaatmadja, 2016). There have also been discussions on the legal aspects (Busro 2017). There has been much less written on Indonesia's intentions to replace IUU fishing with legally-approved fishing and to expand their domestic seafood industry. The development of Indonesia's domestic fishing industry is part of President Joko "Jokowi" Widodo's Maritime Indonesia vision (Hwee Hwee 2016). The intention of this paper is to discuss and prompt more open discussion of how Indonesia should take up the opportunity to replace IUU fishing activity in order to capture the economic benefits for Indonesians. The paper also highlights that there are some significant challenges to address if Indonesia is to sustainably capture more economic benefits from its national fisheries and seafood resources.

2 Fisheries in Indonesia

The nation of Indonesia is a 17,500 island archipelago, with many rural coastal communities representing 60% of the population (MMAF 2017). Indonesia is an archipelagic State and has a large Exclusive Economic Zone (EEZ). The country's tropical location, geography and geology have led to very high species richness, diversity and marine resources productivity. This has been recognised with Indonesia being a major part of the "Coral Triangle" (ADB 2014).

The available data on Indonesia's reported fish catch states that the annual catch in 2014 was 6.02 million tonnes (SOFIA 2016). This makes Indonesia the second-highest capture fish producer in the world and confirms the country's importance in global food supply chains. The population of Indonesia has high fish protein dependence, with over 50% of protein coming from fish and seafood (MMAF 2017). An estimated 85% of the Indonesia fish catch remains in Indonesia (MMAF 2017).

Indonesia has in excess of 560,000 national boats of different sizes and operational capacities fishing in its Indonesian waters. Different sizes operate at various distances from land as reported in Box 1 below.

The term "offshore" is not generally used to describe Indonesian fisheries; hence, the operational areas measured by distance from shore indicated in Box 1 are indicative only. This is in part because Indonesian fishers have a long history of sailing small vessels over long distances and it is not unusual to find small vessels of less than 5 GT with minimal safety equipment in the outer EEZ. Here, we use the term "offshore" to collectively refer to large, commercial, industrial fishing vessels with the capacity to operate in the outer areas of the EEZ for many days or weeks, similar to the vessels that have been found to be IUU fishing.

Box 1 Description of the Indonesia fishing fleet by size class

Artisanal fleet: ~ 500,000 small vessels (<5 Gross Tonnes (GT), operating in area 0–4 nautical miles (nm) from the coast and taking ~40% of the catch), unlicensed.

'Midwater': ~ 50,000 medium vessels (~20 GT, operating 4–12 nm from the coast and taking ~30% of the catch), licensed *'Offshore'*: ~10,000 larger vessels (30GT+ operating in archipelagic or EEZ and taking ~20% of the catch), nationally licensed.

Industrial sector: ~10% of the catch (foreign/joint venture tuna, prawn), nationally licensed in the EEZ.

SOURCE: MMAF DATA

Domestically, Indonesian fisheries are managed by the Central government in Jakarta, through the Ministry of Marine Affairs and Fisheries (MMAF). Commercial fisheries are managed through both national and provincial government licensing systems, with licences renewed annually. Commercial fishing licences are issued by MMAF for access to the national archipelagic and Exclusive Economic Zone fisheries. Provincial governments manage inshore fisheries adjacent to their provinces and can issue fishing licences. District governments manage local marine resources, but no longer issue fishing licenses.

Since 2010, MMAF have been developing a national approach to strengthened fisheries management through the creation of 11 Fishery Management Areas (FMAs) across the Indonesian Archipelago. In 2016, Fishery Management Plans were developed for each of these areas. In 2017, Fishery Management Councils for each FMA began to be appointed. The Fishery Management Plans and Fishery Management Councils' framework operate national fisheries management in the 11 national FMAs.

Despite this movement towards strengthened and coordinated fisheries management frameworks across Indonesia's substantial seascape, no total caps on fishing effort currently exist for fisheries in Indonesian waters; guideline total allowable catches are recommended for each Fisheries Management Area but, critically, current regulations do not restrict fishing effort at the individual fishing vessel level. This lack of quantitative limits has obvious implications for the longer-term sustainability of Indonesian fisheries and for the sustainability of potential fisheries industry investments in turn. To date, the Indonesian fishing industry has not had sufficient port and transport infrastructure to address some of the product spoilage losses arising from poor handling and seafood quality issues (Hwee Hwee 2016).

Internationally, Indonesia has made commitments to sustainable fisheries management, as seen in their adoption of the Law of the Sea Convention (LOSC), UN Fish Stocks Convention (UNFSA) and the UN Port State Measures (PSM) agreement. Indonesia participates in three Regional Fishery Management Organisations (RFMOs) for tuna and highly migratory species (Indian Ocean Tuna Commission-IOTC; The Western Central Pacific Fisheries Commissions-WCPFC; and The Convention for the Conservation of Southern Bluefin Tuna-CCSBT). Indonesia is also part of regional fishery organisations (RFOs), such as the Southeast Asian Fisheries Development Centre (SEAFDEC).

As part of its international commitments to sustainable fisheries management, Indonesia has certain information reporting requirements. However, information on fish stock levels and fishing effort in Indonesian fisheries is currently dated and uncertain (MMAF 2017). Some of this uncertainty exists because IUU is inherently unknown and therefore not counted in fishing catch or effort data. The risk of overfishing and overfished stocks is therefore increased because of this information uncertainty; this is particularly concerning for the health and status of threatened species (e.g. sharks, juvenile tunas), and vulnerable habitats (e.g. coral reefs) that are relatively abundant in productive Indonesian waters. Again, because of the uncertainties that IUU activities create, sustainable fishery management processes are undermined because it makes accurate stock assessment and effective effort control-setting difficult. IUU activities also reduce or eliminate potential domestic development opportunities by undermining the reliability of information on fishery investment returns. Cumulatively, IUU threatens the contributions by licensed fishing to Indonesia's national long-term food supply and hence, social stability.

Lastly, IUU fishing is also in contravention of international law as well as of obligations for sustainability under Indonesia's fishery legislation (Busro 2017). It also breaches the conservation and management measures (CMMs) in place under the three regional Fishery Management Organisations. Concerns about the number of overfished stocks in Indonesia were raised in the report by the main scientific organisation national fishery profile (LIPI, 2011). The report measured the status of the fish stocks nationally, and reflected the impacts of both licensed and illegal fishing activity.

3 IUU and National Sovereignty

In 2014, the new "Jokowi" government announced a vision to build a "Maritime Indonesia" (Hwee Hwee 2016). Part of this was getting tough with IUU fishing in Indonesian waters. The new Minister of Fisheries, Ms. Susi Pudjiastuti, was from a fishing industry background in eastern Indonesia and announced a policy to apprehend and sink IUU vessels. After an initial set of vessel sinkings she declared, "If the perpetrators of IUU fishing are not affiliated with any legal entity, we will sink them immediately ... That's the new rule, the national consensus" (Toh 2016). As of May 2017 there have been 315 vessels sunk under this policy (Soeriaatmadja 2017).

Prior to these changes, the court processes to hear IUU fishing offences were considered to be slow and punishments ineffective (Toh 2016). However, by 2016, waiting times for trials had been reduced to several weeks (Toh 2016). In addition, after court sessions and verdicts, the retained vessels found guilty of IUU fishing were being publicly sunk—an event popular with the Indonesian public. Detained IUU vessels have often been found to originate from nations adjacent to Indonesia, while others were under flags of convenience, including use of false flags (IOM 2016; Busro 2017).

In addition to considering it a fisheries crime, Indonesia also sees IUU as a form of transnational crime (Busro 2017). Detained vessels have also been found to be involved in a range of additional illegal activities and practices, including human trafficking and slavery (Alford 2015; IOM 2016). The following additional problems with IUU vessels were described by the IOM (2016) report:
- corruption;
- money laundering;
- tax violation;
- customs related crime;
- immigration related crime;
- illicit drugs trafficking; and
- human rights violations (human trafficking, forced labour and child labour) (IOM 2016).

Indonesia's "sinkings policy" has strained bilateral diplomatic relations with neighbouring countries (Soeriaatmadja, 2016). Protests lodged with the Indonesian government have nevertheless been met with a firm resolve: "…there's no way IUU fishing is a part of a good bilateral relationship" (Minister Pudjiastuti in Soeriaatmadja, 2016). In contrast, there has been general support for the sinkings from the domestic fishing industry. However, the banning of foreign fishers and transport ships involved in transhipping has reduced the raw fish supply to Indonesian processors (Hwee Hwee 2016). This has led to criticisms that the banning of transhipment policy was too radical as it hurt local canning businesses, a point contested by the Minister (Soeriaatmadja, 2017). This illustrates a tension between politics and industry in developing a "maritime Indonesia".

The national prioritisation of the maritime direction of the Jokowi government has been endorsed by the maritime industry: "God created Indonesia as an archipelago and it is God's will that Indonesia will develop into a maritime nation" (Pak Eddy Loban, Chair of Iperindo, Indonesian Shipbuilding and Offshore Industries Association, quoted in Hwee Hwee 2016). Beliefs and nationalism have been invoked, but there has not been a transparent economic analysis of the development opportunity to date.

4 Foreign Fishing, IUU and Domestic Fishery Development

Foreign fishing is a legitimate part of UN LOSC (Art 62), which gives the coastal State (CS) the duty to determine the Total Allowable Catch (TAC) available, to assess the national capacity to take the resources, and to determine the difference—i.e., "the catch surplus". This surplus can be made available to foreign fishing entities through "agreements or other arrangements" (LOSC Art

62) acceptable to the CS. Generally, this involves negotiation of access fees, which result in government revenue, or other benefits as agreed with the CS. The CS can benefit from importing foreign fishing services to harvest available fish resources (World Bank 2014).

In negotiating these services, the foreign fishing entity provides the vessel capital, pays the CS a "rental fee" for access to CS waters under sovereignty, and takes the operational and financial risk of catching a certain quantity of fish. For example, foreign fishing entities participating in the Western and Central Pacific Parties to the Nauru Agreement (PNA) "Vessel Day Scheme" for tropical tunas can buy "vessel days" in order to fish in PNA party country waters. This generates "rental income" for these nations, which has increased from USD 64 million annually in 2010, to USD 350 million in 2016 (PNA 2016).

Indonesia fishery policy has not prioritised admitting foreign fishing entities under access fee arrangements, preferring joint ventures. This may have contributed to Indonesia experiencing IUU since before 2008 (Sodik 2009).

Conceptually, the fact that "foreign fishing" exists indicates that a foreign fishing entity may have a comparative advantage relative to the CS and that foreign fishing acts as a substitute for trade (Munro 1985; World bank 2014). The World Bank (2014) notes the following factors that may give a foreign fishing entity a comparative advantage:

> A comparative advantage could arise for many reasons throughout the fish value chain. For instance, foreigners may have comparative advantage in the following operations:
> - *Fishing, due to skill or access to capital enabling them to operate more efficiently in capital—intensive fisheries;*
> - *Processing, due to the scale of operations; and*
> - *Marketing, due to market access, proximity to customers, and product branding* (World Bank 2014)

However, a CS generally aspires to use and obtain benefits from their "own" fish resources. Determining comparative advantage requires an examination of fish stocks (a type of land resource), capital (fishing and processing plant), labour, and enterprise resources (Munro 1985). Here, we consider each of these for Indonesia.

Indonesia has abundant fish resources and likely has a comparative advantage due to being adjacent to these fish stocks. It can also maintain the sustainability of these stocks through sound fisheries management.

As a developing nation, Indonesia does not currently have adequate port infrastructure, fishing vessels, or food processing technologies. This lack of capacity restricts the quality and supply of fish processing and exports. Furthermore, Indonesian fishing industry businesses generally have lower access to capital for investment compared to neighbouring countries. These national financing and banking policies reflect government policies.

Comparative advantage in the context of labour requires a comparison with the labour force of other nations. Labour skills for vessel masters, fishing skippers, fishing mates, and crew have to be considered separately as they are different types of labour. Fishing offshore can involve long trips, with low pay and unsocial conditions away from family. Many Indonesians are seafaring people, but may not desire to be commercial fishing all year-round and a long way from home. The Indonesian government has found that some of the IUU vessels are not only using very low costs labour, but also some crew are people who have been trafficked (IOM 2016).

Enterprise is also part of comparative advantage. Enterprise includes competence in implementing fishing operations, seafood processing, business skills and marketing. Indonesia has potential in these areas, but the seafood industry is currently less competent and experienced than industry in some of its neighbouring and competing countries, such as Thailand.

From this short assessment it appears that Indonesia has a comparative advantage in having fish resources, but is less competitive in terms of infrastructure, capital plant, technology, and skilled staff than some of its more established seafood industry neighbours. However, thirty years ago, Thailand did not have a successful seafood industry; this illustrates that the development of a competitive fishing and seafood processing industry in Indonesia will require time, effort, and persistence. The Indonesian government is indicating that investment and policies will be used to promote this development (Soeriaatmadja 2016).

5 Indonesian Fisheries—The Economic Loss from IUU and Related Activities

There are various estimates of the economic loss to Indonesia from IUU fishing, ranging from US $2 billion to $5 billion as an annual value of lost catch (Almuttaqi 2014; Busro 2017). In addition to IUU fishing, the inclusion of illegal transhipment in an analysis of economic losses, led the Minister to claim that IUU was responsible for annual losses of USD $20 billion to Indonesia's fishery sector (Alford 2015; Soeriaatmadja 2017; Busro 2017).

However, there is limited public information on how these estimates were calculated. Nevertheless, there is clearly a significant economic cost to Indonesia from IUU fishing. What could the Indonesian economy gain from this recent removal of foreign IUU fishing capacity? The opportunity that this reduction represents to the fishing industry, the secondary sector, and for increasing domestic food supply, has yet to be fully scoped and evaluated as a national economic opportunity.

6 The Opportunity

The removal of illegal foreign fishing capacity from Indonesia's waters provides an opportunity for the domestic development of the country's fishing, processing, marketing and exports industries. The greatest potential exists for offshore fisheries beyond 12 nm from shore, where Indonesian industrial fishing can replace the fishing effort/capacity that was previously generated by IUU fishing. The removal of IUU fishing effort from offshore areas will generate opportunities for fishing vessels to fish, employment and associated seafood processing benefits in many Indonesian regions. The removal of IUU fishing may also generate benefits to smaller-scale artisanal coastal fishers in inshore waters due to stock recovery offshore, as claimed by the Minister (Soeriaatmadja 2016). The secondary sector also has the potential to benefit in terms of revenue and employment from expanded processing, marketing, and exporting activities.

New Indonesian fish production and processing activities will generate an increase in the supply of services on the input side of the economy. The increased output from both primary and secondary sectors has the potential to generate employment and economic benefits in regional Indonesia where job creation reduces urban drift to the cities for work. There are not only economic opportunities from increasing the volume of exports, but also from increasing the export price received for higher quality export species achieved through improved fish cold supply chain handling in production, processing and export marketing.

The potential size of the national development opportunity is unknown and requires an economic evaluation study. The cost to the national economy and investors to obtain these new revenues, profits, and export incomes is important information for investors seeking finance. Amidst all of this potential, however, maintaining the sustainability of the fishery resource through responsible fisheries management must remain a critical underpinning of sectoral investment and development plans if these opportunities are to be effectively realised into the future.

7 The Challenges

The current plans to develop the fishing and processing industries to take up the opportunity will require investment. The fishing industry and processors will require access to capital via the banking system to invest in fisheries, and secondary sector developments that will benefit the whole Indonesian fish sector and offshore fishing. The banking sector in Indonesia has apparently been less willing to invest in the fishing industry than in neighbouring countries where government and finance sector policies and practices are more aligned, and therefore helped fund IUU fishing in Indonesian waters (Rhein 2016). In April 2016, an economist workshop gathered many banks in the Indonesian finance sector together with government and industry to discuss the way forward. The banking sector indicated its willingness to lend to the fishing and seafood sector, and government (Rhein 2017).

There is also a need for government to assist with the development of core industry infrastructure. Key fisheries landing sites need to be identified and provided with fishing port services, refrigeration, ice production, and clean water as well as roads and airfields to fly the products to national and export markets. These infrastructure needs have generally been provided by national government in the past.

In the processing and marketing sectors, cold supply chains need to be developed by industry. The involvement of government in providing this infrastructure to gain food benefits for home consumers and access to export marketing needs to be considered. Government provisioning of low interest loans to industry may be an option, but this loan scheme would need to be developed and trialed before it is rolled out more widely. Additionally, the development of international trade standards over time now means that export supply chains have high hygiene criteria. Indonesia has a record of achieving lower export prices than its neighbouring countries trading the same fish species (MMAF 2017). It is apparent that exports can gain higher product prices if rigorous cold supply chain processes are in place and able to meet international trade requirements.

There is also a need to regard the maintenance of sustainable fish stocks as a form of investment. In the current absence of fishing effort restrictions, the Indonesian fishery sector risks replaying the over-investment of capital seen in the Indonesian forestry industry, which did not adequately consider sustainability, and led to resource depletion and poor returns on investments (Rhein 2017). If Indonesia implements limits to total catch or a total effort limit, invested fishing capital will not be unlikely to exceed the sustainable resource available, hence leading to long term sustainability and profitability (Costello et al. 2016). For example, issuing only a limited number of licences can be used to restrict effort levels and also give those in the fishing industry an incentive for licence renewal if they both follow regulations and provide log-book data for management.

8 Domestic Industry Development

Developing the domestic fishing industry to replace IUU fishing activities will take time as investment and transition occurs. Domestic fishing industry development assumes that the Indonesian fishing and processing industry will see all areas of potential development as commercially advantageous. There can also be constraints in accessing the capital required to make key investments. It is proposed that going forward there may be four stages to precautionary industry development, each with some time estimates:

1. An initial stimulus period: with government assistance to get new fishing operators "out there" (0–3 years);
2. A period for on-shore infrastructure development: that will help both the fish catching and the secondary processing and exporting sector (0–10 years);
3. A transitional fishery management period: to develop fisheries management as a resource investment framework (0–10 years); and
4. Following the 10 years of development as above, a more stable sustainable fishery management regime is envisaged that underpins the resource and sustainable economic investment.

The government has commenced stimulus measures to increase fishing and secondary processing sector developments. The development of fishery management has started with the formation of Fishery Management Areas, Fisheries Management Plans, and Fishery Management Councils. However, further specific resource management restrictions are needed to enable development of fish resources to be sustainable in the long term.

To achieve the intentions above, capacity development is required in government and industry sectors, and at many levels, for example, for individual staff in government fishery management agencies. There is also an industry training need in management and business skills for leadership in the fishing and processing industry. The expansion of the fishing industry will require trained employees over the next five years. Capacity development in the marine sector is a major national challenge for Indonesia and requires government and donors to create a strong enabling environment to facilitate substantial changes in the next decade.

9 Government Steps to Go Forward

The national government announced a series of steps in 2016 to assist small-scale fishers to benefit from the removal of IUU fishing effort. The government proposed to provide 3,500 new boats to fishermen in stages until 2019, with an estimated cost of USD $300m (Hwee Hwee 2016). Many of these are smaller inshore vessels (5-10GT) will be given to socio-economically disadvantaged communities in remote areas to develop fishing and create local employment. Other new fishing vessels are over 30GT and can operate offshore.

Consultations with local fishers have also led to a commitment by government to provide work insurance for 1,000,000 fishers from 2016 to 2019 (Soeriaatmadja 2016). The lack of insurance for fishers has been identified as a factor that has been limiting the willingness of fishers to increase their fishing activity. The potential risk from loss of life while fishing has high social costs on remaining family members, and hence insurance is both appealing and required (Soeriaatmadja 2016).

The national government has also committed USD $175m to improve the quality of fishery products, marketing and supply chain management in the industry to promote exports (Soeriaatmadja 2016). The Indonesian government also sees the national benefits in the economy being realised through having more fish produced by aquaculture and has proposed USD $165m to support fish farm development across the country. This will potentially further add to fish supply and marketing opportunities.

A government-endorsed "IUU task force" is also progressing additional measures to address IUU fishing, and steps are being taken to improve the use of Vessel Monitoring Systems (VMS) and local fishing vessels to improve the information sources available for management decision-making.

One of the less noted challenges is dealing with what the Indonesian Minister has termed "internal IUU". This refers to Indonesian owned and operated vessels that are fishing illegally or not reporting catches; this is an issue that has not been strongly addressed to date.

10 Discussion

In addressing IUU fishing and then replacing it, experience from international trade economics shows that comparative advantage can be mixed between trading parties. For example, it may benefit Indonesia to develop its domestic fishery only to a certain point. If there is a surplus of "catchable", commercially valuable fish, then UN LOSC Article 62 requires the surplus to be offered under acceptable terms to other nations. As Indonesia is not admitting foreign fishing vessels, there may be a need to revisit the national policy on joint ventures if the Indonesian private sector finds some fishing is not commercially attractive. There has been little open discussion about future joint ventures and use of foreign expertise in the seafood sector in Indonesia.

The extent of Indonesia's domestic fishery development is an issue requiring further strategic planning and management. Domestic fishing investment can encourage the increase of fishing effort to a point where it may exceed the sustainable catch. There are examples worldwide of the costs of not establishing and maintaining sustainable domestic fishing controls in the ensuing vacuum created by vacating foreign fishing effort. In the 1990's in Canada, too much fishing effort was applied in the Atlantic Cod fishery when the Canadian domestic fleet replaced foreign fishing effort, exceeded the available catch, and collapsed the fishery. This in turn led to major socio-economic costs for Canada's Maritime Atlantic provinces, which were heavily reliant on the cod fisheries (Hannesson 1996). Indonesia should be aware of this over fishing and over-investment risk.

Long-term sustainable fisheries management requires restrictions on effort and catch levels to maintain the fish stock. These restrictions provide a sound investment framework to develop a sustainable fishery, as the quantitative limits on catch limit the amount of fishing effort and investment applied to the fishery. Indonesia is currently part of the Convention for the Conservation of Southern Bluefin Tuna (CCSBT) and has a quota limit in place for the catch of this species under the CCSBT management arrangements. However, there are no other restricted catch quota limited fisheries currently in Indonesia. Setting a quota at a sustainable level requires good data, scientific analysis, and skilled staff to advise on precautionary management decisions. Staff have to be trained through capacity development activities to be able to implement and maintain more restrictive fishery management regimes.

The Australian National Centre for Ocean Resources and Security (ANCORS) at the University of Wollongong, and the Indonesian Ministry of Marine Affairs and Fisheries (MMAF) have had an agreement to promote capacity development in Indonesian fisheries management since 2010. This has led to several fisheries management training programs funded by the Australia Government Department of Foreign Affairs and Trade (DFAT). However, this is just a small part of a diverse training need for both government and industry in a range of areas.

11 Conclusions

In confronting the IUU fishing that has occurred in Indonesia's offshore fisheries, there is now a national opportunity to responsibly develop the Indonesian fishing and processing sectors. This undertaking requires significant planning, coordination, and management across many government agencies, as well as coordination with industry and discussions with fishing communities. However, the path to realising sustainable benefits needs to be more comprehensively evaluated in order to avoid over-investment and potential fishery resource over-exploitation.

A combination of government-funded fisheries infrastructure and improved access to capital for private sector fishing enterprises is needed to promote industry development in Indonesia. More restrictive fisheries management approaches need to be implemented by government to provide a much-needed, economically-sustainable investment framework in fishing and processing sectors. There is also a need to keep addressing both foreign and domestic IUU fishing through good enforcement and to encourage Indonesian vessels to fish in areas previously occupied by iuu fishing. There are a range of knowledge and skill developments requiring improved fishery data for analysis and capacity development across a range of industry, government and community levels. This capacity development challenge is in need of more open discussion and planned action if the aspirations of the Indonesian fishery and seafood processing sector are to be met in the next decade.

References

ADB, State of the Coral Triangle: Indonesia. A report by the Asian Development Bank, Mandaluyong City, Philippines, 2014.

Alford, P. "Jakarta to ban foreign trawlers and transport ships in Indonesia". *The Australian* October 31, 2015.

Almuttaqi, I. "Indonesia and the problem of illegal fishing". *The Jakarta Post*, Jakarta, April 5, 2014.

Busro, Z.M. "Burning and/or Sinking Foreign Fishing Vessels Conducting Illegal Fishing in Indonesia: Some Obligations and Loopholes". *Asia-Pacific Journal of Ocean Law and Policy* 2, 2017, 174–179.

Costello *et al.* Reform, investment, and risk in Indonesian fisheries. MMAF (KKP) information brochure. Jakarta 2016.

Hannesson, R., *Fisheries (mis)management: The case of the north Atlantic cod.* Oxford, England: Fishing News Books, 1996.

Hwee Hwee, T. "Transforming Indonesia into a maritime power". *The Business Times*, International Enterprise, Singapore (IES) 2016.

IOM. Report on Human Trafficking, Forced Labour and Fisheries Crime in the Indonesian Fishing Industry. A joint research effort by the Indonesian Ministry of Marine Affairs and Fisheries (KKP), Indonesia Presidential Task Force to Combat Illegal Fishing, the International Organization for Migration (IOM) Indonesia, and Coventry University 2016.

LIPI, The Ministry of Marine Affairs and Fisheries Decision, No. KEP.45/MEN/2011 on Estimation of Fish Stock in Fisheries Management Area of Republic of Indonesia 2011.

MMAF Indonesia marine and fisheries business and investment opportunities. Ministry of Marine Affairs and Fisheries, Republic of Indonesia, Jakarta, 21 February 2017.

Munro, G.R. "Coastal States, Distant Water Fleets and EFJ: Some Long-Run Considerations". *Marine Policy* 9: 2–15, 1985.

PNA. PNA CEO urges members to plan for tuna revenue declines. www.pnatuna.com. https://pnatuna.com/node/271 2016 (accessed 19/10/2017).

Rhein, M. "Historical patterns and current perceptions of growth in Indonesia's Fishery sector. Implications for policy, governance and management". A report to David and Lucille Packard Foundation, January 2017.

Sodik, D.M. "IUU Fishing and Indonesia's Legal Framework for Vessel Registration and Fishing Vessel Licensing", *Ocean Development & International Law*, 40:3, 249–267, 2009, DOI: 10.1080/00908320903076797.

Soeriaatmadja, W. "Curbs on illegal fishing deliver 'amazing' results, Some legal fishermen's catches have doubled: Indonesian Minister". *The Straits Times*, April 16, 2016.

Soeriaatmadja, W. "Indonesia's fishery sector loses bite—Government curbs on illegal fishing lead to shortage of boats to bring in enough fish stocks for export", *The Straits Times*, May 2017.

SOFIA State of the World Fisheries and Aquaculture (SOFIA): Opportunities and Challenges 2016. Fisheries and Aquaculture Department, Food and Agriculture Organization (FAO), 2016.

Toh, M. "Indonesia's Explosive Response to Illegal Fishing Could Be Saving the Industry", Fortune.com. Sep 19, 2016. http://fortune.com/2016/09/19/indonesia-illegal-fishing-explosion/.

World Bank Trade in fishing services: Emerging perspectives on foreign fishing arrangements. Environment and natural resources global practice discussion paper #01, World Bank report number 92622-glb, December 2014.

PART 2

Biodiversity beyond National Jurisdiction

CHAPTER 5

Bridging the Ocean, Water and Climate Action Goals under the 2030 Agenda on Sustainable Development

Ronán Long and Mariamalia Rodriguez Chaves***

Abstract

The paper reviews the science, policy and law, as well as the wider contextual considerations that are shaping the implementation of the 2030 Agenda on Sustainable Development, with a view to deducing elements that are applicable to the aquatic and marine environments. In addition to Goal 14, brief reference is made to Goal 3 (good health and well-being for all, at all ages), Goal 6 (ensuring water and sanitation for all), Goal 11 (making cities and human settlement inclusive, safe, resilient and sustainable), and Goal 13 (climate action). Special attention is paid to the scientific evidence and the regulatory background underpinning the concept of sustainability. The discussion touches various aspects of the outcome of three parallel multilateral negotiation processes, namely: the Paris Agreement on Climate Change, the Addis Ababa Agenda on Financing for Development; and the 2030 Agenda for Sustainable Development. The paper highlights the importance of human rights elements of the 2030 Agenda and the requirement of restoring natural capital in the marine environment. The paper outlines the approach adopted under the Paris Agreement and the push towards the twin pillars of fiscal and environmental sustainability in many instruments, along with the enhancement of the role that educational institutes can play as agents of change. The paper concludes with a short discussion of some of challenges faced by cities and the parallel implementation processes for delivering tangible results under the 2030 Agenda.

1 Introduction

Peace, prosperity and living in harmony with the environment are the hallmarks of stable and inclusive societies. As we approach the third decade of

* Ronán Long is Director of the Sasakawa Global Ocean Institute, World Maritime University, Malmö, Sweden.

** Mariamalia Rodriguez Chaves is an alumna United Nations - The Nippon Foundation of Japan Fellowship Program.

 | DOI:10.1163/9789004366619_010

the 21st century, the international community appears to be falling well short of these laudatory objectives and the world is increasingly characterized by political instability, social inequality,[1] human displacement, extreme weather events, along with the loss of natural capital and the resources that it supports.[2] Urbanization, globalization, along with unprecedented levels of natural resource extraction, production and consumption are the order of the day. In parallel, the international legal system is under stress and longstanding approaches to the observance of treaty obligations are progressively challenged by the advent of new political perspectives on the role of international law in international affairs, along with the diminishing commitment of states in multilateral relations, as typified by the United States withdrawal from both the 2015 Paris Climate Change Agreement and the Iran Nuclear Agreement.[3]

The waning influence of the international order is also evident in the maritime domain with some notable examples of excessive maritime claims apparent in south-east Asia and elsewhere, many of which run counter to the 1982 United Nations Convention on the Law of the Sea (LOS Convention).[4] There is also a pressing need to bolster the work of the United Nations and other multilateral agencies in reasserting the rule of law as a means to arrest the decline of biodiversity, to halt illegal access to fisheries, and to combat the effects of climate change. Transnational threats to the environment demand closer engagement with a broader range of participants and the establishment of more effective mechanisms for coordination and cooperation between States and international bodies.[5] Moreover, complex problems in modern law of the sea generally require sophisticated solutions. This is very much the case in relation to the ocean, where the politics, economics and law of sustainability are closely inter-twined and have a strong bearing on ocean governance decisions and the design of regulatory systems.

---

1 World Inequality Laboratory, World Inequality Report 2018 Executive Summary (Paris: WIL, 2017), available at: http://wir2018.wid.world/files/download/wir2018-summary-english.pdf.

2 United Nations Development Programme, Human Development Report 2016 (New York: United Nations, 2016), at 51 and 156.

3 Paris Agreement, Dec. 1/CP.21, Annex, UN Doc. FCCC/CP/2015/10/Add.1, at 21. On the United States withdrawal, see, S. Mulligan, "Withdrawal from International Agreements: Legal Framework, the Paris Agreement, and the Iran Nuclear Agreement, Congressional Research Service 7-5700," www.crs.gov, R44761. Available at: https://fas.org/sgp/crs/row/R44761.pdf. On the other hand, there has been nearly universal acceptance of the Paris Agreement, with 148 Parties ratifying the treaty by 8 January 2018.

4 J.A. Roach, R. Smith, *Excessive Maritime Claims*, 3rd ed. (Boston/Leiden: Brill, 2012), *passim*.

5 M. Shaw, *International Law*, 8th ed. (Cambridge: Cambridge University Press, 2017), at 641, and 652–653.

The riposte comes in the form of a United Nations General Assembly resolution on the 2030 Agenda for Sustainable Development (the 2030 Agenda) setting out a 15 year blueprint for international action between 2016–2030.[6] The theme of the 2017 Center for Oceans Law and Policy's Conference in Indonesia is therefore very timely and provides an ideal forum to explore contemporary issues in the regulation of the marine environment in light of Sustainable Development Goal (SDG) 14 (life below water). The latter is a suitable *leitmotif* for the conference, as it describes the mission to tackle some of humanity's most difficult challenges and the search to find practical and innovative solutions that serve the needs of current and future generations. This is particularly true in Indonesia, an archipelagic State located at the intersection between two oceans and several regional seas, which has adopted its own distinctive and cutting-edge policy that informs national efforts to use marine resources optimally and sustainably.[7] Moreover, it is important to note that the SDGs are indivisible in the struggle to balance economic development with social and environmental imperatives.[8] Accordingly, the 2030 Agenda will only have a lasting impact if the principle of sustainability is incorporated into a broad range of national policies worldwide that have a bearing on the health of the ocean.[9] Moreover, improving accountability, transparency and oversight are also central to the successful advancement of sustainability as a legal concept that has practical implications at different levels. For instance, redressing marine environmental impairment demands greater flag, coastal and port State responsibility in discharging legal obligations under the LOS Convention, as highlighted by the International Tribunal for the Law of the Sea (ITLOS) in its first two advisory opinions.[10]

In light of the aforementioned, the focus of this paper is to touch upon some of the wider contextual considerations that informed and continue to shape the implementation of the 2030 Agenda with a view to deducing elements that are applicable to the management of limited natural resources in the aquatic

6 UN. Doc. A/RES/70/1—Transforming our world: the 2030 Agenda for Sustainable Development, 21 October 2015.

7 Presidential Regulation (Peraturan Presiden) No. 16 of 2017 on Indonesian Sea Policy. English version, Indonesian Ocean Policy. Available at: https://maritim.go.id/konten/unggahan/2017/07/offset_lengkap_KKI_eng-vers.pdf.

8 UN Doc. A/RES/70/1, at 3/35.

9 Ibid.

10 Responsibilities and Obligations of States Sponsoring Persons and Entities with Respect to Activities in the Area, Advisory Opinion, ITLOS Case No 17, [2011] ITLOS Rep 10. For commentary see David Freestone (2011) 105 *American Journal of International Law*, pp. 755–761. Request for an Advisory Opinion Submitted by the Sub-regional Fisheries Commission (SRFC), ITLOS Case No. 21, 2 April 2015.

and marine environments. In addition to Goal 14, brief reference is made to Goal 3 (good health and well-being for all, at all ages), Goal 6 (ensuring water and sanitation for all), Goal 11 (making cities and human settlement inclusive, safe, resilient and sustainable), and Goal 13 (climate action). In relation to the latter, underpinning all of the SDGs is the requirement for effective measures addressing the effects of climate change: a topic that will undoubtedly define the legacy of this generation as responsible environmental actors. In bridging the sustainability gap between the management of water, oceans, seas and marine resources, as well as the international response to climate change under the 2030 Agenda, it must be borne in mind that there is a spectrum of other SDGs that are clearly germane to ocean related matters including energy (Goal 7), growth and employment (Goal 8), ecosystems and biodiversity (Goal 15) and partnerships (Goal 17). Mention is also made of Goal 4 (quality education and lifelong learning) in light of the important capacity-building efforts undertaken at the Rhodes Academy of Oceans Law and Policy, the Yeosu Academy of Law of the Sea, the World Maritime University in Sweden, as well as a number of initiatives under the auspices of the Nippon Foundation.

First, the scientific evidence underpinning the ocean, climate, and water goals under the 2030 Agenda should be mentioned.

2 Preliminary Matters: Unequivocal Evidence

In line with many other international instruments, the formulation and implementation of the SDGs is evidence-based and largely derived from the work undertaken by specialist scientific advisory bodies that provide independent advice and authoritative research on environmental matters.[11] In this context, much of the damage to the ocean environment was predicted and is predictable. Specifically, the expert evidence underpinning Goals 14 (life below water), Goal 13 (climate action) and Goal 6 (availability and sustainable management of water for all) is captured in several international scientific reports that present compelling findings regarding the deplorable status of the environment

11 See, for example, the reports produced by the Group of Experts on the Scientific Aspects of Marine Environmental Protection (GESAMP) that advises the UN system on the scientific aspects of marine environmental protection, along with the annual and regular specialist reports published by the FAO, the International Council for the Exploration of the Seas (ICES) and the regional seas bodies. Many other scientific advisory bodies including the IMO, UNEP, UNESCO and the WHO also make an important contribution on matters of marine environmental importance.

and the catastrophic consequences of poor management of natural resources, including the aquatic and marine environments.[12]

Most notably, the *United Nations First Global Ocean Assessment* published in 2015 provides a definitive and unparalleled overview and seeks to answer five fundamental questions, namely: (1) What is the overall state of the world's oceans and seas?; (2) Are marine ecosystems around the world improving or declining?; (3) What benefits do we get from the world's oceans and seas, and how are they distributed?; (4) How can we measure the state of the oceans and seas?; and (5) And what threatens them?[13] The answers provided in the 600-page report are unequivocal in so far as the oceans' capacity to withstand anthropogenic impacts was deemed near or at its limit and that urgent action on a global scale was "needed to protect the world's oceans from the many pressures they face."[14] All of the world's oceans and seas are affected to a greater or lesser degree by pollution, depleted fisheries, and the loss of biodiversity. This is having catastrophic consequences because in excess of three billion people, the majority of whom live in developing States, depend on ocean resources and marine biodiversity for their well-being. The findings of the First Global Ocean Assessment are crucial and play an important role in the policy process by informing the formulation and implementation of the 2030 Agenda, including especially its ocean related goals.

In understanding the importance of the oceans for ending poverty and hunger (SDGs 1 and 2), as well as the attainment of food security and improved nutrition, it is important to recall that marine fisheries employ over 200 million people worldwide. Moreover, communities in Africa, Latin America, and the Caribbean depend upon the ocean as a vital source of protein. In this regard, according to the FAO (Food and Agriculture Organization), fish provided more than 3.1 billion people with almost 20 per cent of their average per capita intake of animal protein.[15] Thus, it is unsurprising to note that the *FAO State of World Fisheries and Aquaculture 2016* report points out that one of the world's biggest challenges is "how to feed more than 9 billion people by 2050 in a context of climate change, economic and financial uncertainty, and

12 See *inter alia*: United Nations Economic and Social Council, Progress towards the Sustainable Development Goals, E/2017/66, 11 May 2017; United Nations, *The Sustainable Development Goals Report 2017*. Available at: https://unstats.un.org/sdgs/files/report/2017/TheSustainableDevelopmentGoalsReport2017.pdf.

13 United Nations (Division on Ocean Affairs and the Law of the Sea), *The First Global Integrated Marine Assessment* (Cambridge, Cambridge University Press, 2017).

14 Foreword by the Secretary-General of the United Nations, Ban Ki-moon. Available at: http://www.un.org/Depts/los/global_reporting/WOA_RPROC/Foreword.pdf.

15 FAO, *State of World Fisheries and Aquaculture 2016* (Rome, FAO, 2017).

growing competition for natural resources."[16] The FAO Report states that 31.4 per cent of fish stocks are overfished; 58.1 per cent of the stocks assessed were deemed fully fished; and a paltry 10.5 per cent of stocks were deemed to be under-fished.[17] This problem has escalated over the past four decades with the 2017 SDG Report noting that stocks fished at biologically unsustainable levels have increased from 10 per cent in 1974 to 31 per cent in 2013.[18] The scale of this disaster can only be appreciated in the context of the crucial role that oceans and inland waters play in providing food, nutrition and employment to current and future generations and in meeting commitments under the 2030 Agenda. The problem is compounded when one considers that illegal, unregulated and unreported fishing accounts for up to 26 million tons of fish per year, or more than 15 per cent of the world's total annual capture fisheries output. The World Bank/FAO estimate that the economic loss to the global community is in the order of $83 billion a year, hence the title of their report on the crises in world fisheries: *The Sunken Billions*.[19] Furthermore, coastal waters are subject to widespread pollution and eutrophication, which is anticipated to increase significantly by 2050.[20]

Underlying all of the SDGs are concerns about human induced climate change. The 2030 Agenda on climate action must be read in light of several definitive reports of technical working groups that informed the *IPCC's Fifth Assessment Report,* including: *Climate Change 2013, The Physical Science Basis*; *Climate Change 2014, Impacts Adaptation and Vulnerability*; *Climate Change 2014, Mitigation of Climate Change*.[21] The principal conclusion of the summary synthesis report for policy-makers is that anthropogenic greenhouse

16 Ibid., at 2. Also see: http://www.fao.org/fileadmin/templates/wsfs/docs/expert_paper/How_to_Feed_the_World_in_2050.pdf.

17 FAO, *State of World Fisheries and Aquaculture 2016*, at 151–152.

18 SDG Report 2017, *op cit.*, note 12, at 10.

19 World Bank, *The Sunken Billions Revisited: Progress and Challenges in Global Marine Fisheries. Environment and Development* (Washington, DC: World Bank, 2017).

20 According to the United Nations, coastal eutrophication will increase in over 20 percent of the 63 large marine ecosystems that are subject to assessment under the Transboundary Waters Assessment Programme, SDG Report 2017, *op cit.*, note 12, at 10.

21 IPCC, *Climate Change 2013: The Physical Science Basis. Contribution of Working Group I to the Fifth Assessment Report of the Intergovernmental Panel on Climate Change* (Cambridge: Cambridge University Press, 2013); IPCC, *Climate Change 2014: Synthesis Report. Contribution of Working Groups I, II and III to the Fifth Assessment Report of the Intergovernmental Panel on Climate Change* [Core Writing Team, R.K. Pachauri and L.A. Meyer (eds.)] (Geneva: IPCC, 2014): *Climate Change 2014: Impacts, Adaptation, and Vulnerability. Part A: Global and Sectoral Aspects. Contribution of Working Group II to the Fifth Assessment Report of the Intergovernmental Panel on Climate Change* (Cambridge: Cambridge University Press, 2014); IPCC, *Climate Change 2014: Mitigation of Climate Change. Contribution of Working*

gas emissions "are extremely likely to have been the dominant cause of the observed warming since the mid-20th century."[22] The oceans have warmed, acidified and sea-levels have risen by 19 centimeters (cm) since 1901.[23] The future of many coastal communities is threatened by rising sea levels with the average sea level rises predicted as in the order of 24 to 30 cm by 2065, and an incredulous 40 to 63 cm by 2100.[24] The ongoing melting of ice at polar latitudes is impacting navigation routes and the spatial distribution of marine species. The spatial extent of global sea ice fell to 4.14 million square kilometers in 2016, with experts pointing out that this is the second lowest level since records began.[25] Moreover, "most aspects of climate change will persist for many centuries even if emissions are stopped."[26] This is a major concern as global emissions of carbon dioxide (CO_2) have increased by almost 50 per cent since 1990.[27] Emissions grew more quickly between 2000 and 2010 than in each of the previous three decades.[28] The oceans act as a sink or reservoir and have absorbed over 90 per cent of human-induced greenhouse gases.[29]

Sea-level rise, extreme weather events and the loss of natural barriers in the marine environment are impacting human migration. Annual special reports by the American Meteorological Society tabulate how human induced climate change is impacting upon weather systems worldwide including the heating of the polar ice caps, heat waves in the ocean adjacent to Alaska and Australia, the increased severity of the El Niño, and coral bleaching on the Great Barrier Reef in Australia.[30] Explaining and understanding the links between extreme weather events and their impact on the status and distribution of marine

Group III to the Fifth Assessment Report of the Intergovernmental Panel on Climate Change (Cambridge: Cambridge University Press, 2014).

22 Intergovernmental Panel on Climate Change, "Climate Change 2014 Synthesis Report: Summary for Policymakers" (2014) 2–20.

23 Ibid.

24 Ibid.

25 SDG Report 2017, *op cit.*, note 12, at 9.

26 Summary for Policymakers. In: *Climate Change 2013: The Physical Science Basis. Contribution of Working Group I to the Fifth Assessment Report of the Intergovernmental Panel on Climate Change* (Cambridge: Cambridge University Press, 2013), 1–30.

27 Ibid.

28 Ibid.

29 Ibid.

30 S. Herring, A. Christidis, J. Hoell, C. Kossin, C.J. Schreck III, and P.A. Stott, 2018 Eds., "Explaining Extreme Events of 2016 from a Climate Perspective." *Bull. Amer. Meteor. Soc.*, 99 (1), S1–S157. Citing: "A multifactor analysis of the record 2016 Great Barrier Reef bleaching" (Lewis and Mallela 2018); "Ecological impacts of the 2015/16 El Niño in the central equatorial Pacific" (Brainard *et al.* 2018); and "Multiyear extreme ocean temperatures with impacts on living marine resources off the U.S. West Coast during 2016" (Jacox *et al.* 2018).

living resources is crucial to the future implementation of the UN Fish Stocks Agreement and related treaties,[31] as well as in devising strategies on how best to discharge obligations under the 2030 Agenda. The scientific evidence is building on the socio-economic and environmental impacts of climate change on natural resources with the American Meteorological Society noting that 65 per cent of the papers examined in their annual report *Explaining Extreme Events from a Climate Perspective* have identified a role for human induced climate change.[32] The 2016 report reviews a range of phenomena including the impacts of extreme ocean conditions on the distribution and conservation of marine living resources.[33]

Similar to other natural resources, the availability and status of freshwater is influenced by the adverse effects of climate change including drought, flooding and atypical weather patterns described above. Everyone accepts that fresh water and sanitation are essential services and are vital for human well-being and survival.[34] Yet at a global level, the findings of the *UN-Water World Water Development Report 2017* and related reports are uncompromising in relation to the provision of this basic right, and the most fundamental of services for the developing and the least-developed world.[35] The figures speak for themselves and record the following: 663 million people are still without access to a source of clean water; at least 1.8 billion people globally use a source of drinking water that is contaminated; over 1.7 billion people are currently living in river basins where water use exceeds recharge; 2.4 billion people lack access to basic sanitation services, such as toilets or latrines; more than 80 per cent of wastewater resulting from human activities is discharged into rivers or sea without the removal of pollutants; each day, nearly 1,000 children die due to preventable water and sanitation-related diseases; and floods and other water-related disasters account for 70 per cent of all deaths related to natural disasters.[36] The World Health Organization estimates that over one third of

---

31 Agreement for the Implementation of the Provisions of the United Nations Convention on the Law of the Sea of 10 December 1982 Relating to the Conservation and Management of Straddling Fish Stocks and Highly Migratory Fish Stocks, 4 August 1995, 2167 UNTS 88.

32 *Op cit.*, note 28, at i.

33 R. Webb, F. "Werner Explaining Extreme Ocean Conditions Impacting Living Marine Resources" in S. Herring *et al.*, Eds., *Explaining Extreme Events of 2016 from a Climate Perspective, op. cit.* note 28.

34 S. Solomon, *Water: The Epic Struggle for Wealth, Power and Civilisation* (New York: Harper Perennial, 2010).

35 United Nations, *The United Nations World Water Development Report 2017: Wastewater, The Untapped Resource* (Paris: UNESCO, 2017).

36 United Nations Economic and Social Council, Progress towards the Sustainable Development Goals, E/2017/66, 11 May 2017, at 24.

the global population is living with water scarcity.[37] Over 2 billion people are living in countries with water stress,[38] mainly in Africa and Asia, and are thus without sustainable water resources and sanitation.

In summary, the evidence presented in the independent scientific assessment reports demonstrates that the world is facing the consequences of over 100 years of industrialization and social inequality accompanied by environmental damage and the depletion of resources at an unprecedented scale. Ultimately, responsibility for redressing the problems identified in the reports and in responding to scientific advice rests with a broad spectrum of political and regulatory actors, operating at different levels. Furthermore, in taking action, *laissez faire* is not an option in addressing the sustainability deficit and ensuring the well-being of humanity and the conservation of natural resources. There is considerable room for improvement in the search for workable solutions, many of which hinge on applying the concept of sustainability as a normative construct in decision-making and regulatory action.

3 Sustainability Journey Thus Far

"Rome wasn't built in a day" is a well-worn maxim that attests to the need for time and patience when building great things. Hence, it is important to appreciate the temporal scale and protracted nature of the negotiation process that delivered agreement on the SDGs and the 2030 Agenda. If we look back, then it is clearly evident with the benefit of hindsight that the policy and law of sustainable development has not evolved in a vacuum but has advanced in slow incremental steps over time. The roots of the SDGs are firmly embedded in soft law initiatives that evolved in the policy landscape over the past five decades. Viewed in this light, the story of the birth of the SDGs is a multi-faceted one and is directly linked to Scandinavia in so far as it traces its origins back to the adoption of the Declaration of the United Nations Conference on the Human Environment in Stockholm in 1972.[39] The subsequent report published by the World Commission on Environment and Development also had a strong

37 World Health Organization, *Progress on drinking water, sanitation and hygiene: 2017 update and SDG baselines* (Geneva: World Health Organization, 2017).

38 SDG Report 2017, *op cit.*, note 12, at 10. Stress is defined by the United Nations as the ratio of total freshwater withdrawn to total renewable freshwater resources above a threshold of 25 per cent: United Nations Economic and Social Council, Progress towards the Sustainable Development Goals, E/2017/66, 11 May 2017 at 24/35.

39 See Report of the United Nations Conference on the Human Environment, Stockholm, 5–16 June 1972, UN Doc. A/Conf 48/14/Rev. 1 PC p, 13 and 17.

Scandinavian link because it is commonly known as the Brundtland Report on Sustainable Development, named after its chairwoman the former Prime Minister of Norway, Dr. Gro Harlem Brundtland.[40] The report called for international collaborative action in the form of the United Nations Programme of Action for Sustainable Development and defined the term as "development that meets the needs of present without compromising the ability of future generations to meet their own needs".[41] Crucially, it articulated the fundamental norm that has informed the sustainability discourse ever since, namely: that economic development must go hand-in-hand with environmental and social development.

The ground-breaking Brundtland Report in turn led to the convening of the United Nations Conference on Environment and Development in 1992 (the Rio-Earth Summit) and the adoption of innovative instruments that have since shaped international law and policy on sustainability irrevocably eversince, including: The Rio Declaration on Environment and Development;[42] Agenda 21;[43] the United Nations Framework Convention on Climate Change;[44] the Convention on Biological Diversity;[45] as well as the Statement of Forest Principles.[46] The 1990s also witnessed the convening of important international summits and the adoption of action plans on the issues of common concern to the international community pertaining to sustainability such as children, population, gender, human settlement and food.[47]

The concept of sustainable development received the imprimatur of the International Court of Justice in important cases concerning natural resources,[48] and from the World Trade Appellate Body in the *Shrimp Turtle* case.[49] Although

40 World Commission on Environment and Development, *Our Common Future* (Oxford: OUP, 1987).

41 Ibid., at 43.

42 See in particular, principles 3, 4 and 27 Rio Declaration.

43 Paras. 8 and 39 Agenda 21.

44 Article 3(4), United Nations Framework Convention on Climate Change.

45 Article 3(4) of the Convention on Biological Diversity refers to "sustainable use."

46 UNCED Report A/CONF.151/26 (Vol. III). Non-legally binding authoritative statement of principles for a global consensus on the management, conservation and sustainable development of all types of forests. 14 August 1992.

47 Children (New York, 1990), Population and Development (Cairo, 1994), Social Development (Copenhagen 1995), Women (Beijing, 1995), Human Settlements (Istanbul, 1996), and Food (Rome, 1996).

48 Gabcikovo Nagymaros Project (*Hungary v. Slovakia*) [1997] ICJ Rep 7, 78; Pulp Mills on the River Uruguay (*Argentina v. Uruguay*), Merits, Judgment of 20 April 2010, I.C.J. Reports 2010, 14, 48–49.

49 WTO Appellate Body, 38 ILM 199 at 121, para. 129.

the concept is codified expressly in the Fish Stocks Agreement,[50] scholars have questioned its precise normative value in circumscribing State practice or as a general rule of international law open to adjudication.[51] Nonetheless, the progressive development of international policy reached a ground-breaking marker with the adoption of the non-binding Millennium Development Goals,[52] which set down ambitious targets that were aimed at meeting the needs of human population by 2015.[53] This was followed by several major international initiatives to add substance to the concept of sustainable development including, *inter alia*: the Programme for the Further Implementation of Agenda 21;[54] the Plan of Implementation of the World Summit on Sustainable Development (Johannesburg Plan of Implementation);[55] along with the Johannesburg Declaration on Sustainable Development, the Programme of Action for the Sustainable Development of Small Island Developing States (Barbados Programme of Action);[56] and the Mauritius Strategy for the Further Implementation of the Programme of Action for the Sustainable Development of Small Island Developing States.[57]

Progress in achieving the Millennium Development Goals was reviewed by the United Nations General Assemblies in 2005, 2010 and again in 2013. The path has not been an easy one and there is little doubt that the concept of sustainable development suffered a major blow with the global financial crises of 2008, which led to the adoption of fiscal austerity programmes and the curtailment of economic, environmental and social agendas and spending worldwide. A new strategic shift was adopted at the Rio+20 Conference in June 2012, which marked the 40th anniversary of the adoption of the Stockholm

50 Art 2. Fish Stocks Agreement.

51 See *inter alia*: Y. Tanaka, *The International Law of the Sea*, 2nd Ed. (Cambridge, Cambridge University Press, 2015), at 248–250 citing A.V. Lowe "Sustainable and Unsustainable Arguments" in A. Boyle and D. Freestone (eds.) *International Law and Sustainable Development: Past Achievements and Future Challenges* (Oxford, Oxford University Press, 1999) at 26.

52 United Nations Millennium Declaration. UNGA Resolution A/RES/55/2. September 18, 2000. See also: United Nations Summit on the Millennium Development Goals. UNGA Resolution A/65/L.1. September 17, 2010.

53 Ibid., Para. 2, 19.

54 UNGA Resolution A/RES/S-19/2. 19 September 1997.

55 United Nations. Plan of Implementation of the World Summit on Sustainable Development. Available at: http://www.un.org/esa/sustdev/documents/WSSD_POI_PD/English/WSSD_PlanImpl.pdf.

56 UNGA Resolution A/RES/47/189. 22 December 1992. See also: UNGA Report A/CONF.167/9. October, 1994. Available at: http://www.un.org/esa/dsd/dsd_aofw_sids/sids_pdfs/BPOA.pdf.

57 UNGA Resolution A/CONF.207/CRP.7. 13 January 2005.

Declaration. The Conference recognized the need for a new framework for development as a successor to the Millennium Development Goals in 2015. Much of the preparatory work in this regard was entrusted to the United Nations General Assembly Open Working Group of Sustainable Development,[58] which commenced two and half years of sterling work in December 2012.

4 Three Parallel Processes Converge in 2015

The negotiation of the SDGs should not be viewed solely as a discrete initiative under the auspices of the United Nations General Assembly, as the international community has sought to step up to the mark by providing leadership and forging diplomatic agreement on how best to address several of the complex problems faced by humanity in the 21st century. Specifically, the year 2015 was remarkable in this regard as it saw the conclusion of three significant processes, namely: the Paris Agreement on Climate Change;[59] the Addis Ababa Agenda on Financing for Development;[60] and the 2030 Agenda for Sustainable Development.[61]

Each instrument forms a separate strand in the overall approach to living in greater harmony with the natural environment and to meeting the needs of developing States. The central fulcrum text is the 2030 Agenda with its embedded plan of action for the five "Ps": people, planet, prosperity, peace and partnership.[62] The Agenda is highly ambitious and is founded on an altruistic vision of the future world that is "free of poverty, hunger, disease and want, where all life can thrive."[63] The Agenda consists of 17 SDGs addressing poverty, hunger, education, gender, water, energy, industrialization, inequality, consumption and production, climate change, oceans, ecosystems, institutions and partnerships.[64] The broad thematic areas are further elaborated by 169 ambitious targets that are indivisible, integrated and universally applicable, with the overall objectives of addressing the scourge of hunger, poverty and oppression.[65] In balancing the three limbs of sustainable development, economic, social and

58 UNGA Decision A/67/L.48/Rev.1. January 15, 2013.
59 United Nations. Framework Convention on Climate Change. Report FCCC/CP/2015/10/Add.1. 29 January 2016.
60 UNGA Resolution 69/313. 27 July 2015.
61 UNGA Resolution A/RES/70/1. 25 September 2015.
62 UNGA, A/RES/70/1, at 2/35.
63 Ibid., at 3/35.
64 Ibid., at 15–37/35.
65 Ibid., at 23/35.

environmental, countries and stakeholders more generally are committed to achieving a broad and universal policy agenda over the fifteen-year period between 2016 and 2030.

The 2030 Agenda is underpinned by international treaty and customary law, including the Charter of the United Nations. The LOS Convention and related instruments add substance to the specific targets in Goal 14 including: the prevention and reduction of pollution by 2025; the sustainable management and protection of marine ecosystems and further action for their restoration by 2020; by the same date, the minimization of ocean acidification; the effective regulation of illegal, unreported and unregulated fishing; the conservation of at least 10 per cent of marine areas; the ending of fisheries subsidies, along with increasing economic benefits to Small Island Developing States.[66] Other targets relate to the enhancement of scientific knowledge and the transfer of marine technology, as well as for improved access for small-scale artisanal fishers to marine resources. Shoring up all of the targets is the call for effective implementation of international law and the LOS Convention specifically.

With the exception of the Group of Least Developed Countries, there was considerable reluctance on the part of States at the United Nations to introduce a specific goal on climate change into the SDGs.[67] This was only achieved at the twelfth hour in the last sessions and perhaps explains why the targets on climate action set out in Goal 13 are less prescriptive than those pertaining to the ocean.[68] In general, they take a holistic approach to regulatory matters and are predicated on improving climate change resilience and adaptive capacity to hazards and natural disasters in all countries. The latter is very pertinent to coastal States and the small island developing States (SIDS) in temperate zones, which are particularly susceptible to damage from intense tropical storms, as became evident in 2017 with storms such as Irma, Maria and Jose in the Caribbean, along with over a dozen typhoons in the Asia-Pacific region. Other targets in Goal 13 include the integration of climate change measures into planning and policies at the national level, as well as the adoption of mitigation, adaptation, impact reduction and early warning measures. There is a call for the use of the Green Climate Fund as a means to address the climate related needs of developing countries and the SIDS. Again, considerable emphasis is placed on giving greater effect to international law with the United

66 Ibid., at 23–24/35.

67 F. Dodds, D. Donoghue and J. Leiva Roesch, *Negotiating the Sustainable Development Goals: A Transformational Agenda for an Insecure World* (London/New York: Routledge, 2017) at 39.

68 UNGA, A/RES/70/1, at 23 /35.

Nations Framework Convention on Climate Change highlighted as the primary intergovernmental forum for negotiating the global response to climate change.

Similar to Goal 14, the targets on water in Goal 6 are formulated around achieving specific objectives within tight deadlines.[69] For instance, they require the attainment of "universal and equitable access to safe and affordable drinking water for all" by 2030.[70] They also require the reduction of pollution and the fostering of cooperation in relation to the management of trans-boundary resources. Similar to the ocean, there is an emphasis on the restoration of natural ecosystems, along with the expansion of international cooperation and capacity building, as well as the strengthening of local community management of water resources. The importance of achieving Goal 6 with respect to the marine environment is best appreciated when it is considered that some estimates suggest that in excess of 80 per cent of wastewater is discharged into the ocean without treatment to remove pollutants.[71]

At its core, the 2030 Agenda recognizes that social and economic development are predicated upon the sustainable use and conservation of natural resources including ocean and freshwater resources.[72] Importantly, the SDGs and their constituent targets are implemented and reviewed by using a set of global indicators for measuring progress, which have been developed by the Inter-Agency and Expert Group on Sustainable Development Goal Indicator.[73] Reflecting the need for bottom-up support and transparency, they are complemented by indicators developed by Member States at national and regional levels, including indicators that measure the performance of cities.[74]

5 Raising the Profile of Transnational Solutions

The 2030 Agenda aims to raise the normative profile of sustainability in the regulation and management of a broad spectrum of human activities. Several of the SDGs are concerned directly with environmental sustainability.[75] Pointedly, all of the SDGs call for environmentally sustainable practices in different fields of human endeavor. The overall approach blends environmental, social,

69 Ibid., at 18/35.

70 Target 6.1, Goal 6.

71 UNESCO. United Nations World Water Assessment Programme. "United Nations World Water Development Report 2017," at 2.

72 UNGA, A/RES/70/1 at 2–4/35, and especially para. 133 at 9/35.

73 Ibid., para. 75 at 32/35. Discussed further *infra*.

74 Ibid.

75 Goals 6 (Water), 14 (Oceans) and 15 (Ecosystems).

governance and economic considerations into one seamless political agenda that recognizes sustainable development as the key for delivering sustained, green and inclusive economic growth. This is attractive on a number of levels as it allows for a textured and sophisticated response to global challenges. Indeed, a quick glance at the 2030 Agenda reveals that several of the targets set out in the SDGs are aspirational in nature but others require concrete responses and the implementation of specific legislative measures. Contrast for example the targets under Goal 13 relating to the climate action, which are set at a high level of abstraction with the more express quantifiable targets with deadlines set out in Goal 14 concerning oceans and seas.[76] Moreover, nearly all of the targets concerning the sustainable management of oceans, fresh water and the effects of climate change are beyond the exclusive remit of individual countries to resolve at national or local levels. They thus demand transnational solutions and international cooperation and coordination in addressing acute problems pertaining to the degradation of the marine environment.

The successful implementation of Goal 14 is contingent upon the successful implementation of the law of the sea. This in itself is commendable but it may have to be admitted that Goal 14 adds little to existing commitments under international law, apart from shining the spotlight on the wider political backdrop to sustainability. There are positive and negative aspects to this approach. On the one hand, it is unlikely to be used on its own by international courts and tribunals as a normative standard to review the validity of state practice in relation to the use of natural resources or the impairment of the environment. In specific instances, on the other hand, the 2030 Agenda will undoubtedly be useful in bolstering the economic, social and environmental factors that need to be taken into consideration and balanced to some degree in decision-making and in the application of legal instruments that entail binding obligations on conservation and sustainability. In this regard, an example that could be envisioned within target 14.C.1 on the strengthening of the law of the sea framework is the upcoming negotiation process for the conservation and sustainable use of marine biodiversity in areas beyond national jurisdiction.[77] Additionally, regulatory and management responses should thus not be viewed in isolation, as the SDGs are part of a highly inter-connected multilateral agenda that is focused firmly on resolving issues that are vital for the collective future and prosperity of humanity.

76 UNGA, A/RES/70/1, at 23–24/35.

77 UNGA, A/72/L.7, 15 November 2017.

6 Taking Ownership of "an Agenda for the People, by the People and with the People"

The 2030 Agenda is remarkable due to its universal and participatory nature. This is reflected in the support it has received from 193 Member States of the United Nations, international organizations and agencies, as well as from representatives of civil society including the business community, cities, regions and many other public and private actors. The broad constituency bodes well for the difficult tasks associated with the successful attainment of sustainability. Conspicuously, many of the tasks associated with the specific targets to realize Goal 3 (good health and well-being), Goal 6 (clean water and sanitation), Goal 13 (climate action) and Goal 14 (life below water) can only be achieved through action and cooperation at local, national, regional and international levels. In the latter context, one recurrent theme that permeates any discussion of sustainability is the need to build inclusive communities, respect for human rights and the rule of law. The 2030 Agenda is thus informed by the Charter of the United Nations, the Universal Declaration of Human Rights,[78] the Millennium Declaration,[79] the 2005 World Summit Outcome,[80] along with human rights treaties. Viewed from this perspective, the SDGs are first and foremost about people and their rights to a clean environment, clean water, inclusive education, gender equality, as well as the absence of hunger and poverty at a global level. As noted by Mary Robinson, Former United Nations High Commissioner for Human Rights and the UN Secretary-General's Special Envoy on Climate Change, the 2030 Agenda is "an agenda for the people, by the people and with the people."[81] Accordingly, the SDGs should not be viewed solely in terms of realizing individual or specific goals but about transforming peoples' lives to ensure that no one is left behind in the attainment of sustainable development in the interest of the common good.

In line with contemporary developments in human rights law, sustainable development is closely related to social justice.[82] The social dimension of climate related rights is not free from controversy and increasingly subject

---

78 UNGA Resolution 217 A (III).

79 *Op. cit.* note 50.

80 UNGA Resolution 60/1.

81 Mary Robinson in Felix Dodds, Ambassador David Donoghue and Jimena Leiva Roesch, *Negotiating the Sustainable Development Goals: A Transformational Agenda for an Insecure World* (London/New York: Routledge, 2017) at XV.

82 *Op. cit.* note 1, at 9. Also see I. Khan, "Pathways to Justice: Rule of Law, Human Rights, Sustainable Development and Human Security." Background paper for Human Development Report 2016. (New York: United Nations Development Programme, Human Development Report Office, 2016).

to litigation in the context of health and human well-being. As far back as 2007, the United States Supreme Court upheld the Environmental Protection Agency's right to regulate greenhouse gas emissions that may reasonably be anticipated to endanger public health or welfare.[83] In designing regulatory responses to climate change, particularly in the United States and Canada, the social cost of carbon is a useful regulatory tool, as it captures in one fiscal number the long-term damage done by a ton of carbon dioxide (CO_2) emissions in a given year, with some estimates quantifying it at $47 per metric tonne of CO_2 in 2017.[84] Society bears the burden of climate change and the social cost of carbon helps government measure the costs and benefits of policies and investments related to climate change in the economic modeling of climate change mitigation and adaptation measures. Among the socials costs are changes to agricultural productivity, sea levels, extreme weather and risks to human health. In the United States, the reduction of the cost of carbon to $3.50 per mt$CO_2$ will make it far more difficult to propose policies that substantially lower emissions.[85] This shift in federal policy is predicated on taking into account only damages within national borders, along with a discounted rate for future damage. Other countries have sought to reduce the human impact on the environment and to lower the socials cost by developing new technologies enhancing carbon sinks. In June 2017, for example, the Swedish Parliament adopted a law that aims to ensure no net emissions of greenhouse gases by 2045.[86] The law foresees carbon removal from the atmosphere as a key element in delivering the Paris Agreement and SDG targets. The ocean can play a vital role in carbon removal through the restoration of marine ecosystems.

7 Name of the Game: Restoring Natural Capital and Bridging the SDGs

How can the international community respond to the catastrophic picture painted above about the current and future status of the marine, aquatic and human environments? Clearly, action is required at all levels! All life began in

83 *Massachusetts v. Environmental Protection Agency*, 549 U.S. 497 (2007).

84 The United States EPA uses a different figure, see: https://19january2017snapshot.epa.gov/climatechange/social-cost-carbon_.html.

85 See Environmental Protection Agency: https://19january2017snapshot.epa.gov/climatechange/social-cost-carbon_.html. Also, W. Nordhaus *et al.*, *Social Sciences—Sustainability Science—Physical Sciences—Earth, Atmospheric, and Planetary Sciences: Revisiting the social cost of carbon* (2017) 114 (7) PNAS 1518–1523.

86 Swedish Climate Act (into force 1 January 2018). Press Release: http://www.government.se/press-releases/2018/01/as-of-today-sweden-has-a-new-climate-act/.

the ocean and ultimately the fate of humanity is directly linked to how human impacts on the largest ecosystem on the planet are managed.[87] As the ocean is the planet's largest life support system, health to the ocean means health for over three billion people depend on marine and coastal biodiversity for their livelihoods. An obvious solution to the conundrum posed by the findings of the scientific reports outlined above is to work on the restoration of aquatic and marine ecosystems. This course of action can contribute directly to the realization of many of the SDGs, including the improvement of human well-being and making cities and human settlements more sustainable. Viewed in this light, one obvious means to protect and restore natural capital is through the utilization of spatial management tools and this is the focus of the SDGs. Specifically, target 14.5 of Goal 14 requires the conservation of at least 10 per cent of coastal and marine areas by 2020.[88] Progress in this regard is uneven with the 2017 SDG Report recording that protected areas cover 13.2 per cent of areas under national jurisdiction and a paltry 0.25 per cent of areas beyond national jurisdiction, amounting to a combined 5.3 per cent of the ocean.[89] In other words, half ways towards the 2020 target.

In the context of combatting the effects of climate change, there are some estimates that suggest the ocean could be close to "150 per cent more acidic by 2100."[90] Ocean fertilization and sequestration are particularly topical as potential remedies to combat the effects of climate change. There is however considerable concern about the development of biological means to remove and sequester CO2 from the atmosphere, as well as the impacts on marine ecosystems. This is particularly true in relation to CO2 extraction and storage technologies that come within the scope of the London Convention on the Prevention of Marine Pollution by Dumping of Wastes and Other Material. As noted in UNEP's Emissions Gap Report 2017, any climate change strategy that uses the ocean as a common-pool resource for the purpose of carbon removal or sequestration will require coordination at multilateral and regional levels in line with the requirements of the London Convention and its 1996 Protocol.[91] The UNEP Report recommends that there needs to be further scientific assessment on CO2 removal in achieving the goals of the Paris Agreement.[92]

87 Sylvia Earle, *The World Is Blue: How Our Fate and the Ocean's Are One* (Washington, D.C.: National Geographic, 2010).

88 UN Doc. A/RES/70/1, at 24/35.

89 SDG Report 2017, *op cit.*, note 12, at 10.

90 Ibid.

91 See, UNEP, The Emissions Gap Report 2017. United Nations Environment Programme (Nairobi: UNEP, 2017).

92 Ibid., at 65.

Moreover, studies need to focus on carbon dioxide removal approaches in commercial and governmental endeavours, along with an examination of how to appropriately stabilize atmospheric carbon dioxide concentrations and how to manage common interests. According to UNEP, this approach will also help meet other targets pursuant to SDGs related to "water quality, ecosystem restoration, biodiversity preservation, food and nutrition security, job creation and improved crop yields".[93]

The objectives set down by the SDGs in relation to the ocean are clear-cut and aimed at conserving and sustainably using the oceans, seas and marine resources for sustainable development. Ultimately, they aim to restore the natural capital base underpinning ocean ecosystems. The biological removal of carbon through better management of the marine environment along with the restoration of degraded marine ecosystems can also significantly contribute to the attainment of the objectives of the Paris Agreements, as well as several other targets under the 2030 Agenda. Many of the targets are time specific such as Target 14.3 that aims to ensure that by 2025 we prevent and significantly reduce marine pollution of all kinds, in particular from land-based activities, including marine debris and nutrient pollution. Pollution from land-based sources and cities in particular is the single biggest source of pollutants of the marine environment. Again, this directly impinges upon human welfare and health (Goal 3).

Goal 6.3 aims to ensure availability and sustainable management of water and sanitation for all. There is an ambitious deadline of 2030 to ensure that water quality is improved by reducing pollution, eliminating dumping and minimizing the release of hazardous chemicals and materials, halving the proportion of untreated wastewater and substantially increasing recycling and safe reuse globally. In achieving this target, one should not forget that the correct functioning of the hydrological cycle is linked with the health of humans, sustainable development, as well as the avoidance of conflict, natural disasters and human migration. Accordingly, bridging troubled water by successfully implementing Goals 14, 6 and 4 will help deliver sustainable development of vital natural resources in the marine and aquatic environments to address degraded marine ecosystems, as well as water scarcity and thereby also address the effects of climate change.

In 2017, the UN General Assembly pointed out the importance of ocean-related education in promoting ocean literacy and a culture of conservation, restoration and sustainable use of the ocean.[94] In particular, capacity building

93 Ibid., at 66.

94 United Nations, A/RES/71/312, 17 July 2017.

and training programmes, such as those on offer at specialist academies at Rhodes, Yeosu, and at the World Maritime University can help develop knowledge and key skills concerning how best to ensure the conservation and sustainable use of marine and aquatic environments. The up-skilling of regulators and managers is crucial for developing countries, and the Small Island Developing States in particular, in order to build the practice and culture of sustainable development.

In summary, apart from Goal 14, restoring natural capital in the marine environment is linked with the attainment of other global SDGs including the protection of biodiversity, reduction of pollution, improved water quality, institutional building and lifelong-learning.

8 All Roads Lead to Paris

The scientific evidence adduced by international bodies demonstrates that anthropogenic induced climate change is happening and will continue to influence human well-being, along with the world's ability to implement the ambitious targets set out in the 2030 Agenda.[95] The rapidly escalating effects of human induced climate change remain the single greatest threat to the future of humanity and Goal 13 mainstreams climate action into the concept of sustainable development. This accords with Article 3 (2) of the UNFCCC (United Nations Framework Convention on Climate Change), which codifies the right to sustainable development and goes on to state that climate change policies should be appropriate and integrated into national development programmes.

Cities and human settlements are most at risk and are at the forefront of efforts to mitigate and adapt to the effects of climate change. The OECD (Organisation for Economic Co-operation and Development) estimate that climate change combined with rapid population increases, economic growth and land subsidence could lead to more than a "9-fold increase in the global risk of floods in large port cities between now and 2050".[96] This finding is highly significant as more than a billion people live in low-lying coastal regions, the majority of whom are living in flood plains and coastal areas in Asia. Moreover, the five cities with the projected largest estimated increase in flood risk in 2050 are Alexandria, Egypt; Barranquilla, Colombia; Naples, Italy; Sapporo, Japan;

95 *Op. cit.*, note 19.

96 Stephane Hallegatte, Colin Green, Robert J. Nicholls, Jan Corfee-Morlot, "Future flood losses in major coastal cities," *Nature Climate Change* 3, 802–806 (2013).

and Santo Domingo, Dominican Republic.[97] The OECD report also notes that large increases in port city flood risk may occur in locations that are not vulnerable today, catching many cities and countries off-guard.[98]

The IPPC's Fifth Assessment Report for policy-makers cautions that "continued emissions ... will cause further warming and long-lasting changes ... increasing the likelihood of severe, pervasive and irreversible impacts."[99] Furthermore, that "limiting climate change would require substantial and sustained reductions in greenhouse gas emissions".[100] The UNFCCC provides a framework for taking urgent action to address the threat of climate change. Underpinning all of the SDGs is the core premise, set out in Goal 13, that sustainable development can only be achieved if the international community is successful in achieving the object of the Paris Agreement in holding the increase in the global average temperature to well below 2° Celsius above pre-industrial levels and to pursue efforts to limit the temperature increase to 1.5° Celsius.

Many aspects of the 2030 Agenda mirror the Paris Agreement, which provides a framework for national mitigation efforts in the form of Nationally Determined Contributions. Both the SDGs and the Paris Agreement are informed by the principle of common but differentiated responsibilities and the special needs of developing countries in alleviating hunger and poverty (Goals 1 and 2). The Paris Agreement is the first international treaty to address the duties and needs of both developed and developing countries in a unified framework with a particular focus on nationally determined contributions, so that all States do their best within their respective capacities.

The Paris Agreement also aims to safeguard food security by protecting food production systems from the adverse impacts of climate change. The preamble of the Paris Agreement notes the importance of human rights in addressing climate change, including the concept of "climate justice", as well as "gender equality, empowerment of women and intergenerational equity."[101] Significantly, from the point of view of the ocean, the Paris Agreement recognizes the importance of greenhouse gas sinks and calls for the adoption of measures that ensure the integrity of ecosystems and the protection of biodiversity, including the oceans. Similar to the SDGs, the Paris Agreement places particular

97 Ibid.

98 Ibid.

99 12 IPCC, "Climate Change 2014 Synthesis Report: Summary for Policymakers" (2014) 2–20 <http://www.ipcc.ch/pdf/assessment-report/ar5/syr/AR5_SYR_FINAL_SPM.pdf>.

100 Ibid.

101 *Op. cit.* note 57. Recital 11.

emphasis on education, public participation and access to information, along with stakeholder engagement.

In light of the all-embracing consequences of climate change, it is notable from the point of the specialist training undertaken at the Rhodes Academy, the Yeosu Academy and at the World Maritime University that one of the Goal 13 targets relates specifically to promoting "mechanisms for raising capacity for effective climate change-related planning and management in least developed countries and small island developing States, including focusing on women, youth and local and marginalized communities."[102] Surely, nothing could be more central to advancing human well-being and the concept of climate justice. Moreover, the human rights dimension to climate change is paramount. At a conference on oceans and climate change at the World Maritime University, which was hosted in conjunction with Berkeley Law and the Korean Institute of Ocean Science and Technology, Emeritus Professor Roger Zetter of the Refugees Studies Centre at Oxford University stated that new instruments and approaches are required, which focus on population displacement at global, regional and local scales.[103] Furthermore, he articulated the need to adopt a "three Rs" approach, namely: a model that advances "rights," "resilience" and "resettlement."[104] In his view, this lends to the establishment of a more proactive and comprehensive framework for responding to the impacts of climate change and environmental degradation, along with the challenges of displacement and migration.

A key feature of the 2030 Agenda is the objective of integrating climate action into all of the SDGs and Goal 14 in particular. Clearly in this regard, oceans play a central role in the climate system, helping to moderate the effects of climate change. Conversely, climate change is having and will continue to have major impacts on the health of the oceans and marine ecosystems most particularly. Nonetheless, one major criticism of the international regulatory regime codified in the Paris Agreement and its parent treaty (the UNFCCC) is that they ignore ocean related issues. Although the two legal regimes pertaining to the ocean and the climate have evolved in the separate bi-polar worlds of international diplomacy and as a result there is an absence of a bridge between key provisions including measures on capacity building. Professor Bodansky pointed out at the WMU conference in August 2017 that much of the criticism

102 Target 13.b, SDGS.

103 See *inter alia*: R. Zetter (2017) "The Politics of Rights Protection for Environmentally Displaced People." Migration Institute of Finland. Vol. 44 (5–12).

104 See *inter alia*: R. Zetter (2017) "Why They Are Not Refugees—Climate Change, Environmental Degradation and Population Displacement." Migration Institute of Finland. Vol. 44, at 28.

is unwarranted on several grounds, namely: if UN climate regime succeeds, the health of the oceans will benefit greatly; the climate regime approach does take account of ocean-based renewables; the UNFCCC recognizes the importance of the ocean as a carbon sink; and there is very good cause to assign vessel-source emissions to an international multilateral body in that regard, the International Maritime Organization (IMO).[105] Importantly, the latter body is actively reducing marine pollution by the shipping sector and dumping at sea, as well as enabling the delivery of food and energy cost-effectively, along with promoting gender equality in employment.[106] All together these facts, demonstrate that sustainable maritime transport contributes to a broad range of the SDGs and climate action most particularly.

9 Twin Pillars: Fiscal and Environmental Sustainability

Sustainable lifestyles and stable patterns of consumption and production are fundamental objectives of the Paris Agreement. They are also the focus of Goal 12. The approach in the Paris Agreement is bottom-up and a broad range of stakeholders can contribute to global emissions reductions and climate resilience. One transnational initiative merits a special mention because of its novel approach to ensuring that the global financial system is stable and sustainable in light of climate risk. Specifically, the financial service industry is beginning to play a crucial role in climate action in the form of a campaign, "Climate Action 100+," an investor initiative in partnership with UNEP Finance Initiative and the UN Global Compact, which aims "to engage with the world's largest corporate greenhouse gas emitters to curb emissions, strengthen climate-related financial disclosures and improve governance on climate change."[107] Part of the campaign is to make the corporate boards of carbon intensive companies more climate-accountable and to ensure that climate risk becomes part of their fiduciary responsibilities to shareholders.[108] The line taken is firmly

105 D. Bodansky, Ocean and Climate Change: Introducing the Issues, unpublished presentation, Law of the Sea Institute, Malmö, Sweden. 14 August 2017. (copy with the author).

106 IMO, IMO and the Sustainable Development Goals (London: IMO, 2017). Measures include binding GHG emission controls for the shipping sector, mandatory data collection system for fuel oil consumption, data analysis and decision making on energy efficiency, as well as a road map for developing a comprehensive strategy for GHG emissions from shipping. See, Initial IMO strategy on the reduction of GHG emissions from ships (IMO, July 2018).

107 See press report: https://www.unpri.org/about.

108 See J. Plender, "Investors are demanding corporate action on climate change," *Financial Times*, 17 December 2017, at 9.

rooted in sustainable investment and with positioning companies for a low-carbon economy in line with the objectives of the Paris Agreement. Separately, many companies including ExxonMobil, the world's largest hydrocarbon company, are engaged in reducing greenhouse gas emissions and in taking greater climate responsibility including disclosure on risks to its business.[109]

10 Value of Normative Principles in Setting Priorities and Meeting Deadlines

Normative principles and approaches are crosscutting in ambit and should therefore inform all aspects of the SDGs.[110] They are thus a viable means to integrate sustainability into social and economic policies, plans and programmes including those pertaining to Goals 3, 4, 6, 13 and 14. Indeed, one of the features of the SDGs that distinguish them from the Millennium Development Goals is that the 17 Goals setting down 169 ambitious targets of the latter are founded upon the twin normative pillars of universality and indivisibility. This is particularly important for developing countries and goes to the heart of the debate on capacity building and is closed linked to the principle of common and differentiated which is mentioned in the Preamble of the SDGs. Furthermore, the normative pillars of universality and indivisibility are crucial in setting priorities and meeting deadlines in relation to specific targets including those pertaining to health and well-being, clean water, sustainable use of the marine environment, as well as climate action. As such, overproduction in sectors such as fisheries is primarily a responsibility of developed economies in the North Hemisphere but the impacts are largely felt in the developing world because it threatens food security and the viable existence of coastal communities.

11 Parallel Implementation Mechanisms

The Preamble of the LOS Convention stipulates that "the problems of ocean space are closely interrelated and need to be considered as a whole."[111] Similar

109 See, ExxonMobil statement at COP 21: http://corporate.exxonmobil.com/en/current-issues/climate-policy/climate-perspectives/statement-on-cop-21.

110 P. Sands, J. Peel, R. MacKenzie, *Principles of International Environmental Law* 3rd Ed. (Cambridge: CUP, 2012).

111 United Nations on the Law of the Sea. 1833 UNTS. 3/21 ILM 1261 (1982). Entered into force on 16 November 1994. Recital 3.

to the Law of the Sea Convention, the SDGs are built upon the fundamental duty of cooperation between public and private actors: States, international organizations, industry and civil society. The institutions for the delivery of the SDGs involve collaboration with global, regional, national and sub-national bodies. In particular, they rely upon UNEP's regional seas programmes to take action on environmental protection.[112] Additional complexity arises in the task of implementation due to various bodies that exercise mandates in relation to the stewardship of the living and non-living resources of the ocean, as well as in relation to specific sector activities such as the safety of navigation (the IMO), and scientific research (the Intergovernmental Oceanographic Commission (IOC/UNESCO)). Very often, as can be seen in the context of the SDGs, there is a requirement for parallel implementing mechanisms that entail moving from soft to hard law to ensure the realization of specific targets, such as increasing the economic benefits to SIDS and least developed countries from the sustainable use of marine resources, including through sustainable management of fisheries, aquaculture and tourism.[113] Furthermore, there is a critical requirement to ensure that there is a high degree of convergence between the regulatory and policy regimes to achieve many of the targets. These include the frameworks established under UNCLOS, the UNFCCC and the Paris Agreement, as well as other water-related instruments. In this regard, capacity building and the role of academic institutions can play a significant role.

12 Academic Institutions as Agents of Change

The SDGs and Agenda 2030 provide a compelling and coherent policy backdrop that adds new momentum to international efforts to engender social well-being, peace and prosperity, as well as sustainable economic growth in harmony with a clean environment. The educational sector bears significant responsibilities with Target 5.7 under Goal 4 aimed at ensuring that "all learners acquire the knowledge and skills needed to promote sustainable development, including, among others, through education for sustainable development and sustainable lifestyles, human rights, gender equality, promotion of a culture of peace and non-violence, global citizenship and appreciation of cultural diversity and of culture's contribution to sustainable development."

112 UNEP (2017) Regional Seas Programme. Available at: http://web.unep.org/regionalseas/who-we-are/regional-seas-programmes.

113 Goal 14.7, SDGS.

In promoting lifelong learning for all, the role of the World Maritime University is directly linked to the SDGs through the delivery of postgraduate education, professional training and research that promotes sustainable development. Likewise, the Rhodes Academy of Ocean Law and the Yeosu Academy Law of the Sea make a substantial and cutting-edge contribution to education and research in line with Goals, 4 13 and 14 by up-skilling international lawyers and other mid-career experts in modern international law of the sea. They also play a significant role in fostering effective, accountable and transparent institutions, as well as strengthening the capacity of participants in playing a constructive role in the work of institutions of global governance, in accordance with Targets 16.6 and 16.8 of Goal 16. A major development in this regard is the establishment of WMU-Sasakawa Global Ocean Institute at the World Maritime University through a generous endowment from the Nippon Foundation. The Institute will act as an independent focal point for ocean related research and dialogue.

The United Nations General Assembly has also acknowledged the important contribution of the United Nations-Nippon Foundation of Japan Fellowship Programme. The Programme has awarded 120 fellowships to individuals from 70 Member States since 2004. This fosters human resource development for developing Member States in the field of ocean affairs and the law of the sea and related disciplines, as well as the fostering of global interlinkages and continuing capacity development through the alumni programme.[114]

13 Transformative Role of Cities

Close to two-thirds of the world's population live in cities, which can act as a catalyst in moving the 2030 Agenda from words to action on sustainability. In undertaking this task, it might be pertinent to recall that Charles Dickens drew attention to the possibility of human transformation and resurrection in *A Tale of Two Cities* (1859) in order to move forward to a better society. On a similar note, Goal 4 aims to make cities inclusive, safe, resilient and sustainable. This is particularly problematic in relation to the growth of cities in close proximity to the sea, which is a worldwide phenomenon and has many negative effects by contributing to the loss of coastal ecosystems, land-based pollution of the marine environment, along with increased greenhouse gas emissions,

114 A/RES/70/235 on "Oceans and the law of the sea" (adopted on 23 December 2015) paragraph 43. United Nations—Nippon Foundation Fellowship Alumni. Available at: http://www.unfalumni.org.

making up a staggering total of 75 per cent of all emissions in 2017.[115] The 2030 Agenda thus needs to act as a pivot point in the global battle to meet development challenges and respond to climate change in urban areas. In particular, the loss of natural barriers to extreme weather events and the destruction of the quality of the coastal environment is impacting upon living space, as evidenced by the loss of Miami beach and downtown flooding in hurricane Irma. Mangroves and dune systems provide the best line of defense to storm surges and the destructive impacts of tropical storms. They also act as a carbon sink and concerns have been raised by their catastrophic loss worldwide. In particular, some estimates suggest that they are disappearing at a rate of up to two per cent a year in Indonesia, faster than anywhere else in the world. The scale of the problem may be appreciated when it is considered that one-third of the world's total coastal carbon content is soaked up in Indonesia.[116] There have been a number of efforts by Indonesia to monetize coastal ecosystem services, similar to tropical rainforests, with a view to giving communities access to international carbon markets.[117] In many parts of the world, cities are also plagued by poor water quality, the absence of suitable infrastructure, and the privatization of public space, along with the lack of planning integration.[118] Importantly, cities in the United States, such as Pittsburg, are stepping up their efforts and beginning to play a major role in combatting climate change and implementing commitments under the Paris Agreement.[119] Indeed, the unofficial US delegation at COP23 in Bonn was made-up of 20 states, 110 cities and 1,400 businesses, thus demonstrating the scale of support for the climate change regime.[120]

---

115 Report of the Secretary-General, "Progress towards the Sustainable Development Goals," E/2017/66.

116 See, J. Langenheim, "Hope for Indonesia's valuable but threatened mangroves," *The Guardian*, 15 June 2017.

117 Ibid.

118 Å. Johannessen, A. Rosemarin, F. Thomalla, Å. Gerger Swartling, T. Stenström, G. Vulturius (2014) "Strategies for building resilience to hazards in water, sanitation and hygiene (WASH) systems: The role of public private partnerships," *International Journal of Disaster Risk Reduction*, Vol. 10, 102–115. See also: United Nations. "The Habitat Agenda Goals and Principles, Commitments and the Global Plan of Action." Available at: http://www.un.org/en/events/pastevents/pdfs/habitat_agenda.pdf (Para. 8).

119 See: "COP23: Pittsburgh Mayor on why more and more cities are committing to Paris Agreement." Available at: http://www.un.org/sustainabledevelopment/blog/2017/11/cop23-pittsburgh-mayor-on-why-more-and-more-cities-are-committing-to-paris-agreement/.

120 "As Trump administration touts coal at U.N. gathering, U.S. cities and states target climate change," *Los Angeles Times*, 17 November 2017.

14 Measuring Progress and Solving Practical Problems

In discharging collective responsibilities regarding the implementation of the SDGs, the international community is guided by performance indicators and annual reporting.[121] Again, the task of overcoming the practical problems associated with implementation is by fostering greater engagement with stakeholders, including business and industry most particularly. There is a need to harmonize the actions of the various interest groups including most notably the work undertaken by National Councils for Sustainable Development, local authorities, and stakeholder partnerships including those facilitated and supported by the UN. The role of the private sector cannot be over-emphasized as is evident from their sterling contribution to implementing commitments under the Paris Climate Agreement. For instance, the United Nations Global Compact has a membership of over 8,000 companies and 4,000 non-business partners.

The significant role of international fora in the judicial recognition and legal application of the SDGs needs to be stressed. In particular, the international community can draw from the New Delhi Declaration of the Principles of International Law relating to Sustainable Development, the decisions of international courts and tribunals including the jurisprudence of the International Court of Justice and the International Tribunal for the Law of the Sea, as well as decisions of the United Nations and its constituent bodies.

15 Moving Forward

The 2030 Agenda is galvanizing the international community and providing some badly needed political impetus and additional context to the implementation of international law. Additional context is particularly useful when addressing contentious law of the sea issues and in making difficult decisions concerning the regulation and management of offshore activities in the 21st century. In this regard, one of the inherent strengths of the 2030 Agenda is that several of the SDGs focus the spotlight on the need to foster greater stakeholder engagement and accountability regarding the unsustainable use of the oceans and its resources. By doing so, they provide an essential policy perspective that informs the environmental dimension of contemporary international law of the sea and highlights the importance of effective engagement by all actors in bridging the sustainability goals on oceans, water and climate action under the 2030 Agenda.

121 *Op. cit.* note 12.

CHAPTER 6

Realising Biodiversity Conservation and Sustainable Use in Southern Hemisphere Oceans beyond National Jurisdiction: Challenges and Prospects

*Robin Warner**

Abstract

The southern hemisphere is dominated by oceanic influences with 80% of its surface area consisting of ocean space. It hosts abundant marine biodiversity, which is under increasing pressure from activities such as fisheries and shipping as well as multiple sources of marine pollution and climate change impacts. The process initiated by the UN General Assembly (UNGA) Resolution 69/292 to develop the elements of an international legally binding treaty (ILBI) for conservation and sustainable use of marine biodiversity in areas beyond national jurisdiction (ABNJ) has the potential to contribute to a more integrated and cross-sectoral system of oceans governance at a global and regional scale. This chapter examines key features of the existing global and regional ocean governance framework in the southern hemisphere and how the anticipated ILBI might enhance existing ocean governance frameworks.

1 Introduction

The process initiated by the UN General Assembly (UNGA) in UNGA Resolution 69/292[1] to develop the elements of an international legally binding

* Dr. Robin Warner is Professor at the Australian National Centre for Ocean Resources and Security (ANCORS), University of Wollongong, Australia. Her research interests include law of the sea, oceans governance, marine environmental law, climate law, maritime regulation and enforcement. The author's PowerPoint presentation is available at http://www.virginia.edu/colp/pdf/yogya-warner.pdf.

1 United Nations General Assembly, *Development of an International Legally Binding Instrument under the United Nations Convention on the Law of the Sea on the Conservation and Sustainable Use of Marine Biological Diversity of Areas Beyond National Jurisdiction*, resolution

 DOI:10.1163/9789004366619_011

treaty/instrument (ILBI) for conservation and sustainable use of marine biodiversity in areas beyond national jurisdiction (ABNJ) has prompted wide ranging research into existing ocean governance frameworks, and their applicability to conservation and sustainable use of marine biodiversity in ABNJ. UNGA 69/292 provides that negotiations to develop the new ILBI should address the four elements of a package deal agreed to by States in 2011. These elements comprise marine genetic resources including questions on the sharing of benefits, measures such as area-based management tools, including marine protected areas, environmental impact assessments and capacity building, and the transfer of marine technology. The Preparatory Committee meetings held in 2016 identified additional cross-cutting issues for consideration including definitions, scope of the instrument, relationship of the instrument to other instruments and frameworks, institutional arrangements, compliance, responsibility and liability, dispute settlement, and final clauses.[2] UNGA Resolution 69/292 also stipulates that the process to develop the ILBI should not undermine existing relevant legal instruments and frameworks, and relevant global sectoral and regional bodies.[3] This chapter examines key features of the existing global and regional ocean governance framework in the southern hemisphere and its relevance to the conservation and sustainable use of marine biodiversity in ABNJ. It also discusses how existing instruments and institutions address the elements of the package deal referred to in UNGA Resolution 69/292, and how any new governance architecture contained in the ILBI might relate to existing ocean governance frameworks. For the majority of oceanic regions in the southern hemisphere, this analysis reveals significant gaps in regulating conservation and sustainable use of marine biodiversity in ABNJ and potential challenges in forging new relationships between existing ocean governance regimes and the ILBI. The UNGA process to develop the ILBI represents an historic opportunity to remediate these gaps and create a more cross-sectoral

---

adopted by the General Assembly on 19 June 2015, GA Res 69/292, 69th sess. Agenda Item 74(a), A/Res/69/292 (6 July 2015). See also https://documents-dds-ny.un.org/doc/UNDOC/GEN/N15/187/55/PDF/N1518755.pdf?OpenElement.

2 Preparatory Committee established by General Assembly Resolution 69/292: Development of an international legally binding instrument under the United Nations Convention on the Law of the Sea on the conservation and sustainable use of marine biological diversity of areas beyond national jurisdiction, *Chair's indicative suggestions of clusters of issues and questions to assist further discussions in the informal working groups at the second session of the Preparatory Committee*, http://www.un.org/Depts/los/biodiversity/prepcom_files/IWGs_Indictive_Issues_and_Questions.pdf (accessed 31 May 2017).

3 UNGA Res 69/292, above note 2, paragraph 3.

and integrated system of ocean governance which will benefit not only marine biodiversity in ABNJ, but also within national jurisdictions. This paper explores how the UNGA process could lead to a more effective system for conserving and sustainably using marine biodiversity in the southern hemisphere's ABNJ through provisions and implementation mechanisms that support more interconnected ocean governance systems across all elements of the package deal while taking into account cross-cutting issues. In the face of growing threats and pressures on the marine biodiversity in ABNJ, the paper emphasises that it is timely to promote a more coherent system of ocean governance in all regions that draws on modern conservation principles and measures developed under international environmental law.

2 Southern Hemisphere Ocean Areas

The southern hemisphere is dominated by oceanic influences with 80% of its surface area consisting of ocean space as compared to 60.7% of the northern hemisphere.[4] It encompasses the majority of the Pacific Ocean and Indian Ocean as well as a substantial proportion of the Atlantic and the whole of the Southern Ocean. Many of these ocean areas host abundant and unique repositories of marine biodiversity, which are under increasing pressure from commercial scale activities such as fisheries and shipping, as well as land-based sources of marine pollution and climate change impacts.[5]

A recent scientific study identified six marine areas of exceptional biodiversity in the temperate and tropical regions of the Atlantic Ocean, Indian Ocean, and Pacific Ocean in the Southern Hemisphere which have been most affected by climate change pressures and industrial scale fishing.[6] These areas are located in the central-western Pacific waters of Peru and the Galápagos Archipelago, in the southwestern Atlantic Ocean in the Patagonian waters of Argentina and Uruguay, on the western side of the Indian Ocean along the coasts of South Africa, Mozambique, Tanzania, Kenya, and Madagascar, in a large

---

4 Steve Rintoul, Wenju Cai, Helen Cleugh and Gongke Tan, "New Research Centre Focuses on the 'Ocean Hemisphere,'" (22 May 2017) *ECOS* Issue 231 Climate, Global Warming Oceans, https://blogs.csiro.au/ecos/cshor/ (accessed 31 May 2017).

5 E.R. Selig, W.R. Turner, S. Troeng, B.P. Wallace, B.S. Halpern, K. Kaschner, B.G. Lascelles, K.E. Carpenter and R.A. Mittermeier, "Global Priorities for Marine Biodiversity Conservation" (2014) *PLoS ONE* 9(1): e82898. https://doi.org/10.1371/journal.pone.0082898.

6 F. Ramirez, I. Afan, L.S. Davis and A. Chiaradia, "Climate Impacts on Global Hot Spots of Marine Biodiversity" (2017) 3(2) *Science Advances* e1601198 DOI: 10.1126/sciadv.1601198.

area of the central-western Pacific Ocean including water masses surrounding Indonesia, Malaysia, Philippines, Papua New Guinea, Taiwan, and the south of Japan, in waters surrounding New Zealand and Eastern and Southern Australia and in marine areas in Oceania and the central Pacific Ocean. These areas all exhibit local to regional increasing water temperatures, slowing current circulation, and decreasing primary productivity.[7] The study urges the international community "to find solutions that go beyond the interests and borders of sovereign states to conserve the exceptional biodiversity" of these marine hot spots.[8]

The Convention on Biological Diversity (CBD) began a process in 2004 to describe ecologically and biologically significant areas (EBSAs), "in need of protection, in open ocean waters and deep sea habitats."[9] The EBSA designation process has identified a large number of areas within the southern hemisphere oceans whose ecological uniqueness and sensitivity warrant specific conservation and sustainable use measures. Three of the nine workshops convened to identify EBSAs were specifically focused on southern hemisphere oceanic regions comprising the Western South Pacific, the South Eastern Atlantic and the Southern Indian Ocean. The workshops identified 109 EBSAs of which 30 were in ABNJ.[10] In an article analysing the EBSA process, Bax *et al.* acknowledge that in terms of biodiversity conservation, "systematic management of ABNJ is currently lacking in almost all instances" and more effective implementation of existing agreements combined with a new implementing agreement for biodiversity conservation will be needed to improve management of the world's oceans.[11] This paper reviews some of the strengths and weaknesses of the patchwork of agreements and institutions relevant to marine biodiversity conservation and sustainable use in the southern hemisphere. It analyses different approaches to strengthening the fabric of marine biodiversity governance in ABNJ and the role of the ILBI in achieving this objective.

---

7 *Ibid.*

8 *Ibid.*

9 D.C. Dunn, J. Ardron, N. Bax, P. Bernal, J. Cleary, I. Cresswell, B. Donnelly, P. Dunstan, K. Gjerde, D. Johnson, K. Kaschner, B. Lascelles, J. Rice, H. von Nordheim, L. Wood and P.N. Halpin, "The Convention on Biological Diversity's ecologically or biologically significant areas: origins, development and current status" (2014) 49 *Marine Policy* 137–145.

10 N.J. Bax, J. Cleary, B. Donnelly, D.C. Dunn, P.K. Dunstan, M. Fuller and P.N. Halpin, "Results of efforts by the Convention on Biological Diversity to describe ecologically or biologically significant marine areas" (2016) 30(3) *Conservation Biology* 571–581.

11 *Ibid.* at 580.

3 Existing Governance Frameworks for Marine Biodiversity Conservation and Sustainable Use in Southern Hemisphere ABNJ

A diverse array of binding international law agreements and non-binding arrangements at global and regional levels are relevant to the conservation and sustainable use of marine biodiversity in southern hemisphere ABNJ. The majority of these instruments and their associated institutions are focused on specific sectors including fisheries, shipping, and deep seabed mining with conservation of biodiversity as a secondary rather than as a primary objective. In the fisheries sector, the UN Fish Stocks Agreement (UNFSA)[12] provided the first comprehensive blue print for sustainable fisheries conservation and management in marine ABNJ and model provisions for cooperation between coastal States and flag States with high seas fishing fleets. As well as codifying all the relevant international environmental law principles for conservation and management of marine living resources, it provided practical guidance for regional fisheries management organizations (RFMOs) on cooperative compliance and enforcement measures. Since the entry into force of the UNFSA, a complex pattern of high seas fisheries regulation has continued to emerge as existing RFMOs adapt their agreements and institutions to incorporate the new provisions, and as new RFMOs have been established. Across the southern hemisphere there is a wide variation in the extent to which RFMOs have incorporated biodiversity conservation and sustainable use objectives in their agreements and implemented them in the state practice of their members.[13]

In the shipping sector, the principal environmental protection measure applicable to southern hemisphere ABNJ is the designation of Antarctic waters as special areas under Annexes I and II of the International Convention for the Prevention of Pollution from Ships (MARPOL 73/78).[14] This sectoral measure predates the more holistic measures taken by CCAMLR to designate marine protected areas in the Antarctic Ocean but can be viewed as essentially

12 United Nations Agreement for the Implementation of the Provisions of the United Nations Convention on the Law of the Sea of 10 December 1982 Relating to the Conservation and Management of Straddling Fish Stocks and Highly Migratory Fish Stocks (New York, 8 September 1995, in force 11 December 2001) 2167 UNTS 3 ("UNFSA").

13 R. Warner, K.M. Gjerde and D. Freestone, "Regional Governance for Fisheries and Biodiversity" in S.M. Garcia, J. Rice and A. Charles (eds.), *Governance of Marine Fisheries and Biodiversity Conservation* (Wiley Blackwell, Chichester, West Sussex, 2014) 211–224.

14 International Convention for the Prevention of Pollution from Ships (London, 1 June 1978, Annex I in force 2 October 1983, Annex II in force 6 April 1987, Annex III in force 1 July 1992, Annex IV in force 27 September 2003, Annex V in force 31 December 1988, Annex VI in force 19 May 2005) (1978) 17 ILM 546 ("MARPOL 73/78").

complementary to those measures. For the deep seabed mining sector, the only exploration contractor sites in the southern hemisphere are in the Indian Ocean. There has been no environmental management plan developed for this area as yet, but a recent discussion paper and draft environmental regulations issued by the International Seabed Authority in January 2017 have foreshadowed the development of regional environmental management plans for exploration and exploitation sites in the Area.[15]

In addition to the sector-based instruments, there are a variety of instruments and arrangements in the southern hemisphere which are focused on oceanic conservation, but only some of these apply to ABNJ. The geographic scope of regional seas organizations in the southern hemisphere is predominantly limited to marine areas within national jurisdiction, with the one exception being the 1986 Convention for the Protection of the Natural Resources and Environment of the South Pacific Region (Noumea Convention)[16] and the associated South Pacific Regional Environment Programme (SPREP) which covers the high seas areas enclosed by the Exclusive Economic Zones (EEZs) of its member States.[17]

This section will provide an overview of the biodiversity conservation provisions of key ocean governance instruments and arrangements in selected regions of the southern hemisphere, and some of the measures taken to implement those provisions. It will also review the challenges confronted by these regions in implementing these provisions. Finally, it will highlight the interconnections between regional institutions established under these instruments, particularly in relation to biodiversity conservation and sustainable use obligations in ABNJ.

3.1 *South West Pacific*

The South West Pacific region is characterised by vast tracts of ocean space dotted with land masses which range from sizeable island nations such as Australia and New Zealand to small-island developing States such as

15 International Seabed Authority, "Developing a Regulatory Framework for Mineral Exploitation in the Area. A Discussion Paper on the development and drafting of Regulations on Exploitation for Mineral Resources in the Area (Environmental Matters)" (25 January 2017), https://www.isa.org.jm/files/documents/EN/Regs/DraftExpl/DP-EnvRegsDraft25117.pdf.

16 Convention for the Protection of the Natural Resources and Environment of the South Pacific Region (Noumea, 24 November 1986, in force 22 August 1990) (1987) 26 ILM 41 ("Noumea Convention").

17 *Ibid.*, Article 2(a)(ii).

Tuvalu and Niue.[18] The majority of small-island States in the Southwest Pacific have land areas under 700 square kilometres and are heavily dependent on a healthy marine environment for their survival.[19] The region has one of the highest quotients of marine biodiversity in the world, with a large population of rare and endangered species such as dugongs, turtles, and whales.[20] The EEZs and high seas areas of the region contain some of the world's largest stocks of tuna that have been subject to exploitation by distant water fishing nations.[21] The region also hosts a variety of vulnerable marine habitats such as seamounts, hydrothermal vents, and submarine trenches many of which are in ABNJ.[22] The abundant marine biodiversity of the region is subject to multiple stress factors including marine pollution from land and sea based sources, unsustainable fisheries practices, climate change impacts, natural disasters, and alien species invasion.[23] Many of the small island nations in the region have only attained independence in recent decades, and their capacity to fund and manage environmental protection programmes is severely limited.[24] Much of the finance and technical expertise for environmental management is provided by the developed countries in the region and other sources of international aid.[25] Notwithstanding these challenges, the South West Pacific displays a high level of coherence among its regional institutions devoted to ocean governance and there have been ongoing political commitments among Pacific Island States to work towards marine biodiversity conservation and sustainable use within and beyond national jurisdiction.[26] These commitments

18 R. Herr, "Environmental Protection in the South Pacific: the Effectiveness of SPREP and its Conventions" in O.S. Stokke and O.B. Thommessen (eds.), *Yearbook of International Cooperation of Environment and Development 2002/3* (Earthscan Publications Ltd., London, 2002) 41–43.

19 T.I. Tutangata and M. Power, "The Regional Scale of Ocean Governance: Regional Cooperation in the Pacific Islands" (2002) 45(11) *Ocean and Coastal Management* 873–884.

20 SPREP, "Our Ocean Biodiversity: Pacific Conversations with SPREP" http://www.sprep.org/biodiversity-ecosystems-management/our-ocean-biodiversity-pacific-conversations-with-sprep.

21 Herr, above note 18, 43.

22 *Ibid.*

23 SPREP, "Pacific Biodiversity, Including Marine and Coastal Life" http://www.sprep.org/attachments/Publications/FactSheet/Oceans/pacific-biodiversity-including-marine-coastal-life.pdf (accessed 31 May 2017).

24 Herr, above note 18, 43.

25 *Ibid.*

26 R. Mahon, L. Fanning, K.M. Gjerde, O. Young, M. Reid and S. Douglas, "Transboundary Waters Assessment Programme (TWAP) Assessment of Governance Arrangements for the Ocean, Volume 2: Areas Beyond National Jurisdiction" (UNESCO-IOC Technical Series, Paris, 2015), 1–119, xi, 43 and 51.

are underpinned by modern international environmental law principles and management approaches in instruments such as the Noumea Convention and the Convention on the Conservation and Management of Highly Migratory Fish Stocks in the Western and Central Pacific Ocean (WCPFC).[27] Despite these positive regional attributes, key constraints including deficiencies in human, financial, and technical resources and extra-regional influences, continue to militate against realising biodiversity conservation and sustainable use within and beyond national jurisdiction.

The Noumea Convention anticipates the collaboration of its parties in protecting the marine environment of the whole Convention Area including its high seas enclaves. It commits its parties to prevent, reduce, and control pollution of the Convention Area from any source and to ensure sound environmental management of natural resources.[28] Although the Noumea Convention pre-dates the 1992 Convention on Biological Diversity (CBD)[29] and its codification of biodiversity protection, Article 14 of the Noumea Convention reflects some of the key concepts associated with an integrated and ecosystem-based approach to ocean management by providing that parties shall take all appropriate measures to protect and preserve rare and fragile ecosystems and depleted, threatened, and endangered flora and fauna, as well their habitat in the Convention Area. Article 14 also recommends that parties establish protected areas or regulate any activities likely to have adverse effects on the species, ecosystems, or biological processes of such areas. The Noumea Convention and the associated South Pacific Region Environment Programme (SPREP) demonstrate a strong initial expression of political will and the necessary institutional basis for further integrated environmental protection across the South Pacific region, but they have faced substantial resource and capacity challenges in implementation terms.[30]

The WCPFC was the first comprehensive conservation and management regime for highly migratory fish stocks such as tuna in Pacific ABNJ areas. The area of competence of the WCPFC covers a large swathe of the Pacific Ocean in the southern hemisphere, including a large area of high seas lying outside and between the 200-nautical-mile EEZs of its Parties.[31] The WCPFC Commission

27 Convention on the Conservation and Management of Highly Migratory Fish Stocks in the Western and Central Pacific Ocean (Honolulu, 5 September 2000, in force 19 June 2004) (2001) 40 ILM 277 ("WCPFC Convention").

28 Noumea Convention, Article 5(1).

29 Convention on Biological Diversity (Nairobi, 22 May 1992, in force 29 December 1993) (1992) 31 ILM 822 ("CBD").

30 Tutangata and Power, above note 19, 879.

31 WCPFC Convention, Article 3.

is empowered to adopt principles and measures for conservation and management of the highly migratory fish stocks in its area of competence that reflect key international environmental law principles, biodiversity conservation, and sustainable use objectives.[32] The Commission must apply the precautionary principle and determine the impact of fishing activities on target and associated or dependent species and their environment and adopt plans where necessary to ensure the conservation of species and protect habitats of special concern.[33] The conservation measures to be taken by the Commission also include those which protect biodiversity in the marine environment and which assess the impact of fishing activities on other species belonging to the same ecosystem.[34] While the WCPFC provisions are progressive in terms of biodiversity conservation and sustainable use in ABNJ, implementation of relevant conservation and management measures has proven challenging. A conservation and management measure (CMM) proposed by the WCPF Commission in 2008 capped fishing efforts for big eye and yellow fin tuna in the Convention Area, and closed the high seas enclaves in its area of competence to tuna fishing. This proposed CMM faced objections from distant water fishing nation (DWFN) members of WCPFC and was substantially diluted in its final form. Pacific Island parties to the subregional Nauru Agreement Concerning Cooperation in the Management of Fisheries of Common Interest (PNA Agreement)[35] managed to circumvent the inherent weaknesses in this WCPFC CMM by introducing an innovative arrangement to prevent purse seine vessels licensed by PNA member States to fish for tuna in their EEZs from fishing the high seas between 10 degrees N to 20 degrees S of the Western Central Pacific.

In recent years, there have been additional initiatives taken at the Pacific-wide regional level to coalesce political will around oceanic conservation and sustainable use of marine resources. The Pacific Islands Regional Oceans Policy (PIROP)[36] and the associated Framework for a Pacific Oceanscape (FPO)[37] together form the basis for a more integrated and cross-sectoral approach to

32 *Ibid.*, Article 5(a) and (b).

33 *Ibid.*, Articles 5(c) and 6.

34 *Ibid.*, Article 5(f) and (d).

35 Nauru Agreement Concerning Cooperation in the Management of Fisheries of Common Interest (Nauru, 11 February 1982, in force 4 December 1982), latest version available at http://www.pnatuna.com/sites/default/files/Latest%20Nauru%20Agreement_0.pdf ("Nauru Agreement").

36 Pacific Islands Forum Secretariat, Pacific Islands Regional Oceans Policy, http://www.forumsec.org/resources/uploads/attachments/documents/PIROP.pdf.

37 Pacific Islands Forum Secretariat, Framework for a Pacific Oceanscape, http://www.forumsec.org/resources/uploads/embeds/file/Oceanscape.pdf.

regional oceans governance across the Pacific region. These policy instruments have been complemented by the creation of the Office of the Pacific Ocean Commissioner[38] to strengthen policy coordination in regional ocean governance and the Pacific Ocean Alliance,[39] which is a mechanism for more inclusive consultation among States and other regional stakeholders on technical and policy advice concerning regional ocean governance. Successive iterations of the FPO have focused on marine biodiversity conservation and sustainable use within and beyond national jurisdiction, and many of the issues covered have resonated with key elements being discussed in the Preparatory Committee meetings for the ILBI, including representative networks of marine protected areas and environmental impact assessment. The high level of regional coherence among the Pacific Small Island Developing States (PSIDS) has enabled them to play a very active role in the Preparatory Committee meetings for the ILBI.

3.2 *South East Pacific*

The South East Pacific region extends the full length of the Pacific coast of South America from Panama to Cape Horn off the coasts of Chile, Peru, Ecuador, Colombia and Panama with the majority of the region lying in the Southern Hemisphere.[40] It supports one of the world's most productive fisheries with the potential sustainable fisheries yield in 2013 estimated to be 8.9 million tonnes.[41] In addition to its high fisheries quotient, the region also hosts multiple species of shellfish, birds, marine mammals, algae, and reptiles.[42] There are two deep sea trenches, off the coasts of Peru, Chile, and Central America, and off the coasts of Ecuador and Colombia, which host some of the world's most unique species, habitats, and ecosystems.[43] The regional ocean governance framework is comprised of two RFMOs, the Inter-American Tropical Tuna Commission (IATTC), the more recently established South Pacific Regional Fisheries Management Organisation (SPRFMO) and a regional seas organisation, the Comision Permanente del Pacifico Sur (Permanent Commission for

---

38 Pacific Islands Forum Secretariat, Office of the Pacific Ocean Commissioner, http://www.forumsec.org/pages.cfm/strategic-partnerships-coordination/pacific-oceanscape/pacific-ocean-commissioner.html?printerfriendly=true.

39 Pacific Islands Forum Secretariat, Pacific Ocean Alliance, http://www.forumsec.org/resources/uploads/embeds/file/Pacific_Ocean_Alliance_Charter.pdf.

40 United Nations Environment Programme (UNEP) Regional Seas, South East Pacific, http://drustage.unep.org/regionalseas/south-east-pacific.

41 *Ibid.*

42 *Ibid.*

43 *Ibid.*

the South Pacific) (CPPS).[44] The two RFMOs have jurisdiction respectively over high seas tuna and non-tuna fisheries resources in South East Pacific ABNJ. CPPS is an advisory body with no formal jurisdictional responsibility over marine biodiversity in ABNJ, but its member States have committed to developing a broad regional vision on marine policy which would encompass their interests in living and non-living resources in ABNJ.

The Convention on the Conservation and Management of High Seas Living Resources in the South Pacific Ocean (SPRFMO Convention)[45] entered into force in August 2012. The Convention area comprises the region of the South Pacific beyond areas of national jurisdiction encompassing an area of around 59 million square kilometres.[46] Non-highly migratory fisheries species managed by the South Pacific Regional Fisheries Commission (SPRFMO) include pelagic species such as jack mackerel, chub mackerel, jumbo flying squid and other squid species. The marine environment is quite diverse, ranging from the fine muddy sediments of the deep sea floor, which host a variety of microscopic single celled and shelled animals to the seamounts, banks, and ridges at shallower depths that are home to bottom invertebrates such as lobsters and crabs, and bottom dwelling fish, such as orange roughy and alfonsino.[47] The Convention incorporates a number of provisions which reflect a commitment to improved biodiversity conservation and sustainable use of marine living resources in ABNJ. Biodiversity conservation is highlighted in Article 20 (c) and (d) of the SPRFMO Convention which provides that the conservation and management measures adopted by the Commission shall include measures to:

> (c) maintain or restore populations of non-target and associated or dependent species to above levels at which their reproduction may become seriously threatened; and
> (d) protect the habitats and marine ecosystems in which fishery resources and non-target and associated or dependent species occur from the impacts of fishing, including measures to prevent significant adverse impacts on vulnerable marine ecosystems and precautionary measures

44 UNEP-WCMC, *Governance of areas beyond national jurisdiction for biodiversity conservation and sustainable use: Institutional arrangements and cross-sectoral cooperation in the Western Indian Ocean and the South East Pacific* (Cambridge (UK): UN Environment World Conservation Monitoring Centre, 2017), 71–72.

45 Convention on the Conservation and Management of High Seas Fishery Resources in the South Pacific Ocean (Auckland, 14 November 2009, in force 24 August 2012) ATS 28 ("SPRFMO Convention").

46 SPRFMO Convention, Article 5.

47 SPRFMO, *About SPRFMO*, https://www.sprfmo.int/about/.

where it cannot adequately be determined whether vulnerable marine ecosystems are present or whether fishing would cause significant adverse impacts on vulnerable marine ecosystems.

Consistent with its biodiversity conservation obligations, SPRFMO has already adopted a wide range of CMMs in relation to bottom fishing, the management of jack mackerel, the banning of gillnetting, the collection of detailed fishing data, inspections in port and at sea, transhipments, a Vessel Monitoring System, an IUU List, vessels without nationality, minimising the bycatch of seabirds and establishing a compliance and monitoring scheme.[48] It has also begun to establish formal and informal links with other RFMOs in the region and beyond, and has extended observer status to CPPS and a range of relevant IGOs and NGOs.[49] A recent UNEP-WCMC report notes that there is still significant potential for further intra- and cross-sectoral cooperation between SPRFMO and other regional and global bodies.[50]

The IATTC Convention (Antigua Convention) which entered into force in 2010 also contains biodiversity conservation obligations.[51] In particular, it obligates the Commission to adopt as necessary "Conservation and management measures and recommendations to maintain or restore populations of dependent, associated, or same ecosystem species that are likely to be affected by fishing activities."[52] Since the entry into force of the Antigua Convention, IATTC has adopted a range of CMMs relating to the conservation of associated and dependent species, including sharks, seals, turtles, and incidental seabird mortality.[53] It has forged targeted intra-sectoral links with other tuna RFMOs. Its strongest cooperative relationship is with WCPFC, the other tuna RFMO covering the Pacific Ocean but it also has a memorandum of understanding (MOU) with CPPS mainly focused on mutual training opportunities.[54]

48 SPRFMO, *Conservation Measures,* https://www.sprfmo.int/conservation-measures/.

49 UNEP-WCMC report, above note 44, 85.

50 *Ibid.,* 86.

51 Convention for the Strengthening of the Inter-American Tropical Tuna Commission established by the 1949 Convention between the United States of America and the Republic of Costa Rica (Washington, DC, 27 June 2003, in force 27 August 2010), available at https://www.iattc.org/PDFFiles2/Antigua_Convention_Jun_2003.pdf ("IATTC, Antigua Convention").

52 Antigua Convention, Article VII (1)(f).

53 UNEP-WCMC Report, above note 44, 82.

54 *Ibid.,* 83.

The 1981 Convention on the Protection of the Marine Environment and Coastal Area of the South East Pacific (Lima Convention)[55] and the South East Pacific Action Plan were originally adopted by Chile, Ecuador and Peru in 1981 to provide a legal and institutional framework for the marine environmental protection of the region.[56] Panama and Colombia have since joined the Lima Convention. The CPPS is the institutional body implementing the Convention and Action Plan, and acts as a coordinating mechanism designed to harmonise and align the marine policy of its members.[57] The parties to the Lima Convention have consistently evinced an interest in the conservation and sustainable use of marine biodiversity in adjacent ABNJ through a series of declarations, but this has not resulted in concrete conservation and management measures as the CPPS does not have any jurisdictional competence in ABNJ.[58] CPPS has a broad array of cross-sectoral connections with global and regional organizations, including IATTC, SPRFMO, CCAMLR, OSPESCA, CBD, UNEP, FAO, IOC, and WMO, as well as with a relevant group of NGOs.[59] The experience gained by CPPS in coordinating national marine policies as well as collaborating with multiple global and regional organizations on conservation and sustainable use of marine biodiversity places it in a prime position to coordinate implementation of the ILBI in the South East Pacific.[60]

3.3 *Southern Ocean*

The Convention for the Conservation of Antarctic Marine Living Resources (CAMLR)[61] explicitly adopts a precautionary and ecosystem based approach to marine living resource management that recognises the complex interconnections between all parts of the Antarctic ecosystem.[62] Since its inception in 1982, the Commission for the Conservation of Antarctic Marine Living Resources (CCAMLR) has adopted a suite of innovative measures to implement its ecosystem based approach to conservation and sustainable use of marine living resources in the Antarctic treaty area. These include banning destructive

55 Convention on the Protection of the Marine Environment and Coastal Area of the South East Pacific (Lima, 12 November 1981, in force 19 May 1986), 33 *International Digest of Health Legislation* (1982) 96 ("Lima Convention").

56 UNEP-WCMC Report, above note 44, 76–78.

57 *Ibid.*

58 *Ibid.*

59 *Ibid.*, 79–80.

60 *Ibid.*, 80.

61 Convention for the Conservation of Antarctic Marine Living Resources (Canberra, 20 May 1980, in force 7 April 1981) (1980) 19 ILM 837 ("CAMLR").

62 *Ibid.*, Article II(3).

fisheries practices such as bottom trawling, mandating measures to reduce incidental seabird mortality caused by baited hooks in long line fishing and monitoring the effects of fishing on non-target species by collection of data on CCAMLR member state fishing vessels where the risk to by catch species is thought to be too great.[63] In accordance with the UNGA Resolutions 61/105 and 64/72,[64] CCAMLR has also established a number of high seas bottom fisheries closures in the CCAMLR area.[65] Commercial bottom trawling is prohibited within the CCAMLR area and restrictions apply to long line fishing.[66] As a first step in creating a network of MPAs in the CCAMLR area, CCAMLR established an MPA covering the South Orkney Island's southern shelf in 2009.[67] This was followed by the creation in 2016 of the world's largest MPA beyond national jurisdiction in the Ross Sea covering a total area of 1.55 million square kilometres.[68] Two further proposals for the establishment of MPAs in the east and west Antarctic were submitted at the 2016 CCAMLR meeting foreshadowing the first network of representative MPAs in ABNJ, an objective that has consistently been discussed in the BBNJ Working Group and Preparatory Committee meetings.[69] CCAMLR has an extensive network of links with both tuna and non-tuna RFMOs including IATTC, WCPFC, and SPRFMO, and with the CMS subsidiary agreement ACAP. It has also granted observer status to a wide range of global and regional organizations for its annual meetings. The relationship between the Antarctic Treaty Scheme (ATS), including CCAMLR, and the future ILBI is likely to be a

63 CCAMLR, Conservation Measures, https://www.ccamlr.org/en/conservation-and-management/conservation-measures.

64 UNGA Resolution 61/105, Sustainable fisheries, including through the 1995 Agreement for the Implementation of the Provisions of the United Nations Convention on the Law of the Sea of 10 December 1982 relating to the Conservation and Management of Straddling Fish Stocks and Highly Migratory Fish Stocks, and related instruments, A/RES/61/105(2006); UNGA Resolution 64/7, Sustainable fisheries, including through the 1995 Agreement for the Implementation of the Provisions of the United Nations Convention on the Law of the Sea of 10 December 1982 relating to the Conservation and Management of Straddling Fish Stocks and Highly Migratory Fish Stocks, and related instruments, A/RES/64/72 (2009).

65 G. Wright, J. Ardron, K. Gjerde, D. Currie and J. Rochette, "Advancing marine biodiversity protection through regional fisheries management: A review of bottom fisheries closures in areas beyond national jurisdiction" (2015) 61 *Marine Policy* 134–148, 142–144.

66 *Ibid.*, 142.

67 *Ibid.*

68 CCAMLR, Ross Sea Conservation Measure, https://www.ccamlr.org/sites/drupal.ccamlr.org/files/91-05_4.pdf.

69 Department of the Environment and Energy, Australian Antarctic Division, "A proposal for a representative system of Marine Protected Areas in the East Antarctic planning domain" http://www.antarctica.gov.au/law-and-treaty/ccamlr/marine-protected-areas.

delicate one that will involve not straining the fabric of the ATS while maximising the benefits to be gained from an established regime for conservation and sustainable use of marine biodiversity in ABNJ across all the world's oceans.

3.4 *Western Indian Ocean*

The Western Indian Ocean is characterised by the long continental coastlines of South Africa, Kenya, and Somalia on the eastern side of the African continent, and remote small islands and groups of islands such as the Comoros Islands, the Seychelles, the Maldives, the Chagos archipelago, Reunion and Mauritius.[70] This oceanic region is recognised as a centre of biodiversity that hosts an array of distinctive species including dugongs, different species of sea turtles, the coelacanth and endemic molluscs, such as the double harp and the violet spider conch, as well as cetaceans and reef fish.[71] The waters between Madagascar and Africa, have been estimated to contain one of the highest levels of coral diversity worldwide, with 369 coral species identified in a recent study, and more still to be identified.[72] The key threats to the regions' biodiversity are unsustainable fisheries practices and climate change impacts. Pelagic fish stocks, particularly shark and tuna, have been heavily exploited by local and distant water fishing fleets and the region's coral reefs have been subject to a number of coral bleaching events in recent decades.[73] The key elements of the regional ocean governance architecture for the Western Indian Ocean region comprise the Nairobi Convention[74] for the protection, management and development of the coastal and marine environment of the Western Indian Ocean, the Indian Ocean Tuna Commission (IOTC) and the Southern Indian Ocean Fisheries Agreement (SIOFA).[75] A variety of other bodies, including the Indian Ocean Commission, the Consortium for the Conservation of Coastal and Marine Ecosystems in the Western Indian Ocean (WIO-C), the Western Indian Coastal Challenge, and the Western Indian Ocean Marine Science Association, have interests in and conduct research on regional marine

70 WWF, International Corals Initiative, Western Indian Ocean Marine Ecoregion, assets.panda.org/downloads/westernindianocean.pdf.

71 *Ibid.*

72 D. Obura, "The Diversity and Biogeography of Western Indian Ocean Reef-Building Corals" (2012) *PLoS ONE* 7(9): e45013. https://doi.org/10.1371/journal.pone.0045013.

73 WWF WIO Marine Ecoregion, above note 70.

74 Convention for the Protection, Management and Development of the Marine and Coastal Environment of the Eastern African Region (Nairobi, 21 June 1985, in force 29 May 1996) *Official Journal of the European Community* 1986, C253/10 ("Nairobi Convention").

75 FAO, South Indian Ocean Fisheries Agreement (Rome, 2006, in force 2012) http://www.fao.org/fishery/rfb/siofa/en ("SIOFA") (accessed 8 June 2017).

biodiversity.[76] While the Nairobi Convention and a number of the other non-fisheries bodies have begun to focus on conservation and sustainable use of marine biodiversity in ABNJ in recent years, there is no regional mechanism to coordinate their activities.[77] The only clear jurisdictional responsibilities for conservation and sustainable use of marine biodiversity in the Western Indian Ocean ABNJ rest with SIOFA and IOTC.

SIOFA entered into force in June 2012 and covers the majority of high seas areas in the Southern Indian Ocean. The objectives of SIOFA are to ensure the long-term conservation and sustainable use of the fishery resources in its area of responsibility taking into account the needs of developing States bordering the SIOFA area that are Contracting Parties to the Convention and, in particular, the least developed among them, and small-island developing States.[78] SIOFA covers fishery resources including fish, molluscs, crustaceans and other sedentary species within its area of responsibility, but excludes highly migratory species and sedentary species subject to the fishery jurisdiction of coastal States.[79] The agreement adopts the precautionary approach and provides that biodiversity in the marine environment shall be protected.[80] SIOFA is still at an early stage of development and limited in its capacity to pass conservation and management measures which fulfil its biodiversity conservation and sustainable use obligations in ABNJ.[81] Although SIOFA has not yet adopted high seas bottom fisheries closures to protect vulnerable marine ecosystems (VMEs), the Southern Indian Ocean Deep Sea Fishers Association (SIODFA), consisting of the five commercial fishing operators in the region, has declared 13 areas in the Southern Indian Ocean as voluntary benthic protected areas, and the SIOFA Scientific Committee has recommended that three of these areas be closed to fishing.[82] SIOFA has articulated plans to connect with a broad network of ocean governance organizations at global and regional levels.[83]

The Indian Ocean Tuna Commission (IOTC) is a tuna RFMO established under Article XIV of the FAO Constitution that started operation in 1996. Its objective is to ensure, through appropriate management, the conservation and optimal utilisation of tuna and tuna-like species in the Indian Ocean.[84]

76 UNEP-WCMC Report, above note 44, 50.
77 *Ibid.*
78 SIOFA, "Objectives," above note 75.
79 *Ibid.*
80 UNEP-WCMC Report, above note 44, 57.
81 *Ibid.*; Wright *et al.*, above note 65, 144.
82 Wright *et al.*, above note 65, 144.
83 UNEP-WCMC Report, above note 44, 58.
84 IOTC, Objectives, http://www.iotc.org/about-iotc (accessed 8 June 2017).

A UNEP-WCMC report notes that its mandate has recently been expanded *de facto* to include data collection and conservation measures for associated species, including sharks, seabirds, and turtles, but that sustainable development of tuna and tuna-like species in its area of competence remains the primary objective of the organization.[85] While IOTC is beginning to develop some cross-sectoral relationships on issues such as seabird mortality, it is still at an early stage in building cooperative relationships with other global and regional organizations concerning biodiversity conservation and sustainable use in the Western Indian Ocean region.[86]

The key challenges facing the Western Indian Ocean region in BBNJ governance are in further developing its scientific, technical, and resource capacity, and coordinating the policy mandates and efforts of the multiple bodies involved in conservation and sustainable use of biodiversity in the region. Stronger links with global and regional organizations at a more advanced stage in implementing biodiversity conservation and sustainable use measures could foster similar capacities in the region.[87]

3.5 *South Atlantic Ocean*

The South Atlantic Ocean generally refers to the area south of the equator and north of the Antarctic Convergence.[88] Turra *et al.* in Chapter 36B of the First Global Integrated Marine Assessment note that the main topographical feature in the South Atlantic is the Mid-Atlantic Ridge that runs between Africa and South America from approximately 58 degrees south to Iceland in the north.[89] The predominant currents and upwelling processes in the South Atlantic provide varied habitats for pelagic biodiversity and productive fisheries.[90] Benthic communities beyond the continental shelf break in the South Atlantic are not as well understood at this time due to lack of research.[91] The ocean governance framework of the South Atlantic is quite sparse with no regional seas organization covering any part of ABNJ and no non-tuna RFMO covering the South West Atlantic.[92] The 2001 Convention on the Conservation

85 UNEP-WCMC Report, above note 44, 59.
86 *Ibid.*, 60–61.
87 *Ibid.*, 70–71.
88 United Nations, Division for Ocean Affairs and the Law of the Sea (DOALOS), The *First Global Integrated Marine Assessment*, (2016) Chapter 36 B South Atlantic Ocean, 1, http://www.un.org/Depts/los/global_reporting/WOA_RPROC/Chapter_36B.pdf.
89 *Ibid.*
90 *Ibid.*, 2–3.
91 *Ibid.*, 5.
92 See also Wright *et al.*, above note 65, 140.

and Management of the Fisheries Resources in the South East Atlantic Ocean (SEAFO Convention), which entered into force in April 2003, covers a large proportion of the high seas of the South East Atlantic Ocean.[93] The objective of the SEAFO Convention is to address the long-term conservation of straddling fish stocks and discrete high seas fish stocks such as alfonsino, orange roughy, arrowhead, wreck fish, deep water hake, and red crab in the geographic area covered by the Convention. The conservation of highly migratory stocks in this area which are subject to the jurisdiction of the International Commission for the Conservation of Atlantic Tunas (ICCAT) is not covered.[94] Under the general principles contained in Article 3 of the SEAFO Convention, the Contracting Parties must adopt measures based on the best scientific evidence available for the long-term conservation and sustainable use of the fisheries resources covered by the Convention and must apply the precautionary approach. The need to maintain other components of the marine ecosystems to which the fisheries resources belong is also recognised in the General Principles of the Convention. Contracting Parties must take account of the impact of fishing on ecologically related species such as seabirds, cetaceans, and seals, adopt where necessary measures for species belonging to the same ecosystem, protect biodiversity in the marine environment, and ensure that fisheries practices and management measures take account of the need to minimise harmful impacts on marine living resources as a whole.[95] The SEAFO Convention contains other progressive features, including a requirement for Contracting Parties to provide an explanation of their reasons for not accepting conservation and management measures agreed by the majority of Contracting Parties, reciprocal high seas boarding and inspection procedures between Contracting Parties for each other's flag vessels and an observer program.[96] The Convention also provides for port State measures to prohibit the landing and transhipment of catch taken by non-Contracting party vessels in a manner undermining the effectiveness of conservation and management measures established under the SEAFO Convention.[97] SEAFO has closed 11 seamount areas in its area of competence where VMEs are present.[98] As with SPRFMO, SEAFO has delineated its bottom fishing footprint and has also introduced exploratory fishing

93 Convention on the Conservation and Management of the Fisheries Resources in the South East Atlantic Ocean (Windhoek, 20 April 2001, in force 13 April 2003) (2002) 41 ILM 257 ("SEAFO Convention").

94 *Ibid.*, Articles 1(l) and 2.

95 *Ibid.*, Article 3.

96 *Ibid.*, Articles 16 and 23(1)(d).

97 *Ibid.*, Article 15.

98 Wright *et al.*, above note 65, 139.

protocols for previously unexploited areas.[99] Although there is no RFMO in the South West Atlantic, Spain has closed nine areas to bottom fishing for its vessels fishing in the South West Atlantic.[100] Given the significant gaps in the ocean governance framework covering the South Atlantic, the region stands to benefit immensely from the broader scientific, technical, and financial resources that could be directed to biodiversity conservation and sustainable use in ABNJ under the ILBI.

4 The Role of the ILBI in Developing More Integrated and Cross-Sectoral Frameworks for Marine Biodiversity Conservation and Sustainable Use in Southern Hemisphere ABNJ

With the right mix of flexible and supportive provisions tailored to the needs of particular regions and sectors, the ILBI has the potential to enhance and lift the performance of existing governance frameworks for biodiversity conservation and sustainable use in the southern hemisphere. At a minimum, the ILBI could contain international law principles and standards on conservation and sustainable use of marine biodiversity in ABNJ that are generally applicable to all States Parties. The IUCN recommended high seas principles[101] provide a widely accepted starting point for relevant international law norms and could be operationalised in other provisions of the ILBI. The ILBI could also provide an important focal point for global regional and sectoral organisations with responsibilities for conservation and sustainable use of marine biodiversity in ABNJ through establishing a Conference of the Parties (COP) that meets regularly and has the power to take decisions on the different elements of the ILBI. The COP could establish regular as opposed to *ad hoc* cooperation mechanisms between RFMOs and other global and regional organisations such as the IMO, ISA and regional seas organizations. The COP could also provide a collaborative forum and potentially financial support for such organizations to coordinate their efforts on conservation and sustainable use of marine biodiversity in ABNJ. The ILBI has the potential to further coalesce global scientific information on marine biodiversity in ABNJ, drawing on data, monitoring, and assessment from a wider group of global, regional, and sectoral organizations, and to share such information among its States Parties. This broader

99 *Ibid.*, 140.

100 *Ibid.*

101 IUCN, "10 Principles for High Seas Governance" https://www.iucn.org/downloads/10_principles_for_high_seas_governance___final.pdf (accessed 9 June 2017).

collation of scientific information could provide environmental baselines for different regions in ABNJ against which to measure the cumulative impacts of human activities on marine biodiversity in ABNJ over time. The collation of this information has already begun with initiatives such as the EBSA process undertaken by the CBD,[102] the various Global Ocean assessments undertaken by the UN,[103] and the Global Ocean Biodiversity Initiative (GOBI).[104] A broader picture of marine biodiversity in ABNJ could provide a more informed basis for strategic environmental planning and biodiversity conservation action plans across the whole of ABNJ. For example, access to combined information on shipping, fisheries, and climate change impacts on a particular seamount in ABNJ could enable a suite of complementary conservation measures to be developed by the relevant sectoral and regional organizations in that area. A central repository of information on marine biodiversity in ABNJ could also highlight deficiencies in biodiversity conservation across ABNJ, and initiate measures to mitigate adverse impacts on species, habitats, and ecosystems not protected by existing regional and sectoral organizations. Some discrete high seas fish stocks not covered by existing RFMOs and sedentary species would be clear beneficiaries of such measures.

4.1 *Role of the ILBI in South West Pacific*

The South West Pacific is distinguished by the high level of coherence and integration between its regional institutions, including those related to oceans governance and biodiversity conservation.[105] Collective support for the Pacific Islands Forum and the Council of Regional Organisations of the Pacific (CROP) have created the political conditions necessary for cross-sectoral cooperation and coordination of policies across the marine sector. This regional coherence has also become apparent in the Preparatory Committee meetings for the ILBI and the negotiations for the SDGs where the PSIDS have operated as a bloc and pursued consensus positions. With this background in mind, they argue that the ILBI should be framed so as to form

102 Convention on Biological Diversity (CBD), "Background on the EBSA Process," https://www.cbd.int/ebsa/about (accessed 9 June 2017).

103 United Nations World Ocean Assessment, http://www.worldoceanassessment.org/ (accessed 9 June 2017).

104 Global Ocean Biodiversity Initiative (GOBI), "About GOBI," http://gobi.org/ (accessed 9 June 2017).

105 Robin Mahon, Lucia Fanning, Kristina M. Gjerde, Oran Young, Michael Reid, Selicia Douglas, *Transboundary Waters Assessment Programme (TWAP) Assessment of Governance Arrangements for the Ocean, Volume 2: Areas Beyond National Jurisdiction*. (UNESCO-IOC, Paris, 2015). IOC Technical Series, 119, 43.

constructive links with the existing cohesive ocean governance instruments and institutions in the South West Pacific. They suggest that if the IUCN recommended high seas principles were to be included in the ILBI that this would create a strong link between the South West Pacific regional ocean governance system and the ILBI, as these principles are already reflected in the Pacific Islands Regional Oceans Policy (PIROP) and its implementing action plan, the Framework for a Pacific Oceanscape (FPO).[106] They also raise the issue of compatibility between the biodiversity conservation and sustainable use measures already in place in the South West Pacific within the national jurisdiction of Pacific Island States (PICs) and any new measures introduced by the ILBI for ABNJ to ensure that the economic, political and conservation interests of the PICs are not undermined, and high standards of biodiversity conservation are not diminished. The ILBI also has the potential to strengthen and revive regional institutions and initiatives for biodiversity conservation both within and beyond national jurisdiction, such as the South Pacific Regional Environment Program (SPREP) and the South Pacific Whale Sanctuary. Through its area-based management provisions, the ILBI could potentially provide more global legitimacy for existing high seas fisheries closure areas under the Nauru Agreement.

Effective conservation and sustainable use of marine biodiversity in all regions depends on access to relevant scientific data and knowledge to make informed decisions on conservation and management measures. Access to a sufficiently broad scientific knowledge base combined with trained staff, appropriate technology, and financial support is critical for maintaining the long-term integrity of marine biodiversity within and beyond national jurisdiction. The ILBI represents an opportunity to strengthen systems for global scientific capacity development and technology transfer related to marine biodiversity in ABNJ. A new version of technology transfer could be engendered by the ILBI based on open data, innovation, and greater collaboration in the generation, sharing, and application of scientific knowledge related to marine biodiversity in ABNJ.[107] Bax *et al.* advocate the introduction of regional needs assessments among the member States of the ILBI to match existing marine technology resources to current needs in particular regions. They also argue

106 See above notes 36 and 37.

107 See also Nippon Foundation, Nereus Policy Briefs—Thiele, T., Harden-Davies, H. (2016) Technology transfer. http://www.nereusprogram.org/policy-brief-bbnj-technology-transfer/; Global Ocean Trust. Bax, N., Harden-Davies, H., Thiele, T., Halpin, P., Dunn, D. (2016) Open Data: enabling conservation and sustainable use of biodiversity in areas beyond national jurisdiction, http://www.nereusprogram.org/policy-brief-bbnj-open-data/ (accessed 9 June 2017).

that strengthening links between regional institutions, scientists, and technical personnel in the South West Pacific and international organizations such as IOC is critical to elevating scientific and technical capacity among the PSIDS. The ILBI could contain provisions establishing regional marine science and technology clusters and fostering international research collaborations on marine biodiversity in ABNJ. This could alleviate some of the constraints currently facing researchers in the South West Pacific. It could create its own funding mechanism for technology transfer and capacity development purposes based on existing mechanisms such as the Global Environment Facility (GEF) and the UNESCO-IOC Capacity Development Fund, or source funding from these global bodies for specific initiatives.[108]

4.2 *Role of the ILBI in South East Pacific*

A recent UNEP-WCMC report on "Governance of areas beyond national jurisdiction for biodiversity conservation and sustainable use" acknowledges the relatively limited levels of cooperation among the regional organizations with ocean governance responsibilities in the South East Pacific.[109] The complementary nature of the geographical scope and functional mandates of the three key ocean governance organizations in the region can however be used positively to strengthen conservation and sustainable use of marine biodiversity in ABNJ. The UNEP WCMC report on "Governance of areas beyond national jurisdiction for biodiversity conservation and sustainable use" recommends that any cross-sectoral collaboration should begin with specific issues based on the mandates, objectives, and interests of the three key regional organizations and that one of the organizations might be able to lead or champion specific initiatives for cross sectoral collaboration.[110] The negotiation of the ILBI has the potential to add global legitimacy and impetus to such initiatives.

4.3 *Role of the ILBI in Western Indian Ocean*

There are a range of opportunities to coordinate regional action on conservation and sustainable use of marine biodiversity in ABNJ in the Western Indian Ocean region (WIO). The geographical scope of the Nairobi Convention could potentially be extended into ABNJ and modalities developed for more extensive cooperation and coordination with global, sectoral, and regional organizations such as RFMOS, ISA, IMO, and other regional seas organizations. The ILBI could provide the initial impetus for such an extension

108 *Ibid.*

109 UNEP-WCMC Report, above note 44, 90.

110 *Ibid.*, 91.

and ongoing cooperation. As there are currently no high seas bottom fisheries closures in the WIO, a cooperative process between SIOFA, IOTC, and the Nairobi Convention could be initiated to identify VMEs in the region and establish such closures. Once established, the ILBI could support this process by providing scientific and technical information, and fostering consultation among the regional bodies. The ILBI could also provide global endorsement for the thirteen benthic-protected areas that have been established voluntarily by SIODFA and recommend their adoption by SIOFA.[111] From a shipping perspective, the member States of regional bodies such as the Nairobi Convention could evaluate the shipping threats to marine biodiversity in regional ABNJ, identify potential associated protective measures (APMs), and propose possible particularly sensitive sea areas (PSSAs) to IMO. The ILBI could facilitate this process, providing scientific information to support such proposals. There are four exploration contracts issued by the ISA for polymetallic nodules and polymetallic sulphides exploration in the WIO. The repository of scientific information on marine biodiversity in ABNJ developed by the ILBI could assist both the ISA and regional organizations such as the Nairobi Convention in designating areas of particular environmental interest (APEI) and broader conservation measures, such as regional environmental management and biodiversity action plans in the WIO.

The UNEP-WCMC report on "Governance of areas beyond national jurisdiction for biodiversity conservation and sustainable use" has assessed the readiness of regional organizations in the WIO for more cross-sectoral cooperation in terms of their political will and respective capacities. It notes that cross-sectoral cooperation to date has typically been *ad hoc* and issue driven, but that most regional organizations with ocean governance responsibilities are open to greater collaboration and area-based planning.[112] The ILBI could act as a catalyst for further cross-sectoral collaboration in the WIO, particularly if it becomes a comprehensive source of scientific information on biodiversity in ABNJ and is able to provide technical and financial capacity building in area based management, environmental impact assessment, and marine scientific research.

4.4 *The Role of the ILBI in the South Atlantic*

A defining feature of the regional ocean governance architecture of the South Atlantic is its uneven nature.[113] The complete lack of any regional arrangements

111 See above note 82.

112 UNEP-WCMC, above note 44, 70.

113 Mahon *et al.*, above note 105, 37–38.

for marine environmental protection and sustainable use of fisheries resources in the Southwest Atlantic is a significant governance gap. Elsewhere in the South Atlantic, the two regional sea conventions, the Cartagena Convention[114] and the Abidjan Convention[115] only cover marine areas within national jurisdiction. While SEAFO is a post-UNFSA convention with biodiversity conservation objectives and modern environmental principles embedded within its provisions, it has only seven Contracting Parties with very few of those from the South East Atlantic region. As a result of this paucity of regional institutions for marine biodiversity conservation, the role of the ILBI in the South Atlantic could be more fundamental than in other regions, providing the impetus for States to establish the missing components of regional ocean governance and further accession to existing regional instruments, such as SEAFO. The ILBI could also provide a global model for best-practice marine environmental principles and approaches, and a global system for designating MPAs and overseeing EIAs in ABNJ. There is also a role for the ILBI in updating the biodiversity conservation aspects of the high seas fisheries regime. It has the potential to provide greater clarity around the operation of the extended continental shelf regime and the high seas regime for the water column above. This could occur through the ILBI setting out reciprocal principles and procedures ensuring the commitment of coastal States to biodiversity conservation on their extended continental shelves consistent with sustainable use of the resources, as well as the commitment of all States Parties to the ILBI to respecting coastal State measures while exercising their high seas rights in the water column above the shelf. More broadly, the ILBI could be pivotal in creating greater cohesion among the various sectors and regions involved in ocean governance, both within and beyond national jurisdiction. This will be dependent on the ILBI establishing a viable technology transfer and capacity building system to assist developing countries in the South Atlantic and other regions. In effect, the ILBI must act as a vector for the open provision of scientific data and information on marine biodiversity in ABNJ across regions, matching available training and technology with demonstrated needs, particularly in regions with a preponderance of developing States.

114 Convention for the Protection and Development of the Marine Environment of the Wider Caribbean Region (Cartagena de Indias, 24 March 1983, in force 11 October 1986) (1983) 22 ILM 221.

115 Convention for Co-operation in the Protection and Development of the Marine and Coastal Environment of the West and Central African Region (Abidjan, 23 March 1981, in force 5 August 1984), http://abidjanconvention.org/index.php?option=com_content&view=article&id=100&Itemid=200&lang=en (accessed 9 June 2017).

4.5 *The Role of the ILBI in the Southern Ocean*

The Antarctic Treaty system (ATS) has operated predominantly as a self-contained regime since the adoption of the Antarctic Treaty[116] in 1959. Its relationship with the LOSC and other global instruments such as the CBD has been quite ambivalent. Notwithstanding this uncertain relationship, the advent of a new global instrument designed to conserve and sustainably use marine biodiversity in ABNJ could have some benefits for the long-term integrity of the abundant marine biodiversity in the Antarctic. The Southern Ocean surrounding the Antarctic continent has unique biodiversity and the legal status of these waters is still uncertain. Even taking into account the territorial and offshore zones claimed by some Antarctic treaty partners, there are clearly substantial marine areas beyond national jurisdiction within the geographic scope of the Antarctic Treaty and CAMLR. The ATS already deals extensively with the conservation and sustainable use of marine biodiversity in the substantial areas of ABNJ under its responsibility. Conservation and sustainable use of Antarctic living resources are key objectives of the Antarctic Treaty, CAMLR, and the Protocol on Environmental Protection to the Antarctic Treaty.[117] In line with its ecosystem-based approach, CCAMLR has been in the forefront of designing and designating two marine protected areas in the CAMLR area of responsibility that cover vast areas of ABNJ.[118] The designation and management of these areas by an existing treaty body with both global and regional attributes provides a precedent for the area based management element of the BBNJ package. Other regions which are less advanced in designing and managing MPAs in ABNJ could draw on the CCAMLR experience in implementing their own regional plans under the ILBI. In contrast to the designation of MPAs, the ATS system has not yet seriously tackled issues related to access and distribution of marine genetic resources (MGRs) benefits. Any developments in relation to these issues under the ILBI will therefore be of interest to the ATS, and may influence the treatment of MGRs within the ATS. While the potential for close interaction between the ATS and the ILBI seems unlikely in view of the relative autonomy of the ATS to date, some positive cooperation could evolve over time on particular issues such as area-based management and access to and distribution of MGR benefits.

116 Antarctic Treaty (Washington DC, 1 December 1959, in force 23 June 1961), 402 UNTS 71 ("Antarctic Treaty").

117 Protocol on Environmental Protection to the Antarctic Treaty (Madrid, 4 October 1991, in force 14 January 1998) ("Madrid Protocol").

118 See above notes 68 and 69.

5 Conclusion

This chapter focuses on biodiversity conservation initiatives taken by Southern Hemisphere institutions and the degree to which connections have been forged between the relevant institutions with responsibilities for biodiversity beyond national jurisdiction in the most oceanic of the two hemispheres. The above overview of selected instruments, institutions, and arrangements in the Southern Hemisphere reveals that governance frameworks for conservation and sustainable use of marine biodiversity in the Southern Hemisphere vary widely in their biodiversity conservation activities and coherence across sectors and regions. The chapter has also examined the potential relationship of the ILBI with this selection of regional ocean governance arrangements and institutions in the Southern Hemisphere, and the benefits of such interaction for the conservation and sustainable use of marine biodiversity at the regional and global scales. The ILBI represents an historic opportunity to lay the foundations for a more integrated and cross-sectoral system of oceans governance in ABNJ globally and in the southern hemisphere. Given the growing threats and pressures of anthropogenic activities on southern hemisphere ABNJ and its biodiversity, the realisation of this objective is critical to ensuring the long term integrity and productivity of the vast ocean areas south of the equator.

CHAPTER 7

The BBNJ PrepCom and Institutional Arrangements: The Hype about the Hybrid Approach

*Kristine Dalaker Kraabel**

Abstract

The United Nations General Assembly established the Preparatory Committee (PrepCom), prior to convening an intergovernmental conference, for the purpose of recommending substantive elements for the development of an international legally binding instrument (ILBI) under the United Nations Convention on the Law of the Sea on the conservation and sustainable use of marine biodiversity of areas beyond national jurisdiction. Discussions at the PrepCom included possible institutional arrangements that could govern the new legal regime. Three approaches to the institutional arrangements were proposed at the PrepCom: the global and regional and/or sectoral approaches, together with the newly-proposed hybrid approach. The overall aim of this chapter is to examine the context and emergence of the hybrid approach, together with its implications for the ILBI, given the institutional structure and criteria the delegations proposed for the institutional arrangements.

1 Introduction

In 2016 and 2017, delegations from around the world gathered in New York at the United Nations headquarters for four sessions of the Preparatory Committee, which was established by United Nations General Assembly Resolution 69/292 (UNGA Resolution 69/292). The goal was to recommend substantive elements for the development of an international legally binding instrument (ILBI) under the United Nations Convention on the Law of the Sea (UNCLOS)[1]

* Kristine Dalaker Kraabel is a PhD Research Fellow at the KG Jebsen Centre for the Law of the Sea, UiT, The Arctic University of Norway, Tromsø. The author's PowerPoint presentation is available at http://www.virginia.edu/colp/pdf/yogya-kraabel.pdf.

1 United Nations Convention on the Law of the Sea, adopted 10 December 1982, 1833 UNTS 3 (entered into force 16 November 1994) (UNCLOS).

 | DOI:10.1163/9789004366619_012

on the conservation and sustainable use of marine biological diversity[2] of areas beyond national jurisdiction (referred to as BBNJ).[3] By the close of the fourth session, the PrepCom had fulfilled its mandate and adopted a report for submission to the General Assembly that included recommendations addressing the following four elements of what has come to be known as the "package deal":[4] (i) marine genetic resources; (ii) including questions on sharing of benefits, measures such as area-based management tools (ABMTs), including marine protected areas (MPAs); (iii) environmental impact assessments (EIAs); and (iv) capacity building and the transfer of marine technology.[5]

The PrepCom may, as some observers have remarked, appear to be "high brow" and only of interest to those in the oceans circles.[6] A reading (and re-reading) of the introductory sentences to this chapter would likely confirm that impression, and would diminish the importance of what is at stake in this process. More simply, when discussing marine biodiversity in areas beyond national jurisdiction, a number of key questions need to be considered: What is it that we are protecting? Why are we protecting it? What threats are we protecting it from? How can we best conserve and sustainably use marine biodiversity in areas beyond national jurisdiction?

Areas beyond national jurisdiction are home to rich biodiversity and marine resources. These vast marine areas comprise approximately 64 percent of the Earth's surface and contain some of the world's most unique species

2 The terms "marine biological diversity" and "marine biodiversity" are used interchangeably in this chapter.

3 *Development of an international legally-binding instrument under the United Nations Convention on the Law of the Sea on the conservation and sustainable use of marine biological diversity of areas beyond national jurisdiction*, UN GAOR A/Res/69/292 (UNGA Resolution 69/292), accessed 2 November 2017 at <https://documents-dds-ny.un.org/doc/UNDOC/GEN/N15/187/55/PDF/N1518755.pdf?OpenElement>.

4 *Report of the Preparatory Committee established by General Assembly resolution 69/292: Development of an international legally binding instrument under the United Nations Convention on the Law of the Sea on the conservation and sustainable use of marine biological diversity of areas beyond national jurisdiction*, A/AC.287/2017/PC.4/2, dated 31 July 2017 (PrepCom Report), accessed 2 November 2017 at <http://www.un.org/ga/search/view_doc.asp?symbol=A/AC.287/2017/PC.4/2>. The "package deal" was first proposed in the BBNJ Working Group Recommendations (see Section 3 *infra* and BBNJ Working Group Recommendations, *infra* note 5 at 5).

5 *Letter dated 30 June 2011 from the Co-Chairs of the Ad Hoc Open-ended Informal Working Group to the President of the General Assembly*, UN GAOR Doc. A/66/119 (BBNJ Working Group Recommendations), Annex, para. 1(b).

6 Earth Negotiations Bulletin, "Summary of the Third Session of the Preparatory Committee on Marine Biodiversity Beyond Areas of National Jurisdiction: 27 March–7 April 2017", Vol. 25, No. 129 (10 April 2017) (ENB PrepCom3 Summary), at 16.

and habitats, including seamounts, cold seeps, hypothermal vents, deep sea trenches and coral reefs.[7] More specifically, within the context of UNCLOS, the marine areas beyond national jurisdiction include: (i) the "Area", which comprises the seabed and the ocean floor, and subsoil thereof, beyond the limits of national jurisdiction,[8] and (ii) the high seas, which encompass the water column beyond the exclusive economic zones (EEZs) of coastal States and are governed by the traditional principles of the freedoms of the high seas.[9]

Each of these marine areas enjoys a special legal status under UNCLOS. While the Area and its mineral resources are considered the "common heritage of mankind", whereby all activities must be conducted for the benefit of mankind as a whole,[10] the high seas are grounded in the traditional notion of the Grotian Ocean and the desire to secure freedom of navigation.[11] UNCLOS provides a non-exhaustive list of high seas freedoms: (i) freedom of navigation; (ii) freedom of overflight; (iii) freedom to lay submarine cables and pipelines; (iv) freedom to construct artificial islands and other installations permitted under international law; (v) freedom of fishing; and (vi) freedom of scientific research.[12] However, these freedoms are not absolute, and are subject to a number of limitations and corresponding obligations. Furthermore, due to the increasing and competing uses of the oceans and the resulting degradation, these freedoms have been limited over time.[13] Because the Area and the high seas each enjoy a different legal status under UNCLOS, a dichotomy of

7 Global Ocean Commission, *From Decline to Recovery: A Rescue Package for the Global Ocean* (2014) (Global Ocean Commission Report), at 4, accessed 2 November 2017 at <http://www.some.ox.ac.uk/wp-content/uploads/2016/03/GOC_report_2015.July_2.pdf>.

8 UNCLOS, *supra* note 1, Art. 1(1).

9 UNCLOS, *supra* note 1, Part VII.

10 UNCLOS, *supra* note 1, Arts. 136 and 137.

11 "I shall base my argument on the following most unimpeachable axiom of the Law of Nations, called a primary rule or first principle, the spirit of which is self-evident and immutable, to wit: Every nation is free to travel to every other nation, and to trade with it.… For do not the ocean, navigable in every direction with which God has encompassed the earth, and the regular and the occasional winds which blow now from one quarter and now from another, offer sufficient proof that Nature has given to all peoples a right of access to all other peoples? Seneca thinks this is Nature's greatest service, that by the wind she united the widely scattered peoples, and yet did so distribute all her products over the earth that commercial intercourse was a necessity to mankind. Therefore this right belongs equally to all nations". Hugo Grotius, *The Freedom of the Seas* (trans. Ralph Van Deman Magoffin) (Oxford University Press, 1916), at 7, 8.

12 UNCLOS, *supra* note 1, Art. 87.

13 See, e.g., Agreement for the Implementation of the Provisions of the United Nations Convention on the Law of the Sea of 10 December 1982 Relating to the Conservation and Management of Straddling Fish Stocks and Highly Migratory Fish Stocks, adopted 4 December 1995, 2167 UNTS 3 (entered into force 11 December 2001) (UNFSA).

principles has been created: the common heritage of mankind versus high seas freedoms.[14] This dichotomy has carried through much of the BBNJ discussions (particularly with regard to MGRs),[15] and echoes the ideological issues that complicated the negotiation of Part XI of UNCLOS during the Third United Nations Conference on the Law of the Sea.[16]

The importance of marine biodiversity in areas beyond national jurisdiction cannot be overemphasised.[17] It provides a wealth of resources and services, including seafood and raw materials, genetic and medicinal resources, air purification, climate regulation, and habitat and cultural services.[18] The Global Ocean Commission (2014) found that: (i) high seas ecosystems are responsible for almost half of the total biological productivity of the global ocean; (ii) nearly half a billion tons of carbon are captured and stored by high-seas ecosystems annually, a value of USD $74–222 billion; (iii) close to 10 million tons of fish are caught annually on the high seas; and (iv) the majority of global ocean fish harvests are of species captured both in EEZs and in the high seas, suggesting that overfishing on the high seas is likely to have an adverse impact on fishing in areas within national jurisdiction.[19]

While the importance of areas beyond national jurisdiction cannot be overemphasised, the threats to marine biodiversity in these areas cannot be overly diminished. The threats to areas beyond national jurisdiction include: pollution (in all its forms), overfishing and intensified aquaculture, shipping and expansion of global maritime trade, deep seabed mining, ocean warming, ocean acidification, and numerous emerging uses of the oceans, which include bioprospecting, geo-engineering, energy development, and climate

14 See BBNJ Working Group Recommendations, *supra* note 5 at 5.

15 For the purpose of clarity, this author refers to the PrepCom as the formal process that was conducted under the auspices of the United Nations regarding the conservation and sustainable use of BBNJ, and refers to the "BBNJ process" and "BBNJ discussions" as those conversations that are happening (or have happened) on a formal and informal basis amongst stakeholders with regard to the conservation and sustainable use of BBNJ. This point of clarification recognises that the PrepCom and its subsequent iterations may well stall or come to an end, while the BBNJ process and/or discussions may continue in different forums over time.

16 "A Constitution for the Oceans", Remarks by Tommy T.B. Koh of Singapore, President of the Third Conference on the Law of the Sea (adapted from statements by the President on 6 and 11 December 1982 at the final session of the Conference in Montego Bay) (Constitution for the Oceans).

17 Global Ocean Commission Report, *supra* note 7 at 5–6.

18 *Ibid.*

19 *Ibid.*

change mitigation efforts, such as seabed sequestration of carbon monoxide and ocean fertilisation, among others.[20]

Both the importance of, and the threats to, marine biodiversity in areas beyond national jurisdiction first came to the attention of the United Nations in 2003 with the 4th meeting of the United Nations Informal Consultative Process on the Oceans and the Law of the Sea.[21] Those early discussions eventually developed into the PrepCom, the evolution of which is set out in more detail in Section 3 of this chapter. Discussions at the PrepCom included the elements of the package deal, together with possible institutional arrangements that could govern the new regime. Of the cross-cutting issues discussed at the PrepCom, the institutional arrangements were among the most divisive and fell largely along ideological lines between developed and developing nations, following the long-cited tension between high seas freedoms and the common heritage of mankind. The discussion of institutional arrangements was further complicated by the PrepCom's mandate, which provides that the process "should *not undermine* existing relevant legal instruments and frameworks and relevant global, regional and sectoral bodies".[22]

At the first and second sessions of the PrepCom, some delegations, mostly from developing nations, suggested that the effective implementation of the provisions of an ILBI would require the establishment of a global executive body through which Parties could make decisions, undertake coordination efforts, and perform reviews and assessment of implementation and compliance, with implementation taking place at the regional and sectoral levels.[23] Other delegations, mostly from developed nations, suggested that such a global body would undermine other instruments and frameworks, favouring instead a regional and/or sectoral approach.[24] However, at the third session of the PrepCom, a hybrid approach was suggested, which attracted significant

20 See generally Global Ocean Commission Report, *supra* note 7.

21 *Report on the work of the United Nations Open-ended Informal Consultative Process on Oceans and the Law of the Sea, Letter dated 9 June 2003 from the Co-Chairperson of the Consultative Process addressed to the President of the General Assembly*, UN GAOR Doc. A/58/95 (Report of 4th Informal Consultative Process).

22 UNGA Resolution 69/292, *supra* note 3, para. 3 [emphasis added].

23 See Section 5 *infra*. See also "Summary of the First Session of the Preparatory Committee on Marine Biodiversity of Areas Beyond National Jurisdiction: 28 March–8 April 2016" *Earth Negotiations Bulletin*, Vol. 25, No. 106 (IISD Reporting Services, 11 April 2016) (ENB PrepCom1 Summary); "Summary of the Second Session of the Preparatory Committee on Marine Biodiversity of Areas Beyond National Jurisdiction: 26 August–9 September 2016" *Earth Negotiations Bulletin*, Vol. 25, No. 118 (IISD Reporting Services, 12 September 2016) (ENB PrepCom2 Summary).

24 *Ibid.*

interest and discussion among the delegations.[25] The essential institutions of the hybrid approach could include (with some variation) a Conference of the Parties (COP) that would set out internationally agreed standards and obligations at the global level (e.g., those pertaining to ABMTs and MPAs) with regional and sectoral bodies reporting on implementation, and national entities implementing the ILBI.[26] At the fourth session of the PrepCom, although there was little substantive elaboration on the hybrid approach, there was support in the room for continued discussion of the approach in a subsequent intergovernmental conference (IGC).[27] The possible institutional arrangements were included in the PrepCom Report that was submitted to the General Assembly.[28]

The overall aim of this chapter is to examine the context and emergence of the hybrid approach, together with its implications for the ILBI, given the institutional structure and criteria the delegations proposed for the institutional arrangements by exploring the three approaches for the ILBI that were proposed at the PrepCom—the global and regional and/or sectoral approaches, together with the newly-proposed hybrid approach. This aim is achieved by first examining, as set out in Section 2, the historical context of the development of institutional arrangements in the making of the law of the sea and its relevance to the BBNJ process and the PrepCom. Section 3 outlines the existing ocean governance framework and its relevant inadequacies, while Section 4 tells the story of the beginnings of the discussions of the BBNJ process, including the establishment of the PrepCom. Section 5 describes the discussions at the first two sessions of the PrepCom surrounding cross-cutting issues, with a particular focus on institutional arrangements. Section 6 examines the emergence of the hybrid approach at the PrepCom, together with the criteria and overall structure for the institutional arrangements put forward by the

---

25 See Section 6 *infra*. See also "Chair's Overview of the Third Session of the Preparatory Committee", *Preparatory Committee Established by General Assembly Resolution 69/292: Development of an International Legally Binding Instrument under the United Nations Convention on the Law of the Sea on the Conservation and Sustainable Use of Marine Biological Diversity of Areas Beyond National Jurisdiction* (Chair's Overview of PrepCom3), accessed 2 November 2017 at <http://www.un.org/depts/los/biodiversity/prepcom_files/Chair_Overview.pdf>.

26 Chair's Overview of PrepCom3, *supra* note 25 at 27.

27 "Summary of the Fourth Session of the Preparatory Committee on Marine Biodiversity of Areas Beyond National Jurisdiction: 10–21 July 2017" *Earth Negotiations Bulletin*, Vol. 25, No. 141 (IISD Reporting Services, 24 July 2017) (ENB PrepCom4 Summary). See also this author's notes from the fourth session of the PrepCom (on file with author) (Author's PrepCom4 Notes).

28 PrepCom Report, *supra* note 4 at 15–16.

delegations, and provides an explanation for the emergence of the hybrid approach and its implications for the BBNJ process. Concluding remarks are set out in Section 7 of the chapter.

2 Relevant Historical Background of the Development of Institutional Arrangements under the Law of the Sea

In discussing upcoming developments and innovations in the law of the sea, it may seem strange to begin with a historical note, but this author believes that the history of the law of the sea informs the BBNJ process in significant ways. In 1994, after the long and arduous years of negotiations that followed the signing of UNCLOS in Montego Bay, Jamaica in 1982 and the ensuing informal consultations that brought about the Part XI Implementation Agreement,[29] the Center for Oceans Law and Policy at the University of Virginia School of Law held a conference in Rhodes, Greece on the Entry into Force of the Law of the Sea Convention. The output from that conference was a book for which Sir Robert Jennings, a former President of the International Court of Justice, wrote the foreword, of which an excerpt follows:

> [I]t should not be forgotten that Part XII of the [UNCLOS] on the "Protection and Preservation of the Marine Environment", was a major contribution to the making of shaping of this aspect of the law. This is a problem which is much larger than the law of the sea itself. Since such an immense proportion of marine pollution is from land-based sources, the old juridical idea that the "land dominates the sea" has acquired a new and sinister connotation. It is perhaps worth adding, however, *that this again is a problem which calls for management techniques and the evolution of new kinds of international organization.*[30]

Lord Jennings' words—as reinforced by the text of UNGA Resolution 69/292 requiring that the ILBI be *under* UNCLOS—serve as a timely reminder that UNCLOS provides the relevant legal framework that allows for the development of management techniques and the evolution of new kinds of international

29 Agreement Relating to the Implementation of Part XI of the United Nations Convention on the Law of the Sea of 10 December 1982, 28 July 1994, 33 ILM 1309 (1994) (entered into force 28 July 1996) (Part XI Implementation Agreement).

30 R.Y. Jennings, "Foreword" in M.H. Nordquist and J.N. Moore (eds.), *1994 Rhodes Papers: Entry into Force of the Law of the Sea Convention* (Martinus Nijhoff Publishers, 1995), at xvi [emphasis added].

organisation to combat existing and emerging oceans problems. The flexibility of this framework has allowed for the law of the sea to develop and evolve further through the Part XI Implementation Agreement, the United Nations Fish Stocks Agreement (UNFSA),[31] and the ensuing regional fishery management organisation (RFMO) conventions, among others, in addition to the ongoing interpretation of the legal framework through case law and advisory opinions. Similarly, BBNJ is yet another oceans problem that is being addressed within this dynamic framework, illustrating precisely what the PrepCom is about: to find agreement for new and improved international management for the protection of marine biodiversity in areas beyond national jurisdiction in an increasingly fragmented legal and regulatory space.

The history of the law of the sea demonstrates that the rules of governance and decision making in new international management arrangements, and the processes required to negotiate such arrangements, are critical to universal acceptance and successful implementation of such arrangements. For example, at the Third United Nations Conference on the Law of the Sea, both the "Gentlemen's Agreement" and the "package deal",[32] together with the organisation of the Conference's work in informal working groups, proved to be innovative consensus-building devices that brought about UNCLOS[33] and which are now being employed in part by the PrepCom. Similarly, within the context of the BBNJ process, the specific use of the package deal contemplates that a series of small compromises will need to be made in order to reach overall agreement—in other words, "nothing is agreed until everything is agreed".[34]

The history of the law of the sea also serves as a reminder of the interconnectedness of such arrangements and procedures with the substantive matters. As Ambassador Satya N. Nandan (Fiji), Rapporteur of the Second Committee, remarked in his report to his Government on the First Session of the Third United Nations Conference on the Law of the Sea, which took place in December 1973:

31 UNFSA, *supra* note 13.

32 *Third United Nations Conference on the Law of the Sea, Official Records*, Vol. I (Summary Records of Plenary Meetings of the First and Second Sessions, and of Meetings of the General Committee, Second Session), at 51–52 (adoption of the rules of procedure for the Conference, including the gentlemen's agreement).

33 J. Evensen, "Working Methods and Procedures in the Third United Nations Conference on the Law of the Sea", *Collected Courses of The Hague Academy of International Law Volume: 199* (Brill/Nijhoff Publishing, 1986).

34 G.M. Danilenko, *Law-Making in the International Community* (Martinus Nijhoff Publishers, 1993).

> The controversies affecting procedure highlighted the fact that—on the law of the sea—procedure and substance are probably so inter-connected that it is hard to separate the two. Delegates viewed procedural questions in the light of their stand on substantive matters.[35]

Delegations at the PrepCom have often echoed Ambassador Nandan's view, noting on a number of occasions that "form follows substance" and important cross-cutting issues can only be considered after the substantive provisions of the ILBI are sufficiently developed.[36] Whether discussing the procedures regarding governance of an IGC or the establishment of a new legal regime for the oceans—whether in 1973 or in 2017—the inter-connectedness remains the same. The Part XI Implementation Agreement serves as a further example of this inter-connectedness, with the long and difficult negotiations regarding decision making under the Agreement nearly costing UNCLOS universal acceptance.

In the early BBNJ discussions, there was some debate about whether an ILBI would follow the format of the Part XI Implementation Agreement or the UNFSA. Unlike the Part XI Implementation Agreement, which was unique in that it was concluded for the purpose of ensuring universal acceptance of UNCLOS as the Convention neared entry into force by setting out how one part of UNCLOS would be implemented, the UNFSA is a freestanding agreement, which falls under the framework convention of UNCLOS for the purpose of governing a specific aspect of marine biodiversity. As the PrepCom progressed, substantial agreement consolidated around the view that the ILBI could borrow in large part from the UNFSA. In particular, the ILBI would likely follow the UNFSA's provisions for those matters that bear directly on the relationship between UNCLOS and the resulting agreement.[37] However, the Part XI Implementation Agreement still remains relevant for the BBNJ process, with delegations suggesting that the Agreement's establishment of the International

35 S.N. Nandan, "Brief Outline on the First Session of the Conference on the Law of the Sea" (New York, 3–15 December 1973) (*Papers of the United States Conferences on the Law of the Sea at the University of Virginia,* Arthur J. Morris Law Library Special Collections, Box 69 of 204), at 1.

36 "Development of an international legally-binding instrument under UNCLOS on the conservation and sustainable use of marine biological diversity of areas beyond national jurisdiction (BBNJ Process)", Written Submission of the EU and its Member States on Cross-cutting Issues (5 December 2016) (EU Submission of 5 December 2016), at 3.

37 See e.g., Article 4 of the UNFSA states: "Nothing in this Agreement shall prejudice the rights, jurisdiction and duties of States under the Convention. This Agreement shall be interpreted and applied in the context of and in a manner consistent with the Convention".

Seabed Authority (ISA) and its subsidiary bodies using a cost-effective, evolutionary approach to the establishment of institutional arrangements[38] be used as a model or template for the new legal regime.[39]

It is critical that these examples and past lessons remain foremost in the minds of the delegations as they engage in the BBNJ process to find a set of new international management arrangements that will adequately and comprehensively protect the marine environment in areas beyond national jurisdiction.

3 The Existing Ocean Governance Framework for BBNJ

The Global Ocean Commission concluded in its 2014 report that the existing ocean governance framework, which is outlined below, lacks a comprehensive set of overarching governance principles and is characterised by widespread fragmentation.[40]

38 Part XI Implementation Agreement, *supra* note 29, Annex:
1.The International Seabed Authority (the "Authority") is the organization through which States Parties to the Convention shall, in accordance with the regime for the Area established in Part XI and this Agreement, organize and control activities in the Area, particularly with a view to administering the resources of the Area....
2.In order to minimize costs to State Parties, all organs and subsidiary bodies to be established under the Convention and this Agreement *shall be cost-effective*. This principle shall also apply to the frequency, duration and scheduling of meetings.
3.The setting up and the functioning of the organs and subsidiary bodies of the Authority *shall be based on an evolutionary approach*, taking into account the functional needs of the organs and subsidiary bodies concerned in order that they may discharge effectively their respective responsibilities at various stages of development of activities in the Area. [emphasis added].

39 See also this author's notes from the third session of the PrepCom, where on 6 April 2017 Tonga suggested the evolutionary approach used under the Part XI Implementation Agreement as a template for the development of institutional arrangements under the ILBI, beginning with a conference of the parties (on file with author) (Author's PrepCom3 Notes).

40 Global Ocean Commission Report, *supra* note 7, at 4. Several scholars have noted the fragmented nature of the existing ocean governance framework. See, e.g., D. Tladi, "Ocean Governance: A Fragmented Regulatory Framework" (2011), accessed 2 November 2017 at <http://regardssurlaterre.com/en/ocean-governance-fragmented-regulatory-framework>; R. Rayfuse and R. Warner, "Securing a Sustainable Future for the Oceans Beyond National Jurisdiction: The Legal Basis for an Integrated Cross-Sectoral Regime for High Seas Governance for the 21st Century", *The International Journal of Marine and Coastal Law* 23 (2008), 399–421.

3.1 *UNCLOS*

UNCLOS is a framework convention and provides the basic "Constitution for the Oceans",[41] which includes some general environmental duties and obligations for States (primarily set out in Part XII of the Convention), including: (i) the duty to conserve and manage the living resources of the high seas;[42] (ii) the general obligation to protect and preserve the marine environment;[43] (iii) the duty to take all measures necessary to prevent, reduce and control pollution of the marine environment;[44] (iv) the specific duty to take necessary measures to "protect and preserve rare or fragile ecosystems as well as the habitat of depleted, threatened or endangered species and other forms of marine life";[45] and (v) the duty to cooperate with other States, both at the regional and global levels.[46]

3.2 *Convention on Biological Diversity (CBD) and Related Targets and Commitments*

The CBD[47] is closely related to the BBNJ process, with its objectives including: (i) the conservation of biological diversity;[48] (ii) the sustainable use of its components; and (iii) the fair and equitable sharing of the benefits arising out of the utilisation of genetic resources.[49] Article 4 of the CBD limits the jurisdictional scope of the Convention, providing that the rights and obligations of Contracting Parties with respect to the components of biological diversity are divided between those within and beyond the limits of national jurisdiction, with its application beyond the limits of national jurisdiction being limited to jurisdiction over processes and activities concerning the components of biological diversity.[50] Despite the CBD's jurisdictional limitation in areas beyond

41 Constitution for the Oceans, *supra* note 16.

42 UNCLOS, *supra* note 1, Arts. 116–119.

43 *Ibid.*, Art. 192.

44 *Ibid.*, Arts. 194–196.

45 *Ibid.*, Art. 194(5).

46 *Ibid.*, Art. 197.

47 Convention on Biological Diversity, adopted 5 June 1982, 1760 UNTS 79 (entered into force 29 December 1993) (CBD).

48 Article 2 of the CBD defines biological diversity as "the variability among living organisms from all sources including, *inter alia,* terrestrial, marine and other aquatic ecosystems and the ecological complexes of which they are part: this includes diversity within species, between species and of ecosystems". *Ibid.,* Art. 2.

49 *Ibid.*, Art. 1.

50 "Subject to the rights of other States, and except as otherwise expressly provided in this Convention, the provisions of this Convention apply, in relation to each Contracting Party: (a) In the case of components of biological diversity, in areas within the limits of its national jurisdiction; and (b) *In the case of processes and activities, regardless of where*

national jurisdiction, the CBD has been useful to the BBNJ context, with the following significant contributions: one, the development of a process to describe ecologically or biologically significant marine areas (EBSAs),[51] and two, the adoption of voluntary guidelines for the consideration of EIAs and strategic environmental assessments (SEAs) in areas beyond national jurisdiction.[52]

Additional targets and commitments are provided by other initiatives. In 2010, objectives relevant to marine biodiversity in areas beyond national jurisdiction were adopted within the CBD, which are known as the "Aichi Targets".[53] The Rio+20 outcome document, *The Future We Want*, reaffirmed many of these objectives.[54] At the Rio+20 Conference, States also agreed to develop the Sustainable Development Goals (SDGs)[55] to bring together the Millennium

---

their effects occur, carried out under its jurisdiction or control, within the area of its national jurisdiction or beyond the limits of national jurisdiction". *Ibid.*, Art. 4 [emphasis added].

51 G. Wright, J. Rochette, E. Druel, and K. Gjerde, "The Long and Winding Road Continues: Towards a New Agreement on High Seas Governance", IDDRI Study, No. 01 (16 March 2016) (IDDRI Study), at 18 (citing CBD Decision XI/17 on Marine and Coastal Diversity, UNEP/CBD/COP/DEC/XI/17, 5 December 2012, accessed 2 November 2017 at <https://www.cbd.int/doc/decisions/cop-11/cop-11-dec-17-en.pdf)>.

52 IDDRI Study, *supra* note 51 at 18 (citing CBD Decision XI/18 on Marine and Coastal Diversity, UNEP/CBD/COP/DEC/XI/18, 5 December 2012, at 7–8, accessed 2 November 2017 at <https://www.cbd.int/doc/decisions/cop-11/cop-11-dec-18-en.pdf>).

53 Strategic Plan for Biodiversity 2011–2020 and the Aichi Targets (Aichi Targets), accessed 2 November 2017 at <https://www.cbd.int/sp/targets/>. Of special relevance, is Target 11:

> By 2020, at least 17 percent of terrestrial and inland water, and 10 percent of coastal and marine areas, especially areas of particular importance for biodiversity and ecosystem services, are conserved through effectively and equitably managed, ecologically representative and well-connected systems of protected areas and other effective area-based conservation measures, and integrated into the wider landscapes and seascapes. Ibid.

54 *The Future We Want,* Outcome document of the United Nations Conference on Sustainable Development Rio de Janeiro, Brazil, 20–22 June 2012, UN GAOR A/Res/66/288 (*The Future We Want*), as evidenced by States committing to the following:

> We recognize the importance of the conservation and sustainable use of marine biodiversity beyond areas of national jurisdiction. We note the ongoing work under the UN General Assembly of an Ad Hoc Open-ended Informal Working Group to study issues relating to the conservation and sustainable use of marine biological diversity beyond areas of national jurisdiction. Building on the work of the ad hoc working group and before the end of the 69th Session of the United Nations General Assembly we commit to address, on an urgent basis, the issue of the conservation and sustainable use of marine biological diversity of areas beyond national jurisdiction including by taking a decision on the development of an international instrument under UNCLOS. *Ibid.*, para. 162.

55 *Transforming our world: the 2030 Agenda for Sustainable Development*, UN GAOR A/Res/70/1, accessed 2 November 2017 at <http://unctad.org/meetings/en/SessionalDocuments/ares70d1_en.pdf>.

Development Goals and environmental concerns into one comprehensive global development agenda. The SDGs were adopted in 2015, with an Ocean SDG (SDG 14), which includes targets related to the conservation of marine ecosystems and effective regulation of fishing.[56]

3.3 *International Organisations and Instruments with Sectoral Mandates*

Although there are international organisations with sectoral mandates in areas beyond national jurisdiction, no overarching framework exists. Fishing is covered by the mandate of the Food and Agriculture Organization of the United Nations (FAO),[57] which is further complemented on the regional level by regional fisheries management organisations (RFMOs). The exploration and exploitation of mineral resources in the Area is governed by the ISA. Shipping and dumping fall under the governance framework of the International Maritime Organization (IMO). Marine science is dealt with under the auspices of the Intergovernmental Oceanographic Commission (IOC) of the United Nations Educational, Scientific and Cultural Organization (UNESCO). Trade and intellectual property relating to BBNJ are considered in the context of the World Trade Organization (WTO) and the World Intellectual Property Organization (WIPO).

Other sectoral instruments relevant to BBNJ include the International Convention on the Regulation of Whaling[58] and the Convention on the Conservation of Migratory Species of Wild Animals.[59]

56 Most recently, the United Nations Ocean Conference was convened from 5–9 June 2017 at United Nations Headquarters in New York to support SDG 14. The high-level Conference coincided with World Oceans Day and was co-hosted by the Governments of Fiji and Sweden.

57 Fishing is also governed by soft law instruments, see, for example, United Nations Resolution 61/105, which addresses deep sea fisheries. *Sustainable fisheries, including through the 1995 Agreement for the Implementation of the Provisions of the United Nations Convention on the Law of the Sea of 10 December 1982 relating to the Conservation and Management of Straddling Fish Stocks and Highly Migratory Fish Stocks, and related instruments*, UN GAOR A/Res/61/105 (UNGA Resolution 61/105), accessed 2 November 2017 at <https://documents-dds-ny.un.org/doc/UNDOC/GEN/N06/500/73/PDF/N0650073.pdf?OpenElement>.

58 International Convention for the Regulation of Whaling, 364 UNTS 1953 (entered into force 10 November 1948).

59 Convention on the Conservation of Migratory Species of Wild Animals, 1651 UNTS 333 (entered into force 11 January 1983), together with its two supplemental agreements: (i) Agreement on the Conservation of Cetaceans of the Black Sea, Mediterranean Sea and contiguous Atlantic Area (ACCOBAMS) and (ii) the Agreement on the Conservation of Small Cetaceans of the Baltic, North East Atlantic and North Seas (ASCOBANS). The challenge is that these agreements do not have the same parties.

3.4 *Regional Organisations and Instruments*

The United Nations Environmental Programme (UNEP) was created in 1972 to serve as a focal point for environmental action and coordination within the United Nations system.[60] Eighteen regional initiatives are taken through UNEP's Regional Seas Programme (RSP) for the protection and sustainable use of global and marine environments.[61] While UNEP provides overall coordination of the RSP, only seven of these regional initiatives are administered directly by UNEP, seven fall outside of UNEP administration, and four are administered independently.[62] The regional initiatives under the RSP comprise both conventions and action plans, which are tailored to each region's unique environmental, socio-economic and political requirements, but generally include five common elements: (i) economic assessment, (ii) economic management, (iii) economic legislation, (iv) institutional arrangements, and (v) financial arrangements.[63]

Regions that have established measures that impact BBNJ through their respective regional initiatives include the following:[64] the Mediterranean through the Barcelona Convention;[65] the Southern Ocean through the Convention for the Conservation of Antarctic Marine Living Resources (CCLAMR);[66] the North-East Atlantic through the OSPAR Convention,[67] and the South Pacific through the Noumea Convention.[68] Three of these regional

60 United Nations Environmental Programme (UNEP), accessed 9 December 2017 at <https://www.unenvironment.org>.

61 UNEP Regional Seas Programme, "Overview", accessed 9 December 2017 at <http://web.unep.org/regionalseas/who-we-are/overview>.

62 *Ibid.*

63 UNEP Region Seas Programme, "Action Plans and Conventions", accessed 9 December 2017 at <http://web.unep.org/regionalseas/who-we-are/regional-seas-action-plans>.

64 IDDRI Study, *supra* note 51 at 18. UNEP contributed a study to the fourth session of the PrepCom on the regional seas programmes covering areas beyond national jurisdiction, accessed 12 December 2017 at <http://www.un.org/Depts/los/biodiversityworkinggroup/Regional_seas_programmes_ABNJ.pdf>.

65 Convention for the Protection of the Marine Environment and the Coastal Region of the Mediterranean 1102 UNTS 45 (adopted 16 February 1976 and amended 10 June 1995; entered into force 12 February 1978 and 9 July 2004, respectively).

66 Convention for the Conservation of Antarctic Marine Living Resources, 1329 UNTS 48 (entered into force 7 April 1982).

67 Convention for the Protection of the Marine Environment of the North-East Atlantic, 23 LOSB 32 (entered into force 25 March 1998).

68 Convention for the Protection of the Natural Resources and Environment of the South Pacific Region (entered into force 22 August 1990), accessed 2 November 2017 at <https://www.ecolex.org/details/treaty/convention-for-the-protection-of-the-natural-resources-and-environment-of-the-south-pacific-region-tre-000892/>.

initiatives have taken action in areas beyond national jurisdiction through the creation of MPAS.[69]

4 The Beginnings of the BBNJ Discussions and Establishment of the PrepCom

Parties to UNCLOS have focused primarily on areas within national jurisdiction, where coastal States not only have the authority to regulate human activities in marine areas under their national jurisdiction, but also the obligation to protect such areas through the use of a number of biodiversity conservation measures, including EIAS and MPAS, among others. In areas beyond national jurisdiction, however, the freedom of the high seas, ineffective flag State jurisdiction on the high seas, and the inability to bind third parties not party to certain instruments, such as the UNFSA, has led to an ocean governance framework that is widely perceived as inadequate to protect marine biodiversity in areas beyond national jurisdiction and ensure its long-term health and sustainable use.

Questions were first raised at the fourth meeting of the United Nations Informal Consultative Process on the Oceans and the Law of the Sea in 2003 regarding the lack of effective legal and institutional mechanisms for governing areas beyond national jurisdiction. Discussions on these issues led to a recommendation, approved by UNGA A/58/95, to form the *Ad Hoc* Open-Ended Informal Working Group on Marine Biodiversity in Areas Beyond National Jurisdiction (BBNJ Working Group).[70]

4.1 *The BBNJ Working Group's Recommendations*

In the BBNJ Working Group, there was general recognition for the view that "the status quo was not a sustainable situation, nor an acceptable solution".[71] Through the work of the BBNJ Working Group, it was noted that although a number of relevant instruments, bodies and frameworks exist, not all human activities within areas beyond national jurisdiction are adequately regulated and not all regions are covered. The BBNJ Working Group also recognised that modern governance principles were not being employed within the existing

69 IDDRI Study, *supra* note 51 at 18–19.

70 *Report on the Work of the United Nations Open-ended Informal Consultative Process on Oceans and the Law of the Sea*, UN GAOR A/58/95, accessed 2 November 2017 at <http://undocs.org/A/58/95>.

71 BBNJ Working Group Recommendations, *supra* note 5 at 10.

framework, such as the need for decision-making processes to be transparent and open, thereby frustrating the ability to implement a coordinated, cross-sectoral approach.[72] The general view among States was that gaps in the governance framework should be addressed under the auspices of UNCLOS.[73]

The BBNJ Working Group noted in its 2011 recommendations the following gaps, all of which are related to the institutional aspects of ocean governance. First, they noted a lack of legal clarity regarding the status of MGRs in areas beyond national jurisdiction.[74] The starting point for discussion of access and benefit sharing of MGRs in the areas beyond national jurisdiction context has been that MGRs do not fall within the scope of the Nagoya Protocol, a supplementary agreement to the CBD that sets out the implementation of access and benefit sharing under the CBD.[75] Accordingly, the BBNJ Working Group noted the absence of an international body to regulate access to such resources.[76]

Second, the BBNJ Working Group noted an absence of a global framework to establish MPAs[77] in areas beyond national jurisdiction[78]—i.e., where the IMO can identify Particularly Sensitive Sea Areas (PSSAs),[79] the ISA can designate Areas of Particularly Environmental Interest (APEIs)[80] and RFMOs can

72 *Ibid.* at 4–5.

73 *Ibid.* at 4.

74 *Ibid.* at 5–6.

75 *Ibid.* at 6. The Nagoya Protocol on Access to Genetic Resources and the Fair and Equitable Sharing of Benefits Arising from their Utilization applied to genetic resources within the scope of Article 15 of the CBD and to traditional knowledge associated with genetic resources within the scope of the Convention. The CBD entered into force on 12 October 2014.

76 *Ibid.* at 6.

77 A Marine Protected Area can be defined as a clearly defined geographical space, recognised, dedicated and managed, through legal or other effective means, to achieve the long-term conservation of nature with associated ecosystem services and cultural values. CBD Subsidiary Body On Scientific, Technical And Technological Advice, "Report of the Ad Hoc Technical Expert Group on Marine and Coastal Protected Areas, Note by the Executive Secretary", UNEP/CBD/STSTTA/8/INF/7 (13 February 2003), accessed 2 November 2017 at <https://www.cbd.int/doc/meetings/sbstta/sbstta-08/information/sbstta-08-inf-07-en.pdf>.

78 BBNJ Working Group Recommendations, *supra* note 5 at 6–7.

79 IMO, "Revised Guidelines for the Identification and Designation of Particularly Sensitive Sea Area", A 24/Res. 982 (6 February 2006), accessed 2 November 2017 at <http://www.imo.org/en/OurWork/Environment/PSSAs/Documents/A24-Res.982.pdf>.

80 ISA, "Decision of the Council of the International Seabed Authority relating to Amendments to the Regulations on Prospecting and Exploration for Polymetallic Nodules in the Area and Related Matters", ISBA/19/C/17 (22 July 2013), accessed 2 November 2017 at <https://www.isa.org.jm/sites/default/files/files/documents/isba-19c-17_0.pdf>.

designate fishing closures to protect Vulnerable Marine Ecosystems (VMEs).[81] The Working Group emphasised, however, that the establishment of MPAS should not follow a uniform methodology or what they termed a "one-size-fits-all" approach.[82] In addition, the CBD was viewed as a basis for multilateral cooperation for addressing this gap.[83]

Third, a lack of global rules for EIAS and SEAS in areas beyond national jurisdiction was identified by the BBNJ Working Group as a governance gap. Article 206 of UNCLOS does not provide details regarding the events that trigger EIAS, nor does it provide the minimum standards for EIAS, resulting in poor implementation of this provision.[84] The BBNJ Working Group noted that the duty to conduct EIAS beyond areas of national jurisdiction is governed by the general framework of the CBD, together with the ISA regulations and the advisory opinion of the Seabed Disputes Chamber of the International Tribunal for the Law of the Sea on the responsibilities and obligations of States sponsoring persons and entities with respect to activities in the Area.[85] The BBNJ Working Group observed that States have a role in formulating guidelines for the conduct of EIAS, and recognised that other international organisations also have a role, such as the FAO, the IOC, UNEP, and other instruments, such as the CBD, among others.[86] The incorporation of EIA requirements in regional instruments was also encouraged, with specific mention being made to the Protocol on Environmental Protection to the Antarctic Treaty.[87]

Fourth, the BBNJ Working Group noted that several delegations wanted to increase the capacity and participation of developing States in scientific research in the Area, as well as in the conservation and sustainable use of marine biodiversity, noting that this was a gap under the existing framework of Article 268 of UNCLOS.[88] They noted further that this was the area in which UNCLOS "suffered from the gravest implementation gap".[89]

81 See UNGA Resolution 61/105, *supra* note 57.

82 BBNJ Working Group Recommendations, *supra* note 5 at 7.

83 *Ibid.* at 4, 7.

84 UNCLOS, *supra* note 1, Art. 206: "Where States have reasonable grounds for believing that planned activities under their jurisdiction or control may cause substantial pollution of or significant and harmful changes to the marine environment, they shall, as far as practicable, assess the potential effects of such activities on the marine environment and shall communicate reports of the results of such assessments in the manner provided in Article 205".

85 BBNJ Working Group Recommendations, *supra* note 5 at 8.

86 *Ibid.*

87 *Ibid.*

88 *Ibid.* at 9.

89 *Ibid.*

In addition, the BBNJ Working Group noted the following weaknesses: (i) the need for better coherence and coordination of regional and international processes, and (ii) the need for a cross-sectoral approach due to the multiplicity of existing international bodies and forums that dealt with marine biodiversity.[90] Furthermore, given that fisheries are currently the greatest threat to marine biodiversity in areas beyond national jurisdiction, many delegations argued that a new agreement could make improvements to the existing fisheries management framework, at least with respect to the way RFMOs protect biodiversity in the marine environment.[91]

The work of the BBNJ Working Group concluded by recommending to the General Assembly that a process be initiated to assess the adequacy of the legal framework for the conservation and sustainable use of marine biodiversity in areas beyond national jurisdiction "by identifying gaps and ways forward, including through the implementation of existing instruments and the possible development of a [new] multilateral agreement under the United Nations Convention on the Law of the Sea".[92] Furthermore, the BBNJ Working Group recommended that the process consider "together and as a whole" the four key elements of the package deal.[93] These elements gained momentum and political support during the 2012 UN Conference on Sustainable Development (Rio+20),[94] eventually leading to the adoption of UNGA Resolution 69/292 and the establishment and convening of the PrepCom.

4.2 *The Mandate of the PrepCom*

The General Assembly established the PrepCom, prior to convening an IGC, for the purpose of recommending substantive elements for the development of an ILBI under UNCLOS on the conservation and sustainable use of marine biodiversity of areas beyond national jurisdiction. The resolution stipulated that the process set out in paragraph 3 of UNGA Resolution 69/292 "should *not undermine* existing relevant instruments and frameworks and relevant global, regional and sectoral bodies".[95] In the first and second sessions of the PrepCom, discussions on how to foster better coordination to resolve the gaps was coupled with substantial discussions about the language of the mandate,

90 *Ibid.* at 12–13.

91 *Ibid.* at 12.

92 *Ibid.*, para. 1(a).

93 *Ibid.* at 5.

94 *The Future We Want, supra* note 54, para. 162.

95 UNGA Resolution 69/292, *supra* note 3, para. 3 [emphasis added]. See also UNFSA, *supra* note 13, Art. 33.

particularly the "not undermine" requirement.[96] This whole idea of what "not undermine" may mean in practice attracted a number of different views throughout the PrepCom.[97]

5 The Discussion of Cross-Cutting Issues at the First Two Sessions of the PrepCom

5.1 *Considerations for Institutional Arrangements*

The first session of the PrepCom (PrepCom1) was convened on 28 March 2016 at the United Nations headquarters in New York and continued through 8 April 2016 under the chairmanship of Ambassador Eden Charles (Trinidad and Tobago) (PrepCom1). After ten years of discussions within the BBNJ Working Group, the PrepCom began its work with a "constructive spirit".[98] Building on this momentum, the delegates started to "unpack" and "defragment" the elements of the package deal.[99] In doing so, the delegates quickly realised the extent of the inter-linkages between the elements, most notably in the case of ABMTs, including MPAs and EIAs.[100] With that realisation, the challenge for the delegations to construct a new instrument that would "not undermine" the existing instruments and frameworks that had "mushroomed" elsewhere became that much more daunting, and focused the discussions on the PrepCom mandate and the institutional arrangements that may be needed to implement the new legal regime.[101] The outcome of the first session was a roadmap that organised the work of the Prepcom around the package deal with the establishment of five informal working groups: one for each element of the package and one for cross-cutting issues.[102]

During the second session of the PrepCom (PrepCom2), which ran from 26 August to 9 September 2016, it was still unclear to many delegates how existing ocean governance gaps could be addressed without undermining existing relevant instruments and frameworks as per UNGA Resolution 69/292.[103]

96 ENB PrepCom1 Summary, *supra* note 23 at 5. Norway's intervention supported the view that there was no consensus about the meaning of "not undermine", and suggested the use of UNFSA Article 44(1) as a model. ENB PrepCom2 Summary, *supra* note 23 at 16.

97 Chair's Overview of PrepCom3, *supra* note 25 at 26.

98 ENB PrepCom1 Summary, *supra* note 23 at 18.

99 *Ibid.*

100 *Ibid.*

101 *Ibid.*

102 *Ibid.* at 17.

103 ENB PrepCom2 Summary, *supra* note 23 at 19.

Significant focus, discussion and emphasis centred on the strength of the ILBI's mandate to address the gaps. Some delegations argued for an ILBI with a strong mandate that would create a mechanism for increased cooperation and coordination by advising existing institutions, communicating information, and formulating recommendations.[104] Others cautioned against an ILBI with a strong mandate, suggesting that proactive intervention would undermine the mandates of existing instruments and frameworks.[105] For instance, China highlighted that the ILBI should not interfere with the existing mandates of the FAO, IMO, ISA or RFMOs; rather, it should promote coordination and cooperation, and avoid duplication and overlap.[106] In New Zealand's intervention, it was suggested that the new legal regime should build on BBNJ expertise in regional and sectoral organisations, and by creating incentives for the improvement of the existing ocean governance framework.[107] Delegations also suggested that Parties should be called upon to strengthen existing institutions in accordance with the priorities and principles of the resulting ILBI.[108]

For the most part, the discussions at both PrepCom1 and PrepCom2 proceeded along the lines of the dichotomy.[109] Those delegations that supported a global approach also supported the common heritage of mankind principle for marine biodiversity in areas beyond national jurisdiction.[110] Therefore, they called for a global mechanism to fill this gap through a comprehensive approach drawing on existing bodies with expertise and jurisdiction over activities in areas beyond national jurisdiction, particularly relating to the establishment of ABMTs.[111] Those delegations that supported traditional high seas freedoms called for a more decentralised, regional and sectoral approach, favouring reliance on existing regional and sectoral bodies for implementation.[112]

With regard to the discussions on institutional arrangements, there was support at the first two sessions of the PrepCom for a number of different variations, including having a decision-making body (i.e., a COP) that could

104 *Ibid.* at 19–20.

105 *Ibid.* at 16–17, 19–20.

106 ENB PrepCom2 Summary, *supra* note 23 at 16.

107 ENB PrepCom2 Summary, *supra* note 23 at 17. New Zealand referred to Article 8(6) of the UNFSA on consultations with RFMOs as a possible model. *Ibid.*

108 *Ibid.*

109 ENB PrepCom1 Summary, *supra* note 23; ENB PrepCom2 Summary, *supra* note 23.

110 ENB PrepCom1 Summary, *supra* note 23 at 5–6; ENB Prepcom2 Summary, *supra* note 23 at 4.

111 ENB PrepCom1 Summary, *supra* note 23 at 9–10; ENB Prepcom2 Summary, *supra* note 23 at 17, 19–20.

112 ENB PrepCom1 Summary, *supra* note 23 at 5–6; ENB Prepcom2 Summary, *supra* note 23 at 4.

establish subsidiary organs, including a scientific committee.[113] Other suggestions included a clearinghouse mechanism, a periodic review conference, a compliance mechanism, and access and benefit-sharing mechanisms.[114] Some delegations suggested a centralised approach where the effective implementation of the new regime would require the establishment of a new global executive body, (i) either through the establishment of a new body, or (ii) through the expansion of the mandate of an existing organisation, such as the ISA, through which Parties could take decisions, undertake coordination efforts, perform reviews, and assess implementation and compliance.[115] Expansion of the ISA mandate was seen as a possible fix, but this idea has since lost substantial support.[116] The delegates supporting a centralised approach emphasised that a global authority could ensure consistency, coordination, and possible mechanisms for enhancing capacities of existing bodies.[117] Other delegations suggested that such a global body would undermine other instruments and frameworks, favouring instead a regional and/or sectoral approach where implementation would occur at the regional and sectoral levels.[118] These delegations favoured a more decentralised approach with "light arrangements".[119] There were other delegations that expressed openness to a "mixed approach" whereby minimum standards for regional and sectoral management would be set, with significant reliance on the experiences and expertise of existing arrangements.[120] This approach, however, received little traction at the first two sessions of the PrepCom.

There was some discussion on the criteria for the institutional arrangements under the ILBI. For example, some delegations emphasised that the resulting institutional structure must be cost-effective and efficient.[121] This drove some

113 ENB PrepCom1 Summary, *supra* note 23 at 16, 19; ENB Prepcom2 Summary, *supra* note 23 at 17, 19–20.

114 ENB PrepCom1 Summary, *supra* note 23 at 16, 19; ENB Prepcom2 Summary, *supra* note 23 at 17, 19–20.

115 ENB PrepCom1 Summary, *supra* note 23 at 9–10, 16; ENB PrepCom2 Summary, *supra* note 23 at 17.

116 ENB PrepCom1 Summary, *supra* note 23 at 5, 16; ENB PrepCom2 Summary, *supra* note 23 at 20.

117 ENB PrepCom1 Summary, *supra* note 23 at 5, 16, 19; ENB PrepCom2 Summary, *supra* note 23 at 17, 19–20.

118 ENB PrepCom1 Summary, *supra* note 23 at 5, 16, 19; ENB PrepCom2 Summary, *supra* note 23 at 17, 19–20.

119 ENB PrepCom2 Summary, *supra* note 23 at 17, 19–20.

120 ENB PrepCom2 Summary, *supra* note 23 at 19–20.

121 ENB PrepCom1 Summary, *supra* note 23 at 16, 19; ENB Prepcom2 Summary, *supra* note 23 at 17.

of the conversations regarding the possibility of expanding the ISA's mandate, which was raised again at PrepCom2, but as mentioned above retained little momentum in subsequent sessions.[122] Transparency, accountability and inclusiveness were also emphasised as key principles of the ILBI, but there was little convergence around a specific approach.[123]

5.2 *The Chair's Non-Paper on Considerations for Institutional Arrangements*

Ambassador Charles, the Chair of the PrepCom, organised the work of the first two sessions, such that the output following the conclusion of PrepCom2 would be a Chair's non-paper that compiled the ideas and proposals put forward by delegations for the purpose of assisting in their consideration of the issues to be discussed at subsequent sessions of the PrepCom. The non-paper was intended to provide a structured presentation of the issues and ideas reflected in the proposals for elements of a draft text of an ILBI, as well as possible areas of convergence from the Chair's understandings and those issues and ideas that were discussed extensively during PrepCom2.[124]

In his non-paper, the Chair consolidated the views on institutional arrangements in the section on cross-cutting issues,[125] which fell largely along the lines of the global approach[126] and regional and/or sectoral approach.[127] In doing so,

122 ENB PrepCom1 Summary, *supra* note 23 at 5, 16; ENB Prepcom2 Summary, *supra* note 23 at 20; see also "Chair's non-paper on elements of a draft text of an international legally-binding instrument under the United Nations Convention on the Law of the Sea on the conservation and sustainable use of marine biological diversity of areas beyond national jurisdiction" 28 February 2017 (Chair's non-paper), at 95, n. 694 (CARICOM).

123 ENB PrepCom2 Summary, *supra* note 23 at 16, 20.

124 Chair's non-paper, *supra* note 122 at 93, n. 678 (Chair's understanding on cross-cutting issues).

125 Specifically, the Chair's view was that there was support for institutional arrangements at the global level, which could include: (i) a decision-making forum; (ii) a scientific forum; (iii) a clearing house; and (iv) a secretariat. Chair's non-paper, *supra* note 122 at 94, n. 688 (Chair's understanding on cross-cutting issues); EU Submission of 5 December 2016, *supra* note 36 at 3. Other delegations were more specific in suggesting that institutional bodies could include: (i) a secretariat; (ii) a decision-making body such as a COP; (iii) a scientific or technical body with an advisory competence which can play a role in the establishment of ABMTs, including MPAs; (iii) a clearing-house mechanism to promote and facilitate technical and scientific cooperation, knowledge and data sharing; and (iv) a mechanism in charge of access and benefit sharing of marine genetic resources. Chair's non-paper, *supra* note 122 at 94, n. 689 (G77 and China).

126 Chair's non-paper, *supra* note 122 at 93–103.

127 There was support for the view that implementation could occur by States or groups of States at a regional level, supported by regional agreements and institutions responsible for the implementation of conservation and sustainable use objectives, taking into

he also addressed the criteria put forth for such arrangements. Such criteria can be grouped into three main categories, suggesting that there was a desire among the delegations that the resulting arrangements: (i) be "fit-to-purpose", cost-effective and efficient; (ii) promote transparency and accountability; and (iii) foster participation and inclusiveness. With regard to the first category, support existed for the ILBI to be equipped with a "robust, but cost-effective institutional structure".[128] Delegations also emphasised that the arrangements must be "fit-to-purpose" with existing institutional arrangements to be used wherever possible, but recognising that in some cases establishment of new bodies may be required.[129] Another recommendation was that scientific input should take into account all existing processes, with due consideration given to other relevant processes.[130]

With regard to the second category, transparency and accountability, the Chair's non-paper noted that there was support for the view that to ensure transparency, the actions, decisions and decision-making processes under the ILBI would need to be open to an appropriate level of scrutiny by Parties, civil society and, where appropriate, outside institutions.[131] A specific suggestion was made for States and sub-regional or regional biological diversity management organisations and arrangements to ensure transparency in the mechanisms for management and in an agreed decision-making process.[132] It was further suggested that this could be achieved through opening meetings to observers where appropriate, and making information about and distributed at such meetings publicly available.[133]

To ensure that the third category of participation and inclusiveness was achieved, it was suggested in the Chair's non-paper that participation in meetings should be open to non-Contracting Parties, relevant non-governmental organisations, and other stakeholders, in an observer capacity.[134] In addition, there was support for the view that representatives from relevant organisations, both governmental and non-governmental, concerned with BBNJ should be afforded the opportunity to take part in meetings of sub-regional and regional organisations and arrangements as observers or otherwise, as

account the region-specific issues and the different environments, stakeholders and activities. *Ibid.*

128 *Ibid.* at 94, n. 690 (Greenpeace).

129 *Ibid.* at 93–95.

130 *Ibid.* at 94–95, n. 693 (EU).

131 *Ibid.* at 93, n. 680 (New Zealand).

132 *Ibid.* at 93, n. 682 (Fiji).

133 *Ibid.* at 93, n. 684 (Australia).

134 *Ibid.* at 93, n. 681 (New Zealand).

appropriate, in accordance with the procedures of the organisation or arrangement concerned.[135]

Arguably, a fourth category of criteria for the institutional arrangements for the ILBI in the Chair's non-paper was the language of the PrepCom mandate requiring that the process "not undermine" existing instruments, frameworks and bodies. This part of the mandate was discussed at the first two sessions of the PrepCom in the context of the ILBI's relationship with other instruments and frameworks. Specifically, Article 4 of the UNFSA was raised as a means of addressing the relationship of the ILBI to UNCLOS, providing that nothing in the ILBI "shall affect the rights, obligations or jurisdiction of States under the Convention" and that the ILBI shall be "interpreted and applied in a manner consistent with the Convention".[136] There was also support for the suggestion that the inclusion of a "without-prejudice clause" in the ILBI would assist in fulfilling the "not undermine" requirement of the mandate.[137]

The Chair's non-paper was well received by the delegations at PrepCom3 and aided in the emergence of the hybrid approach, as is discussed in more detail below.

6 The Hybrid Approach at the PrepCom

6.1 *The Emergence of the Hybrid Approach*

The "hybrid approach" was born with New Zealand's paper on cross-cutting issues,[138] which was one of several papers submitted between PrepCom2 and PrepCom3 at the Chair's invitation for the purpose of assisting his preparation of the non-paper.[139] New Zealand's submission attracted a great deal of attention in the course of PrepCom3, and was first raised in the discussions of ABMTs, where it was proposed that the ILBI could set out internationally agreed standards and obligations at the global level, a number of which, particularly those pertaining to ABMTs and MPAs, could then be implemented by States

135 *Ibid.* at 93. Such representatives may be given timely access to the records and reports of such meetings, subject to the procedural rules on access to them. *Ibid.* at 93, n. 683 (Fiji).

136 EU Submission of 5 December 2016, *supra* note 36 at 3.

137 *Ibid.*

138 New Zealand's complete set of submissions is available online at PaperSmart, accessed 2 November 2017 at <https://papersmart.unmeetings.org>.

139 The third session of the PrepCom ran from 27 March to 7 April 2017, and was conducted under the chairmanship of Ambassador Carlos Sobral Duarte (Brazil).

through relevant regional and sectoral frameworks.[140] Such a framework, New Zealand proposed, could promote greater global coherence and coordination through a comprehensive set of obligations that would guide actions by States individually, as well as through their participation in relevant regional and sectoral frameworks.[141] Under this framework, States would be required to report their individual activities, including those undertaken regionally and sectorally, to the COP to inform further policy decisions and recommendations and promote broader cooperation.[142] In a subsequent intervention, New Zealand expounded on the hybrid approach, proposing that the ILBI could establish a global body, which would have responsibilities for the following functions: (i) making high level decisions including the identification of priority areas for establishment of ABMTs or MPAs; (ii) articulating when or how activities in areas beyond national jurisdiction trigger the need for an EIA; (iii) establishing processes for cooperation and coordination with existing bodies; (iv) administering a global information repository; and (v) undertaking a regular review of implementation of the ILBI.[143]

The hybrid approach was first discussed in more detail in the Informal Working Group on MPAs at PrepCom3. Within the context of ABMT governance, three approaches were debated: a global approach, establishing a global institution to consider and decide on ABMT proposals;[144] a hybrid approach, reinforcing regional and sectoral organisations' mandates through regional cooperation mechanisms, and providing global guidance and oversight;[145] and a regional and/or sectoral approach, recognising regional and sectoral bodies' authority for decision making, monitoring and review of ABMTs, with the ILBI providing general policy guidance to promote cooperation, without global-level oversight.[146] Later, in the Informal Working Group on cross-cutting

140 "Preparatory Committee on Conservation and Sustainable Use of Marine Biological Diversity of Areas Beyond National Jurisdiction", New Zealand Submission (December 2016) (New Zealand Submission), at 1.

141 *Ibid.*

142 *Ibid.*

143 "New Zealand—Intervention No. 2 IWG on Cross-cutting Issues" (New Zealand Submission No. 2 on Cross-cutting Issues), at 1.

144 Those favouring a global approach to ABMTs included: African Group, EU, Mexico, El Salvador, Iran, Argentina, Pakistan, Costa Rica, Peru, Indonesia and IUCN. ENB PrepCom3 Summary, *supra* note 6 at 6.

145 Those favouring a hybrid approach to ABMTs included: Norway, New Zealand, Australia, Japan, PSIDS, and FSM. *Ibid.* at 6.

146 Those favouring a regional and sectoral approach to ABMTs included: Iceland, Russian Federation, and the International Commission for the Conservation of Atlantic Tunas (ICCAT). *Ibid.*

issues, the hybrid approach again garnered significant discussion. New Zealand, Norway, Chile, and Tonga, among others, elaborated on various aspects of the approach, which could include a COP (a forum for States and stakeholders to exchange views and set recommendations for implementation at regional and sectoral levels), a scientific body, and a Secretariat (possibly through an expanded mandate of DOALOS).[147]

The viability and importance of the hybrid approach was formalised in the Chair's Overview of PrepCom3, together with the "global approach"[148] and the "regional and/or sectoral approach" that had been outlined previously in the Chair's non-paper.[149] The hybrid approach that emerged was essentially a variation on the "mixed approach" referenced earlier, where standards are agreed at the global level and implemented at a regional and/or sectoral level. The Chair's Overview of PrepCom3 elaborated on the hybrid approach as follows:

> [T]he scientific and technical expertise and decision-making mandates of regional and sectoral organisations in the establishment of ABMTs would be relied upon and reinforced, possibly through regional coordination mechanisms. At the same time, at the global level, general guidance and objectives would be developed, building on existing knowledge and the exchange of information between stakeholders, to enhance cooperation

147 *Ibid.* at pp. 12–13. DOALOS is the Division for Ocean Affairs and the Law of the Sea, which provides (i) advice, studies, assistance and research on the implementation of UNCLOS; (ii) secretariat services under UNCLOS; and (iii) functions concerning the deposit of charts and coordinates under certain provisions of UNCLOS.

148 "One approach, which could be termed the 'global model', envisages that a global institution would be given the mandate to consider proposals for ABMTs, including MPAs. It would, with the assistance of a subsidiary scientific or technical body and a mechanism for stakeholder consultations, take decisions on the establishment of such areas, possibly through a conference of the parties; monitor and review implementation by sectoral/regional bodies; and monitor compliance. This approach also foresees the enhancement of cooperation and coordination through consultations with existing regional and sectoral mechanisms. ABMTs adopted by existing regional and sectoral mechanisms could go through a process of recognition by the global mechanism". Chair's Overview of PrepCom3, *supra* note 25 at 12.

149 "A third approach would rely on a 'regional and/or sectoral model', which envisages that general policy guidance to promote cooperation and coordination could be provided under the new instrument at the global level, while recognising the full authority, without oversight from a global mechanism, of regional and sectoral organisations in decision-making, monitoring and review. This approach considers that regional and sectoral bodies represent the most efficient mechanisms to regulate activities in ABNJ, including through ABMTs, as they have the knowledge and expertise necessary for the adoption of measures". *Ibid.*

> and coordination and provide a level of oversight to the decision-making and implementation by regional and/or sectoral mechanisms.[150]

In the Facilitator's Report on Cross-cutting Issues included in the Chair's Overview of PrepCom3, the Facilitator suggested that it would be "beneficial to continue discussing these different approaches and the proposed allocation of roles and responsibilities within them, including how to deal with existing regional and sectoral measures, with the goal of identifying strengths and weaknesses in each approach and building a better understanding of the possible options".[151] In addition, the Chair noted that with regard to the institutional arrangements for ABMTs, references were also made to assisting the implementation of the obligations of States under Part XII of UNCLOS, as well as the commitments under internationally-agreed targets, such as SDG 14 and the CBD Aichi Biodiversity Targets.[152]

6.2 *Criteria and Overall Structure for the Institutional Arrangements as Put Forward by the Delegations*

The Chair's Overview of PrepCom3 encouraged further consideration of the issues identified in the Facilitator's oral report on cross-cutting issues, including on how guiding principles and approaches could be reflected in the ILBI and how they might be applied in the context of the various elements of the package deal, as well as to issues relating to institutional arrangements.[153] The Chair was also of the view that subsequent discussions could be aided by considering the inter-linkages between the elements of the package deal for the purpose of ensuring coherence and identifying how they mutually support one another.[154] Suggesting that "form follows function", the Chair acknowledged that further discussion would be necessary in order to elaborate the substantive elements of the ILBI on a number of the other cross-cutting issues, such as review, monitoring and compliance, responsibility and liability, and dispute settlement.[155]

Noting that there were many commonalities regarding the various institutions that might be needed under the ILBI, as well as the functions that they

150 *Ibid.*

151 *Ibid.* at 13, 27. The Chair concurred with this view, stating "Further discussions may be useful on these various approaches, including their strengths and weaknesses, including as they apply to each element of the package". *Ibid.* at 27.

152 *Ibid.* at 10.

153 *Ibid.* at 32.

154 *Ibid.*

155 *Ibid.*

might perform, the Chair was of the view that there was convergence for the inclusion of an assembly or COP, a scientific and/or technical body, and a secretariat.[156] He also noted that there was significant support for the development of a clearinghouse mechanism.[157] However, differing views were expressed by the delegations regarding whether relevant existing bodies should report or be accountable to an institution established under the ILBI and how to address the situation where no regional or sectoral organisation exists.[158]

The Chair's Overview of PrepCom3 reinforced the criteria set out in the Chair's non-paper, noting that there was convergence around the need to apply principles of good governance to the new ILBI.[159] Specifically, the Chair found that there was support among the delegations that the resulting arrangements: (i) be "fit-to-purpose", cost-effective and efficient; (ii) promote transparency, where there would be exchange of information and data, public participation, and public access to environmental information; (iii) promote accountability through reporting, monitoring and compliance mechanisms; and (iii) foster participation and inclusiveness through openness to non-Contracting Parties.[160] With regard to the PrepCom's mandate that the process "not undermine" existing instruments, frameworks and bodies, the Chair also noted that discussions on institutional arrangements touched upon the issue of the relationship between the institutions established under the ILBI and relevant existing sectoral and regional bodies.[161]

PrepCom3 was followed by a short intersessional period, with the session ending on 7 April 2017, followed closely by the deadline of 24 April 2017 for the final round of written submissions by the delegations to assist the Chair in his preparation of the Overview of PrepCom3.[162] Soon after, the delegations had to contend with the negotiations of the Call for Action that was adopted at the United Nations Ocean Conference in June 2017.[163] The fourth session of the PrepCom convened a few weeks later at the United Nations headquarters in New York on 10 July 2017, concluding with the adoption of the PrepCom Report late in the evening on 21 July 2017 (PrepCom4). Due in part to the tight frame leading up to PrepCom4, expectations for further substantive development of draft elements of the ILBI were low. In addition, further substantive debate

156 *Ibid.* at 27.
157 *Ibid.*
158 *Ibid.* at 26.
159 *Ibid.* at 13.
160 *Ibid.* at 27.
161 *Ibid.*
162 ENB PrepCom4 Summary, *supra* note 27 at 19.
163 *Ibid.*

was hampered by the shift in focus of PrepCom4—from further elaboration of the elements of the package deal to negotiation of the text of the PrepCom Report.[164] To aid in the negotiation of the Report, the Chair prepared a streamlined non-paper prior to the commencement of PrepCom4 and draft texts in the course of the session.[165]

The predominant view of the delegations was that the provisions on institutional arrangements included in the PrepCom Report be non-prescriptive in order to allow for further discussion of the three different approaches to the institutional agreements at the IGC.[166] No one wanted to be locked into any particular position. The United States, supported by Norway, Costa Rica, Australia, New Zealand, Singapore, Monaco, the EU and Iceland, among others, proposed that the language of the PrepCom Report not preempt discussion of the three approaches to the institutional arrangements.[167] Canada supported certain provisions of the text that introduced the hybrid approach, noting that such language served as a good marker to discuss the approach in subsequent discussions.[168] With regard to the process for establishment of ABMTs, Norway favoured language that did not preempt future discussions on global, regional and hybrid approaches, suggesting that a combination of approaches may be workable.[169] Similarly, Singapore, New Zealand, Japan, together with the FAO, favoured non-prescriptive language on potential global/regional/hybrid decision-making approaches in order to allow for flexibility at a later stage in the discussions.[170]

There were, however, instances during PrepCom4 when the substance regarding the three different approaches to institutional arrangements was raised. For example, consistent with the global approach, several delegations aligned with Mexico's proposal to establish a central entity to coordinate

164 *Ibid.*

165 "Chair's streamlined non-paper on elements of a draft text of an international legally-binding instrument under the United Nations Convention on the Law of the Sea on the conservation and sustainable use of marine biological diversity of areas beyond national jurisdiction", accessed 2 November 2017 at <http://www.un.org/depts/los/biodiversity/prepcom files/Chairs streamlined non-paper to delegations.pdf>.

166 ENB PrepCom4 Summary, *supra* note 27 at 16; Author's PrepCom4 Notes, *supra* note 27 at 77, 101.

167 ENB PrepCom4 Summary, *supra* note 27 at 15; Author's PrepCom4 Notes, *supra* note 27 at 28, 29, 53, 54.

168 Author's PrepCom4 Notes, *supra* note 27 at 100.

169 ENB PrepCom4 Summary, *supra* note 27 at 11–12; Author's PrepCom4 Notes, *supra* note 27 at 28.

170 ENB PrepCom4 Summary, *supra* note 27 at 11–12; Author's PrepCom4 Notes, *supra* note 27 at 29, 30.

measures adopted under the ILBI and existing measures established by other instruments.[171] Supporting a decentralised approach, the United States questioned the role and function of a central entity for the establishment of ABMTS, with the Russian Federation dismissing the need for such an entity outright,[172] cautioning that it could lead to duplication, advocating instead that they be established within the context of existing mechanisms and instruments.[173] The lack of substantive elaboration of the institutional arrangements was in no way a dismissal of their importance, but was rather an indication of their critical value to the process. As the delegate from Nauru (acting as the PSIDS representative) said in one of her final interventions at PrepCom4, "The institutional arrangements are the 'glue to our instrument'".[174]

Although the PrepCom Report was widely viewed by the delegations as "less than what was hoped for", most delegations supported the text as a "result that would bring everyone along".[175] The resulting PrepCom Report recommended that the ILBI "would set out institutional arrangements, taking into account the possibility of using existing bodies, institutions and mechanism".[176] Possible institutional arrangements could include: (i) a decision-making body/ forum; (ii) a scientific/technical body; and (iii) a secretariat.[177] The PrepCom Report also recommended that "possible general principles and approaches" could include good faith and good governance principles of public participation, transparency, and availability of information.[178] This recommendation echoed some of the same good governance principles that were included in the Chair's Overview of PrepCom3 as described above and raised in the course

171 ENB PrepCom4 Summary, *supra* note 27 at 11; see also Author's PrepCom4 Notes, *supra* note 27 at 53–57.

172 Author's PrepCom4 Notes, *supra* note 27 at 27.

173 ENB PrepCom4 Summary, *supra* note 27 at 11; Author's PrepCom4 Notes, *supra* note 27 at 27, 48. With regard to the institutional arrangements, the Russian Federation suggested that a "personalised approach" be applied to each type of biodiversity. It was not clear what this approach would entail. Author's PrepCom4 Notes, *supra* note 27 at 54.

174 Author's PrepCom4 Notes, *supra* note 27 at 106.

175 *Ibid.* at 112–117.

176 PrepCom Report, *supra* note 4 at 15.

177 *Ibid.* at 15–16. Bangladesh supported the need for a scientific body to conduct monitoring, while Nauru (PSIDS) supported global oversight for such monitoring. On the other hand, Norway said that it was premature to mandate the establishment of a scientific body. Author's PrepCom4 Notes, *supra* note 27 at 51, 56.

178 PrepCom Report, *supra* note 4 at 9–10. See also Author's PrepCom4 Notes, *supra* note 27 at 24, 27, 48, 53. The EU, Tonga and the Federated States of Micronesia supported the need for fair and cost-effective institutional arrangements. Author's PrepCom4 Notes, *supra* note 27 at 53, 54.

of PrepCom4.[179] The PrepCom Report also made recommendations regarding the ways in which the ILBI could address its relationship to UNCLOS and other existing instruments and frameworks and relevant global, regional and sectoral bodies.[180] Similarly, the PrepCom Report recommended that the ILBI could address the relationship of ABMT measures, including MPAs, to measures under other relevant instruments, frameworks and bodies,[181] but noted that "further discussions are required on the most appropriate decision-making and institutional setup, with a view to enhancing cooperation and coordination, while avoiding undermining existing legal instruments and frameworks and the mandates of regional and/or sectoral bodies".[182] With regard to other cross-cutting issues, the PrepCom Report also recommended that "[f]urther discussions are required on institutional arrangements and the relationship between the institutions established under an international instrument and relevant global, regional and sectoral bodies".[183] Related cross-cutting issues that would require further attention included monitoring, review, compliance, dispute settlement, and responsibility and liability.[184]

6.3 *Explanation for the Emergence of the Hybrid Approach and Its Implications for the BBNJ Process*

In this author's assessment, the emergence of the hybrid approach at PrepCom3 can be explained by three factors: (i) the manner in which the Chair organised and conducted the work of PrepCom3; (ii) the flexibility of the UNCLOS framework and the relevance of the historical development of institutional arrangements under the law of the sea; and (iii) the need for the delegates to look outside the UNCLOS framework—not only because of the crowded regulatory space and the "not undermine" mandate—but for inspiration.

First, with regard to the organisation of the PrepCom's work, the Chair drafted the non-paper in such a way that it looked like an instrument,[185] which allowed for the discussions to focus more on elaborating the elements of an ILBI

179 PrepCom Report, *supra* note 4 at 9. With regard to establishing ABMTs, China said that the procedures must be open and transparent. Author's PrepCom4 Notes, *supra* note 27 at 28.

180 PrepCom Report, *supra* note 4 at 8–9.

181 *Ibid.* at 11.

182 *Ibid.* at 17.

183 *Ibid.* Iceland, with the support of the Russian Federation, Australia and New Zealand, said that the BBNJ process should not undermine existing mechanisms and instruments and that such language should be included in the PrepCom Report. Author's PrepCom4 Notes, *supra* note 27 at 54.

184 PrepCom Report, *supra* note 4 at 17.

185 Chair's non-paper, *supra* note 122; ENB PrepCom3 Summary, *supra* note 6 at 15.

and less on the thorny process of "parking" issues for which there was significant consensus in a written document. In addition, the organisation of the PrepCom3's work moved the discussions away from a fixation on "gaps" and those actions that could potentially "undermine" a new regime and instead began to focus on "inter-linkages".[186] The focus on inter-linkages allowed cross-cutting issues to be discussed during PrepCom3, sometimes in significant depth, in each Informal Working Group. For example, during the discussion on ABMTs the three different approaches—global, hybrid, and regional and/or sectoral—were first discussed at length, rather than at the Informal Working Group on Cross-cutting Issues. In addition, every Informal Working Group's Facilitator's Report identified the inter-linkages to other elements of the package deal.[187]

With regard to the second factor, as previously discussed in Section 2 of this chapter, the language of UNGA Resolution 69/292 and the history of the law of the sea dictate that the ILBI not only falls squarely under the UNCLOS framework, but is also heavily influenced by the Convention's implementing agreements and jurisprudence, particularly the UNFSA. Many delegations in the course of the PrepCom drew upon the provisions of the UNFSA as a model and template for the ILBI. However, not all invocations of the UNFSA were positive. For example, in the discussion of cross-cutting issues at the PrepCom, the *lack* of institutional arrangements under the UNFSA served as a cautionary tale, given the Agreement's lack of universal acceptance and perceived inability to conserve and manage effectively the world's straddling and highly migratory fish stocks.[188]

Therefore, it is important to keep in mind that UNCLOS is a Constitution for the Oceans—a flexible framework that grows and evolves over time—and not a stringent code of rights and obligations stuck in the Cold War of the time at which it was negotiated. Also of note is the fact that the landscape of ocean governance has changed significantly in the years since the adoption of UNCLOS in 1982. There have been previously unthinkable technological advances, power-shifting changes in the political influence of States (e.g., the downfall of the Soviet Union and rise of the BRICs (China, India and Brazil), etc.), increasing and emerging ocean uses, escalating climactic changes and effects, and most importantly with respect to institutional arrangements in the BBNJ

186 "With regard to linkages, going forward, it would also be useful to consider the inter-linkages between measures, such as ABMTs, including MPAs, with other elements of the package of issues that are under consideration by the Preparatory Committee, including environmental impact assessments and capacity-building and the transfer of marine technology". Chair's Overview of PrepCom3, *supra* note 25 at 13. See also *ibid.* at 18.

187 See Chair's Overview of PrepCom3, *supra* note 25.

188 Author's PrepCom4 Notes, *supra* note 27 at 55.

process, the growth and development of an increasingly crowded and regulatory space involving a greater number and variety of stakeholders. As the Global Ocean Commission noted in its 2014 report, "It is crucial that any attempt at driving real and appropriate ocean governance reforms for this century considers the complex economic factors related to ocean services and resources and the role that States, markets, civil society and private industry can play in contributing to sustainable solutions".[189] So while the UNCLOS framework provided structure and context to guide the discussions at the PrepCom, inspiration was sought from other governance frameworks and theories.

The third factor—the need for the delegates to look outside the UNCLOS framework—was influenced in large part by the international community's renewed focus on the oceans, which culminated in the United Nations Ocean Conference. What the Conference may not have delivered in binding commitments and obligations, it delivered in awareness, momentum and inspiration. In June 2017, just weeks prior to the convening of PrepCom4, over 5000 delegates gathered in New York at the United Nations headquarters for the Conference. Of special relevance to the PrepCom was that both the President of the United Nations General Assembly Peter Thomson (Fiji) and the actor, Leonardo DiCaprio, made their case for a new treaty to govern BBNJ, with DiCaprio declaring, "We can't afford to leave 70% of our oceans to unlimited exploitation. We need a Paris Agreement for the ocean with ambitious and measureable goals".[190] The Paris Agreement[191] retained its focus at PrepCom4 with a side event organised by Mission Blue/High Seas Alliance & Ocean Unite entitled, "How the Paris Agreement Was Achieved".[192] In addition, a number of delegations included the United Nations Ocean Conference, the Paris Agreement and the 2030 Agenda in their interventions during PrepCom4, with Costa

189 Global Ocean Commission Report, *supra* note 7 at 6.

190 ENB PrepCom4 Summary, *supra* note 27 at 19.

191 The Paris Agreement was adopted on 12 December 2015 at the twenty first session of the Conference of the Parties to the United Nations Framework Convention on Climate Change held in Paris from 30 November to 13 December 2015. The Paris Agreement entered into force 4 November 2016 (Treaty Registration No. 54113), accessed 12 December 2017 at <https://unfccc.int/resource/docs/2015/cop21/eng/l09r01.pdf>.

192 See Division for Ocean Affairs and the Law of the Sea, "List of Side Events", *Preparatory Committee established by General Assembly resolution 69/292: Development of an international legally binding instrument under the United Nations Convention on the Law of the Sea on the conservation and sustainable use of marine biological diversity of areas beyond national jurisdiction*, accessed 2 November 2017 at <http://www.un.org/depts/los/biodiversity/prepcom.htm>.

Rica referencing the Paris Agreement in its intervention on EIAS;[193] Mexico suggesting the incorporation of references to the 2030 Agenda and the United Nations Ocean Conference in the final clauses of the ILBI;[194] Indonesia referring to the Call for Action adopted at the United Nations Ocean Conference;[195] and Monaco citing the United Nations Ocean Conference, the Rio Agenda and the urgency to address BBNJ issues, among others.[196]

In studying the governance of the commons, Elinor Ostrom, a renowned political scientist, "studied systems that were government owned and managed, co-managed, community managed, or not managed at all".[197] Her theory of polycentric governance gained traction in emerging and evolving theories of post-sovereign governance and climate change governance. The former theory "examines as one aspect of the emergent structure a mode of *hybrid* problem-solving governance in which sovereign states and non-state parties actively collaborate, roughly as equal partners, to address certain kinds of highly complex problems that appear to be beyond the capacity of sovereign states alone to solve".[198] Arguably, the Paris Agreement has been interpreted as an example of both post-sovereign governance and polycentric global climate governance.[199] Given the fragmented nature of the regulatory space in BBNJ and the number and diversity of stakeholders, there should, therefore, be no surprise that delegates at the PrepCom looked outside of the UNCLOS framework for the conservation and sustainable use of BBNJ, and consequently, to the Paris Agreement, and its underlying theoretical basis, as a possible "hybrid approach" to institutional arrangements under the ILBI.

7 Conclusions

As with most every development in the making of the law of the sea, the road to an IGC for the conservation and sustainable use of BBNJ has been a long and arduous one. Over a decade ago, the BBNJ Working Group discussed at length

193 Author's PrepCom4 Notes, *supra* note 27 at 34. Fiji noted that it was the first to ratify both UNCLOS and the Paris Agreement. *Ibid.* at 78.

194 *Ibid.* at 63.

195 *Ibid.* at 111.

196 *Ibid.* at 114.

197 E. Ostrom, "A Long Polycentric Journey", *Annual Review of Political Science* 2010.13:1–23, at 16.

198 B.C. Karkkainen, "Post-Sovereign Environmental Governance", *Global Environmental Politics,* 4:1, February 2004, 72–96, at 74, 75 [emphasis added].

199 M.J. Dorsch and C. Flachsland, "A Polycentric Approach to Global Climate Governance" *Global Environmental Politics* 17:2, May 2017.

the inclusion of modern governance principles in the BBNJ process, including those of transparency and openness. Later, discussions at PrepCom1 and PrepCom2 culminated in the Chair's non-paper, in which the Chair noted the need for "robust" institutional arrangements and specified the criteria for such arrangements. At PrepCom3, the negotiations regarding institutional arrangements built on the ten years of discussions in the BBNJ Working Group and the first two sessions of the PrepCom, elaborated on the global and regional/sectoral approaches—and ultimately resulted in the articulation of the hybrid approach. Now, with the conclusion of the PrepCom and the adoption of the PrepCom Report, the importance and centrality of the hybrid approach to the upcoming IGC is clear.

First and foremost, the emergence of the hybrid approach at PrepCom3 demonstrates measurable progress on the thorny cross-cutting issues, specifically with regard to institutional arrangements. The hybrid approach gave welcome relief to the polarised positions that had hampered progress these many years. Second, as an extension of Ostrom's theory of polycentric governance, the hybrid approach—of the three approaches to institutional arrangements—may be best suited not only to provide for a coherent and integrated regulatory and institutional framework for regional and cross-sectoral coordination, regulation and cooperation for BBNJ, but to fulfill the PrepCom's mandate of "not undermining" existing relevant legal instruments and frameworks and relevant global, regional and sectoral bodies.

In subsequent negotiations at the IGC, each element of the package deal will likely be unpacked and a different variation of the hybrid approach will be applied to each element. In this regard, the PrepCom Report "left the door fully open for exploring all possible shades of hybridity at the IGC".[200] However, the risk is that the hybrid approach—in whatever way(s) it is applied to the institutional arrangements—could be used to dilute the overall effectiveness of the resulting ILBI.

In the last hours of PrepCom4 prior to the adoption of the PrepCom Report, Algeria (on behalf of the African Group) invoked the ghost of Arvid Pardo and the weighty history of the law of the sea to emphasise the importance of the work before the PrepCom, with the EU adding that the 19th of July was the eighteenth anniversary of Pardo's death.[201] Algeria's words brought to mind Judge Jennings's call for "new kinds of international management" to be developed under the UNCLOS framework, when at the close of the PrepCom at 10 o'clock on the night of 21 July 2017, General Assembly President Peter Thomson

200 ENB PrepCom4 Summary, *supra* note 27 at 20.

201 Author's PrepCom4 Notes, *supra* note 27 at 79.

expressed the dire warning that "the ocean is in deep trouble"[202] and the hope that the General Assembly would maintain the momentum of the PrepCom and convene an IGC for the negotiation of an ILBI for the conservation and sustainable use of BBNJ.[203] His hope was well founded for on 24 December 2017 the General Assembly agreed in United Nations General Assembly Resolution 72/249 to convene such an IGC, with an organisational session scheduled for April 2018, together with four 10-day substantive sessions, with the first session to commence in 2018, two sessions to be held in 2019 and the last session to be concluded in the first half of 2020.[204] This author will be following the upcoming sessions with great interest, particularly with regard to the implications for the necessary institutional arrangements.

---

202 *Ibid.* at 119.

203 ENB PrepCom4 Summary, *supra* note 27 at 18.

204 *International legally binding instrument under the United Nations Convention on the Law of the Sea on the conservation and sustainable use of marine biological diversity of areas beyond national jurisdiction,* UN GAOR A/Res/72/249, accessed 21 March 2018 at <http://www.un.org/en/ga/search/view_doc.asp?symbol=A/RES/72/249>.

CHAPTER 8

The Legal Framework and Relevant Issues on the Marine Protected Areas in the Areas beyond National Jurisdiction

Su Jin Park and Ki Hyeon Kim***

Abstract

As the pressure on and significant adverse impacts to marine biodiversity areas within the jurisdiction as well as beyond national jurisdiction from the human activities have been gradually increasing, cautious management for conservation and sustainable use has been required accordingly.

In order to address such challenges, the international community has made efforts through United Nations framework—in particular, through Ad Hoc Open-ended Informal Working Group on BBNJ. In 2008 at the 2nd Ad Hoc Open-ended Informal Working Group on BBNJ, the role of area-based management tools including MPAS (Marine Protected Areas) was addressed as one of major issues. Since then, the international community has reached convergence that area-based management tools including MPAS are essential and effective tools as precautionary and ecosystem-based approaches to human activities in the conservation and sustainable use of marine biological diversity beyond national jurisdiction.

Although MPAS are very useful management tools as well as a foundation in conserving marine biodiversity, the actual application of MPAS to ABNJ (Area Beyond National Jurisdiction) has faced challenges under the current framework of UNCLOS. The reason is that the current framework of UNCLOS has divided the rights and obligations of coastal States based on the zones—territorial sea, continental shelf, Exclusive Economic Zone, high seas, and the Areas. The zonal approach is the fundamental cause for controversy over the interpretation of existing international instruments as well as relevant issues of a new international instrument on the designation and management of MPAS at the UN BBNJ meetings over the last decade.

In this regard, this study explores the challenges and future perspectives in addressing the MPAS in ABNJ based on the analysis of the legal issues, recent developments

* Associate Research Fellow, Ph.D., Korea Maritime Institute, Republic of Korea. Her PowerPoint presentation is available at http://www.virginia.edu/colp/pdf/yogya-park.pdf.

** First Secretary, Ph.D., Embassy of the Republic of Korea in Singapore, Ministry of Foreign Affairs, Republic of Korea.

 | DOI:10.1163/9789004366619_013

and future perspectives on the MPAs in the ABNJ including the definition, designation process, monitoring, review process and effectiveness of MPAs and their relation to existing relevant international instruments and frameworks.

I Introduction

Worldwide species are estimated at 1.4 billion, of which only 13%, or about 1.75 million species, are identified and each year new species are discovered.[1] It is estimated that there are about one billion species of marine life in the world. However, to date, only 250,000 species have been documented.[2] Meanwhile, worldwide species and biodiversity are declining rapidly. Between 1970 and 2006, 31% of the world's wild vertebrates declined and about 30% of fish stocks were overfished, creating a major threat to the marine ecosystem.[3]

Given that conservation of biodiversity is the foundation of sustainable social development since it contributes to maintaining human health, wellbeing and economic benefit, conservation of biodiversity should be regarded as a key factor not only from the perspective of environmental conservation but also from that of sustainable social and economic development. As a result, there is a growing recognition that biological diversity is a global asset with tremendous value to present and future generations.[4]

In particular, the adverse impacts and pressures on marine biodiversity in areas beyond national jurisdiction as well as within national jurisdiction continue to increase as a result of human activities. Accordingly, the necessity for management of conservation and sustainable use of marine biodiversity is also increasing.

ABNJ—the High Seas and the Area—is an important habitat for marine life, accounting for 60% of the entire ocean.[5] The international community is committed to protecting the habitat of marine life in the oceans, as it is essential

---

1 UNESCO, data for 2010; Republic of Korea Ministry of Environment, 3rd National Biodiversity Strategy Report, 2014, p. 11.

2 Bevis Fedder, *Marine Genetic Resources, Access and Benefit Sharing: legal and biological perspectives* (New York: Routledge, 2013), p. 3.

3 CBD, Global Biodiversity Outlook 4—A mid-term assessment of progress towards the implementation of the Strategic Plan for Biodiversity 2011–2020, 2014, p. 13. p. 49, p. 57.

4 https://www.cbd.int/history/default.shtml.

5 Tomas Greiber, "Common pools for marine genetic resources: A possible instrument for a future multilateral agreement address in marine biodiversity in areas beyond national jurisdiction" in *Common Pools of Genetic Resources: Equity and innovation in international biodiversity law*, Evanson Chege Kamau and Gerd Winter, eds. (New York: Routledge), pp. 399–400.

to maintain and preserve marine biodiversity for sustainable conservation of abundant marine life.

Establishing Marine Protected Areas (MPAs) is the most effective policy measure taken most frequently as a tool for habitat protection. As part of this effort, the international community adopted the Aichi Biodiversity Goal 11 at the 10th Conference of the Parties to the Convention on Biological Diversity (CBD) held in 2010 at Nagoya in Japan. The goal decided upon was to designate 10% of the world's coasts and oceans by 2020.[6] However, in reality, it will be difficult to achieve the CBD Aichi biodiversity target of designating 10% MPAs among the entire sea area through only designating MPAs within national jurisdictions.

According to the UNEP report in 2016, MPAs take approximately 4.12% of the earth's oceans. Among 4.12%, MPAs located in coastal and marine areas within national jurisdiction accounts for 10.2%, compared to the 8.4% that was reported in 2014. Most of this increase is attributed to the designation of large protected areas in Australia, New Zealand, the United States, the United Kingdom and Spain.[7]

Only 2.8% of marine areas are designated as MPAs throughout the globe. Moreover, 9.7% of the total protected areas are designated within the 12-nautical mile waters, while 4.6% of the total protected areas are designated in the EEZs. The area of MPAs designated in the high seas is only 0.14%.[8]

In order to achieve the Aichi target, the international community has no choice but to discuss the issue of designating MPAs in ABNJ. In this regard, the international community has addressed legal issues relating to the designation and management of MPAs in the ABNJ through two main fora: an "*Ad hoc* open-ended informal working group to study issues relating to the conservation and sustainable use of marine biological diversity beyond areas of national jurisdiction"[9] and the "Preparatory Committee" established by the General Assembly Resolution 69/292.[10]

---

6 In 2010, the 10th Conference of the Parties to the Convention on Biological Diversity held in Nagoya City, Aichi Prefecture, Japan, adopted the Biodiversity Strategic Plan (2011 2020) and Aichi Biodiversity Targets until 2020. https://www.cbd.int/sp/elements/default.shtml, https://www.cbd.int/sp/targets/default.shtml.

7 UNEP-WCPA, IUCN *et al.*, Protected Planet REPORT 2016, UNEP, 2016. pp. 32–33. https://www.protectedplanet.net/c/protected-planet-report-2016.

8 IUCN, UNEP-WCMC, The World Database on Protected Areas (WDPA), Cambridge, UK, 2013.

9 *Ad Hoc* Open-ended Informal Working Group established pursuant to paragraph 73 of General Assembly resolution 59/24 of 17 November 2004.

10 In its resolution 69/292 of 19 June 2015, the General Assembly decided to develop an international legally binding instrument under the United Nations Convention on the Law of

Global interest in the marine biodiversity issue in ABNJ has been increasing. The World Summit on Sustainable Development (WSSD), held in Johannesburg, South Africa in September 2002 emphasized three issues: the conservation of vulnerable marine and coastal biodiversity, including that within ABNJ; designation of MPAs in the high seas; and eradication of ecosystem destructive illegal fishing.[11] Such international attention played a crucial role in the adoption of resolution 59/24 at the UN General Assembly in 2004 and the UN BBNJ Working Group was established and discussed since 2006.[12]

With the start of its first meeting of UN BBNJ Working Group in 2006, a total of nine WG meetings were held by 2015. Afterwards, the Preparatory Committee was held four times for the two years, from 2016 to 2017.

The first BBNJ WG meeting was in February 2006 at the United Nations Headquarters in New York. Attendees discussed a wide range of issues including marine genetic resources, illegal fishing on the high seas, designation of protected areas on the high seas, environmental impact assessments and oceanographic surveys. The key discussion was how to resolve the implementation gap of the current international regime, such as the United Nations Convention on the Law of the Sea (UNCLOS).[13] With regard to MPAs, the second BBNJ WG meeting in 2008 dealt with MPAs, focusing on the role of area-based management tools (ABMTs). Since then, the international community has established a consensus that ABMTs, including MPAs, are effective and ecosystem-based precautionary means to conserve marine biodiversity and ensure sustainable use in ABNJ.

This study analyzes the main discussions of the BBNJ WG meetings under UNCLOS and BBNJ Preparatory Committee for the development of international instruments. Furthermore, it also explores the relevant legal issues including the definition of MPAs in ABNJ, the standards and process for MPAs designation, the relationship with other existing instruments, and the feasibility of MPAs in ABNJ.

the Sea (UNCLOS) on the conservation and sustainable use of marine biological diversity of areas beyond national jurisdiction, https://documents-dds-ny.un.org/doc/UNDOC/GEN/N15/187/55/PDF/N1518755.pdf?OpenElement.

11 UN Plan of Implementation of the World Summit on Sustainable Development, 2002, p. 18, http://www.un.org/esa/sustdev/documents/WSSD_POI_PD/English/WSSD_PlanImpl.pdf.

12 Marine Biological Diversity of Areas Beyond National Jurisdiction is abbreviated BBNJ.

13 The United Nations Open-Ended Informal Consultative Process on Ocean and the Law of the Sea has also been discussed as a major agenda for the exploration of biological resources. In particular, at the eighth meeting held at the UN headquarters from June 25 to 29, 2007, intensive discussions on marine genetic resources were conducted.

II Management of Marine Living Resources in UNCLOS

1 *The Method of Management of Marine Living Resources in UNCLOS*

MPA is a useful means of managing marine biodiversity and marine ecosystems, and each country has protected such areas as important habitats for endangered marine life, marine landscape, coral reefs, and important historical and cultural areas by designating them as MPAs within its jurisdiction.[14] Each state has managed the marine living resources within its national jurisdiction by designating MPAs in accordance with its domestic law and national circumstances without violating the rights and obligations of coastal States stipulated in UNCLOS.

Currently, the management of marine living resources under UNCLOS takes two basic approaches: the zonal management approach and the species-specific approach.[15] UNCLOS applies different management rules for marine living resources in accordance with zones such as territorial sea, continental shelf, exclusive economic zone (EEZ) and the high seas.[16] This is because the coastal States have sovereignty in their internal waters, territorial sea and archipelagic waters, whereas they have sovereign rights over the continental shelf and EEZ. In accordance with Article 87 (1) (e) of the UNCLOS, all states are obligated to preserve biodiversity in the high seas.

On the other hand, UNCLOS has additional provisions relating to conservation and development, utilization of specific fish stocks or species that stipulates direct cooperation between states, cooperation through international organizations and necessary measures. Those specific fish stocks or species include stocks occurring within the EEZ of two or more coastal States, or both within the EEZ and in areas beyond and adjacent to it (Article 63), highly migratory species (Article 64), Anadromous stocks (Article 66), Catadromous species (Article 67) and Sedentary species (Article 68). In addition, Article 65 prescribes the right of a coastal State or the competence of an international

14 MPAs have three main objectives including conservation of marine biodiversity; maintaining the productivity of coastal ecosystems; contribution to the socio economic development and the well-being of human communities. http://www.commissionoceanindien.org/archives/environment.ioconline.org/marine-protected-areas/the-role-of-the-mpas.html.

15 Historically, the development of the legal framework for marine living resources falls into the two distinct phases. The first phases, until the 1970s, was characterized by narrow zones under national jurisdiction, while the second, since the mid-1970s, is marked by the extension of national jurisdiction over marine living resources. Nele Matz-Lück and Johannes Fuches, "Marine Living Resources", in Donald R. Rothwell *et al.*, *The Oxford Handbook of the Law of the Sea*, Oxford, 2015, pp. 497–498.

16 Yoshfumi Tanaka, *The International Law of the Sea*, Cambridge, 2012, p. 222.

organization to prohibit, limit or regulate the exploitation of marine mammals more strictly than provided for in UNCLOS.

2 *Conservation and Management of Marine Living Resources in the High Seas*

The adverse effects of activities such as over-fishing, destructive fishing practices, and pollution of the marine environment have been well recognized. The ocean has also faced various risks including pollution from ships, marine debris and noise, ocean fertilization, CO2 sequestration, construction of artificial islands and pipelines, offshore oil and gas exploration, bioprospecting and marine scientific research, etc.[17] Specifically, fisheries practices in ABNJ have been carried out in an unsustainable way such as using long line, mid-water and bottom trawls to catch highly migratory, straddling and discrete stocks in the high seas.[18] Accordingly, the biodiversity of the high seas has been threatened and it requires protection and management. Therefore, the international community has conserved and managed marine living resources in the high seas through the provisions of UNCLOS Part VII and the adoption of relevant international conventions.

Specifically, according to Article 87 of UNCLOS, the high seas are open to all states regardless of whether they are coastal or landlocked States, and states have the right to fish on the high seas and enjoy the freedoms of the high seas if certain requirements and conditions are met.

In accordance with Article 116 of UNCLOS, all states have the right to fish on the high seas subject to a) their treaty obligations, b) the rights and duties as well as interests of coastal States, and c) the provisions of Section 2 of Part VII.

In addition, freedom of the high seas, which is one of the important principles of international maritime law, is exercised in accordance with the conditions laid down by UNCLOS and other international law. This freedom includes a) freedom of navigation, b) freedom of overflight, c) freedom to lay submarine cables and pipelines, d) freedom to construct artificial islands and other installations permitted under international law, e) freedom of fishing, and f) freedom of scientific research.

States in exercising such rights as freedom of fishing and freedom of navigation should ensure that the high seas should be used for peaceful purposes,

---

17 Rosemary Rayfuse, "The Anthropocene, Autopoiesis and the Disingenuousness of the Genuine Link: Addressing Enforcement Gaps in the Legal Regime for Areas Beyond National Jurisdiction", in Erik J. Molenaar, Alex G. Oude Elferink, eds., *The International Legal Regime of Areas Beyond National Jurisdiction: Current and Future Developments*, Martinus Nijhoff Publishers, pp. 165–166.

18 Daniela Diz Pereira Pinto, *Fisheries Management in Areas beyond National Jurisdiction—the impact of ecosystem based law-making*, Martinus Nijhoff Publisher, 2013, p. 85.

and that all States cooperate and take certain conservation measures for the management and conservation of living resources in the high seas.[19] However, given that UNCLOS stipulates freedom of high seas shall be exercised "with due regard" to the interests of other States in exercising such freedoms, the "due regard" concept shall be considered more precisely.[20]

Specifically, according to Article 118 of UNCLOS, all States are obliged to cooperate with other countries with a view to taking measures necessary for the conservation of the living resources of the high seas. In addition, all States whose nationals exploit living resources concerned in the high seas shall enter into negotiations with a view to taking the necessary measures for the conservation of the living resources, and shall cooperate to establish subregional or regional fisheries organizations.

In addition, in accordance with Article 119 of the UNCLOS, each State shall take measures to maintain or restore populations of harvested species at levels that can produce the maximum sustainable yield in determining the allowable catch and establish other conservation measures. Furthermore, UNCLOS clearly stipulates measures of States for conserving the living resources in the high seas in its Article 119 (2) including available scientific information, catch and fishing effort statistics, and other data relevant to the conservation of fish stocks. In this case, conservation measures and their implementation should ensure that there is no formal or substantive discrimination against fishermen of any state.

In addition to UNCLOS, the international community has adopted a number of international conventions with a view to further elaborating on and facilitating the implementation of the issues relating to the management and conservation of biological resources in the high seas of Section II of Part VII of UNCLOS.[21] Those efforts include the "1995 UN Convention for the Implementation of the Provisions of the United Nations Convention on the Law of the Sea of 10 December 1982 Relating to the Conservation and Management of Straddling Fish Stocks and Highly Migratory Fish Stocks (1995 UN Fish Stock

---

19 Arts. 88, 117 to 119 of the United Nations Convention on the Law of the Sea.

20 Tullio Treves, "Principles and objectives of the legal regime governing areas beyond national jurisdiction", Erik J. Molenaar, Alex G. Oude Elferink, eds., *The International Legal Regime of Areas Beyond National Jurisdiction: Current and Future Developments*, Martinus Nijhoff Publishers, p. 22.

21 In general, pollution vessels are regulated by international law and the flag State, and international fisheries agreements, including certain restrictions on the freedom of fisheries, one of the principles of freedoms of the high seas. Peter Malanczuk, *Akehurst's Modern Introduction to International Law*, Routledge (7th rev. ed.), 2009.

Agreement, UNFSA),"[22] the "1993 FAO Agreement to Promote Compliance with International Conservation and Management Measures by Fishing Vessels on the High Seas", and the FAO "Code of Conduct for Responsible Fisheries".

In the UNFSA, Article 6 stipulates a precautionary approach. Article 7 also requires conservation and management measures established for the high seas and those adopted for areas under national jurisdiction to be compatible in order to ensure conservation and management of the straddling fish stocks and highly migratory fish stocks in their entirety.[23] In order to implement those regulations, the international community has made efforts to guide universal practices and actions of states by adopting UNGA resolutions relating to fisheries. These basic principles and regulatory methods have been further elaborated by the FAO Fisheries Committee. FAO is responsible for the monitoring, control and implementation of fishing activities in the high seas based on UNFSA and FAO Regulations and Guidelines.[24]

III Major Discussions of the UN BBNJ Meeting

1 *UN BBNJ Working Group Meeting*

1) Establishment of the UN BBNJ Working Group

Attention by the international community to the conservation and sustainable use of marine biodiversity in ABNJ increased due to the World Summit of Sustainable Development in 2002. Thereafter, a UN BBNJ Working Group was established for the conservation and sustainable use of diversity at the 59th UN General Assembly resolution in 2004.[25]

---

22 The 1995 UN Fish Stock Agreement entered into force on December 11, 2001, as one of the implementation agreements of the United Nations Convention on the Law of the Sea. The 1995 UN Fish Stock Agreement is based on the basic principles for the conservation and management of migratory fishes and highly migratory fishes on the boundary of the high seas and on the high seas, the joining of local fisheries organizations, and the search for boarding vessels of other countries. Korea entered into force on March 2, 2008.

23 Agreement for the implementation of the provisions of the UNCLOS of 10 December 1982 relating to the conservation and management of straddling fish stocks and highly migratory fish stocks, Article 7.

24 Korean Association of International Law of Sea, *The issues on and response to UNCLOS*, SeChang Publisher, 2012, p. 119.

25 It was established in accordance with UNGA-59/24, paragraph 73 (2004.1.17). Paragraph 73 of A/RES/59/24(2004) reads:

The Working Group had been held from 2006 to 2015 to discuss topics relating to the conservation and sustainable use of marine biodiversity in ABNJ. The 1st WG meeting was held at the UN Headquarters from February 13 to 17, 2006, and the 2nd and 3rd WG meeting was held every two years. Afterwards the 4th, 5th, and 6th WG meetings were held annually for in-depth discussion from 2011 to 2013. In 2014 the 7th and 8th WG meetings were held and in January, 2015 the last 9th WG meeting was held.

The 9th WG meeting agreed to establish an international instrument legally-binding under UNCLOS for the conservation and sustainable use of marine biodiversity in ABNJ. In addition, it also agreed that four preparatory committee meetings would be convened from 2016 and would submit the results of the discussions on elements of international instrument by the end of the 72nd session of the UNGA in 2017. This let the UNGA decide whether to hold an intergovernmental conference (IGC) for the development of the BBNJ international document. Accordingly, based on the chair's recommendation, the outcome of the 9th WG meeting formed the basis for the UNGA resolution 69/292 in 2015.

2) Major Discussions of the UN BBNJ WG Meeting

(1) *From the 1st to 3rd BBNJ WG Meeting*

Prior to discussions on the international instrument through the UN BBNJ WG meeting, there existed relevant norms for the conservation and sustainable use of marine biodiversity in ABNJ. The international community has partially discussed marine living resources and habitats in ABNJ through UNFSA, FAO Code of Conduct and various RFMOs. However, there has been a growing demand for a more comprehensive discussion at global level and for an international regulatory framework.

Decides to establish an Ad Hoc Open-ended Informal Working Group to study issues relating to the conservation and sustainable use of marine biological diversity beyond areas of national jurisdiction:

(a) To survey the past and present activities of the United Nations and other relevant international organizations with regard to the conservation and sustainable use of marine biological diversity beyond areas of national jurisdiction;

(b) To examine the scientific, technical, economic, legal, environmental, socio-economic and other aspects of these issues;

(c) To identify key issues and questions where more detailed background studies would facilitate consideration by States of these issues;

(d) To indicate, where appropriate, possible options and approaches to promote international cooperation and coordination for the conservation and sustainable use of marine biological diversity beyond areas of national jurisdiction;

The initial purpose of the discussions on biological diversity in ABNJ and BBNJ WG meeting was to address trawl fishing by understanding the influence of the trawl fishing on the marine eco-system and marine biological diversity, and to facilitate international cooperation.[26] Afterwards the aim of the BBNJ WG meeting discussion was expanded to the establishment of a comprehensive international instrument including fisheries, MPAs, marine genetic resources, environmental impact assessment, capacity building, and marine science and technology transfer in ABNJ.

At the 1st BBNJ WG meeting, the comprehensive and broad analysis were conducted for past and present activities relating to conservation and sustainable use of marine biological diversity in ABNJ, and scientific, technical, socio-economic, legal and environmental issues were discussed. Since the initial purpose of the BBNJ WG meeting was to address trawl fishing, the discussion focused on these activities in the Area at the 1st WG meeting.

At the 2nd BBNJ WG meeting in 2008, the role of ABMTs including MPAS was discussed as one of the major issues.[27] Representatives from different countries shared the position that coordination and cooperation approaches were significant for ecosystem-based management. Since then, the international community has agreed that ABMTs are essential and effective tools for the conservation and sustainable use of marine biodiversity in ABNJ, and the core function of MPAs is to provide management measures to govern human activities as an ecosystem-based approach and a preventative measure.[28]

At the 3rd BBNJ WG meeting in 2010, the future direction of BBNJ WG meetings was a key issue. Developing countries emphasized that a new implementing instrument should adopt the "package deal" approach, including marine genetic resources, MPAs and environmental impact assessments. On the other hand, developed countries strongly objected to the package deal approach and stressed that it is necessary to discuss separately those issues within the current UNCLOS system.

26 UNGA, Report of the *Ad Hoc* Open-ended Informal Working Group to study issues relating to the conservation and sustainable use of marine biological diversity beyond areas of national jurisdiction, 2006.4.20, p. 2. UNGA/A/61/65, https://documents-dds-ny.un.org/doc/UNDOC/GEN/N06/277/50/PDF/N0627750.pdf?OpenElement.

27 Regarding the definition of ABMTs, UN BBNJ WG meeting and Preparatory Committee could not reach an agreement, but there are opinions that IMO PSSAs, FAO VMEs, CBD EBSA, and RMFOs management tools should be taken into consideration.

28 UNGA, Report of the *Ad Hoc* Open-ended Informal Working Group to study issues relating to the conservation and sustainable use of marine biological diversity beyond areas of national jurisdiction, A/63/79, 2008.5.16, p. 7, https://documents-dds-ny.un.org/doc/UNDOC/GEN/N08/344/16/PDF/ N0834416.pdf? Open Element.

(2) *From the 4th to 6th BBNJ WG Meeting*

In 2011, the 4th BBNJ WG meeting pointed out that it was necessary to develop a common methodology for designating sea areas that need protection based on existing standards in order to select ecologically or biologically important marine areas. That is to say, MPAs should be based on scientific evidence and information, comply with international law, and activate regional-based management measures.[29]

At the 5th BBNJ WG meeting in 2012, comprehensive discussions were held on the necessity and designation of MPAs, but more discussions were conducted as to the form of the new legal instrument in ABNJ at the meeting.[30]

At the 6th BBNJ WG meeting in 2013, the friends of Co-chair actively discussed how to bring about the specific form of the new legal instrument for the conservation of marine biodiversity. Based on the draft text proposed by various states, more active discussions were held on the basis of their interests. While the EU and developing countries insisted that the new international regime should be introduced in the form of new "implementing agreements", the United States, Iceland and Republic of Korea emphasized that the process for a new international regime should respect and use existing international norms.[31] In the end, states agreed that the new international regime should be the form of an international instrument under UNCLOS and decided to proceed with the discussion of the new international instrument from the 7th BBNJ WG meeting.

(3) *From the 7th to 9th BBNJ WG Meeting*

At the 7th BBNJ WG meeting held in April 2014, full-scale negotiations on the international legal instrument in ABNJ under UNCLOS were conducted. Developed states and developing countries continued their sharp controversy, arguing for key issues including the common heritage of mankind principle, the principle of freedoms of the high seas, and the gap analysis between the new international norms and existing international norms.

At the 8th BBNJ WG meeting held in June 2014, more specific discussion on the scope, the parameters, and the feasibility of the new international

29 Please refer to the 4th United Nations Conference on the Agenda of the UN General Assembly (UNGA A / 66/70, April 22, 2011).

30 A/RES/66/288(2012).

31 Developing countries used the term of "implementing agreement" but developed countries used the term of "international regime" under the UNCLOS. On the other hand, the Russian Federation considered it too early to discuss a new implementing agreement under UNCLOS. IISD, Summary of the sixth meeting of the working group on marine biodiversity beyond areas of national jurisdiction: 19–23 August 2013, p. 2. http://enb.iisd.org/oceans/marinebiodiv6/brief/brief_marinebiodiv6e.pdf.

instrument were conducted than at the 7th WG meeting. Many countries argued that geographic coverage of the new international instrument should include the high seas and the Area. Australia and New Zealand, which had been considered as somewhat neutral states in relation to the BBNJ, expressed their positions to support those of the EU, G77+ China to require the new implementing agreement from the 8th BBNJ WG meeting.[32]

The 9th BBNJ WG meeting adopted a recommendation for the 69th session of the UNGA, which contained provisions on the establishment and operation of the preparatory committee. To be specific, it included that a legally-binding instrument for conservation and sustainable use of marine biodiversity in ABNJ under UNCLOS would be established. Four preparatory committee meetings would be convened for the next two years from 2016 to 2017; the elements of the international instrument would be discussed and the results submitted to the UNGA; and the decision recommended to the UNGA whether intergovernmental conferences (IGC) would be convened based on the outcomes of preparatory committee meetings by the end of the 72nd UNGA.

2 *UN BBNJ Preparatory Committee*

1) Establishment of UN BBNJ Preparatory Committee

In 2015 it was decided in UNGA Resolution 69/292 to establish the Preparatory Committee for the preparation of and development of an international legally binding instrument under the UNCLOS on the conservation and sustainable use of marine biodiversity in areas beyond national jurisdiction.

The first preparatory committee was held at the United Nations Headquarters in New York in March 2016, the second committee met in August 2016, the third and fourth preparatory committee meetings were held in April and July 2017 respectively. The UN BBNJ Preparatory Committees through its four meetings discussed the elements of the draft of the international legally binding instrument under UNCLOS. This was supposed to be submitted to the UNGA of 2017, based on the results of the UN BBNJ *Ad Hoc* Working Group Meetings and that of UN General Assembly Resolution 69/292.

2) Main Discussion at the UN BBNJ Preparatory Committee

(1) *From the 1st BBNJ Preparatory Committee to the 2nd BBNJ Preparatory Committee*

At the 1st BBNJ Preparatory Committee held in March of 2016, the international community had exchanged opinions on the scope of the international

32 IISD, Summary of the eighth meeting of the working group on marine biodiversity beyond areas of national jurisdiction: 16–19 June 2014, pp. 3–7, http://enb.iisd.org/oceans/marinebiodiv8/brief/brief_marinebiodiv8.pdf.

instrument, the relations with the existing international documents, principles and approaches, the details of the components of the package deal agreed on in the year 2004 and the position of each country. At this meeting, states had agreed that a comprehensive approach was required for conservation and sustainable use of marine biodiversity, use of marine genetic resources and benefit sharing, ABMTs including MPAs, environmental impact assessment, capacity building, and marine science transfer.

In particular, discussions had taken place on the definition, concepts, objectives, principles, standards and procedures of the ABMTs, including the MPAs that centered on the necessity of complementing the current international regulatory system including UNCLOS and RFMOs, the usefulness, guidelines and principles of ABMTs, and the balance between conservation and sustainable use, including the maintenance of ecosystem services.[33]

Most countries shared a common opinion that the ABMTs are useful tools for achieving the objectives of conservation and sustainable use of marine biodiversity in ABNJ, but they took different positions on the fisheries sector. In other words, the United States, South Africa, and Venezuela had argued that fisheries should be included in the process of discussing MPAs, while Norway stated that additional discussions on how fisheries would be handled in international documents were needed.

Many countries mentioned that the FAO guideline, the IUCN guideline, the Aichi Biodiversity target 11, the UN SDGs 14, and the Biological and Ecological Significant Areas (EBSAs) could be guidelines for MPAs. They emphasized that ABMTs should be based on the precautionary principle, polluter-pay principle, ecosystem approach principle, scientific-based approach and transparency principle. They also suggested that participation of all stakeholders such as civil society should be guaranteed.[34]

At the second preparatory committee, opinions were exchanged on possible areas of convergence of views and possible issues for further discussion. In relation to ABMTs, including MPAs, possible areas of convergence of views were: a number of principles and approaches to be taken in the establishment of ABMTs, including MPAs, such as transparency, ecosystem approach, and science-based approach; states have the obligation to protect and preserve the marine environment; and ABMTs, including MPAs, should collectively contribute to the objective of conservation and sustainable use of marine biological diversity of areas beyond national jurisdiction. On the other hand,

33 http://www.iisd.ca/vol25/enb25104e.html.

34 Summary of the first session of the preparatory committee on marine biodiversity areas beyond national jurisdiction: 28 March-8 April, 2016, pp. 9–10, http://www.iisd.ca/oceans/bbnj/prepcom1/.

there were still many possible issues requiring further discussion: whether ABMTs, including MPAs, should contribute to rehabilitation and restoration of ocean ecosystems and health; whether definitions of/use of terms related to ABMTs, including MPAs, should be based on existing definitions, adapted to the context of marine biodiversity of areas beyond national jurisdiction; the possible need to include a definition of marine reserves; further discussion on what combination of elements, including vertical, horizontal, top-down, and bottom-up approaches would be most effective in delivering on the objectives of the mandate; a new mechanism/process/global framework/ instrument would provide for a consultative, integrated, transparent and inclusive approach to ABMTs, including MPAs; the decision to designate MPAs, especially in areas which adjoin areas under national jurisdiction, should be taken with the consent of neighboring coastal States and after a consultation process which seeks to take into consideration the views and concerns of all stakeholders.[35]

On the other hand, no concrete discussions or agreements were made on specific definitions, designation standards, procedures and management methods of MPAs.

(2) *From the 3rd BBNJ Preparatory Committee to the 4th BBNJ Preparatory Committee*

The 3rd Preparatory Committee, which was held at the United Nations Headquarters in New York from March 27 to April 7, 2017 discussed the elements for the draft of an international legally binding instrument under UNCLOS to be submitted to the UNGA in 2017 on the basis of the results of the 1st and 2nd preparatory committees, nine *ad hoc* working group meetings from 2006 to 2015 and the outcome of UNGA 69/292 in 2015.[36]

In the context of MPAs, discussions were held on purpose, definition, designation procedures, principles and approaches. There was a certain level of convergence of opinion that the ultimate goal of ABMTs should include MPAs for the conservation and sustainable use of marine biodiversity in areas beyond national jurisdiction. However, AOSIS suggested that contribution to the adaptation to climate change, the restoration of marine ecosystems, and resilience should be included in the purpose of ABMTs, while the EU argued that long-term conservation needed to be specified. On this issue, the United States

35 Chairs overview of the second session of the Preparatory Committee, http://www.un.org/Depts/los/biodiversity/prepcom_files/Prep_Com_II_Chair_overview_to_MS.pdf.

36 Chairs overview of the third session of the Preparatory Committee, http://www.un.org/Depts/los/biodiversity/prepcom_files/Chair_Overview.pdf.

suggested that ABMTs should be set up in a customized way, taking into consideration the characteristics of each sea area.

In terms of the definition of the MPAs, the opinions were divided. Many countries argued that MPAs should be defined based on the definition of Protected Areas from Article 2 of CBD and the definition of MPAs of IUCN, but no consensus was reached. Norway stated that it enacted a relevant domestic legislation and implemented without defining, and it was not necessary to have a definition of MPAs.[37]

Regarding the MPAs designation process, the Caribbean community proposed recommendations which relate to MPAs designation decisions, reviews and monitoring procedures. The EU proposed identification, designation proposals, consultation, scientific evaluation, decision-making, implementation, reporting and review process.[38]

With regard to the subject of designation of MPAs, there were conflicts of opinions among states. Countries such as Japan, the EU, Argentina, China and Switzerland proposed that States should propose MPAs. Meanwhile, AOSIS (Maldives), Peru, and Mexico expressed the position that relevant organizations could propose other than states. The EU added that states could propose MPAs designation individually or collectively.

As for the decision-making method on the designation of MPAs, the EU, Japan, and Switzerland expressed their opinions that the Parties should follow consensus. Developing countries stated that countries should make efforts for consensus but needed to adopt a majority method if they could not reach consensus.

With regard to relations with existing organizations, Japan, Canada, and China wanted to clarify relations with existing organizations, to consult with them, and not to undermine the role of the existing organizations under the principle of cooperation and respect for existing organizations. Australia emphasized the need for regional organizations to participate in decision-making processes, even if global designation criteria would be subsequently set up. Argentina argued that since protecting and conserving marine biodiversity were not the main objectives of RFMOs, it was opposed to strengthening or expanding the mandate of RFMOs. This was contrary to Japan's argument on the emphasis of a role of RFMOs.

37 IISD, Summary of the third session of the preparatory committee on marine biodiversity beyond areas of national jurisdiction; 27 March–7 April, 2017, p. 5, http://enb.iisd.org/download/pdf/enb25129e.pdf.

38 Ibid.

At the 4th Preparatory Committee similar conflicts of opinions among States continued during discussion of the following issues: the purpose of MPAS; the guiding principles and approaches; the process for the establishment including the identification of the site, proposal, consultation, proposal evaluation and decision-making; compliance; capacity building and technology transfer; monitoring and review.

Regarding "the principle of adjacency" proposed since the 3rd Preparatory Committee, the difference of opinions was revealed. This principle is to emphasize the consideration of the rights of the coastal States adjacent to the high seas in designating MPAS in ABNJ. On this principle, States did not specify, but tried to seek a compromise in the way that neighboring countries participate in the process of designation. As for the argument of the EU and developing states which emphasizes the adjacent principle, the United States and Japan proposed amendments to the "interested States" which was a category still broader than the specific expression of considering "the adjacent coastal States". They emphasized, together with China, the need to discuss the end of the designation period with the achievement of the ABMTS objectives.

The outcome of the 4th Preparatory Committee's meeting was adopted as a report to submit to the UNGA, which included: the establishment and discussions of the UN BBNJ Preparatory Committee; the schedule of the Preparatory Committee and the election of the bureau; the main results of the four Preparatory Committee meetings and the recommendations of the Preparatory Committee.[39]

IV Major Legal Issues on MPAS

Modern international law governing the high seas is based on the rule that the high seas, whether individual or collective, are not subject to acquisition from national occupation.[40] Grotius argued in the *Mare Liberum* (1609) that the oceans are not possible to own and that international trade through the sea is a recognized right in natural law. Within national jurisdiction, the coastal States have stronger intention to conserve and sustainably use marine ecosystems

39 Advance, unedited version of the Report of the Preparatory Committee established by General Assembly resolution 69/292: Development of an international legally binding instrument under the United Nations Convention on the Law of the Sea on the conservation and sustainable use of marine biological diversity of areas beyond national jurisdiction, 2017.7.21, http://www.un.org/Depts/los/biodiversity/prepcom.htm.

40 Park Chan Ho, Kim Han Taek, *International Law of the Sea, Seoul Economy and Management*, 2011, p. 138.

and marine biological resources than ABNJ. However, in ABNJ the freedom of the high seas has been more emphasized and greater focus is on the use of marine biological resources rather than protection of such resources through MPAS. In this regard, it is relatively easy to establish domestic MPAS within national jurisdiction, but to designate and manage MPAS in areas beyond national jurisdiction is a more difficult challenge.

1 *Geographical Scope and Definition of MPAS*

At the 4th UN BBNJ Preparatory Committee, the Committee had reached some level of agreement that the geographical scope of the waters designated as MPAS, in accordance with the new international instrument, is the area beyond national jurisdiction.

However, there was a conflict on the details of the definition of MPAS. Regarding the definition of MPAS, Kelleher (1999) states that "MPAS are any area of intertidal or sub-tidal terrain, together with its overlying waters and associated flora, fauna, historical and cultural features, which has been reserved by law or other effective means to protect part or all of the enclosed environment".[41]

The definitions of MPAS can be defined by using the CBD Article 2 and the IUCN Guidelines, but it is necessary to consider the legal nature and the regulatory intensity after designation. It is also necessary to take into consideration the specific purpose and the limitations of activities when defining MPAS as area-based management methods. This is because the current area-based management tools such as management areas of RFMOS, the protected area of the CBD, and the Particularly Sensitive Sea Areas (PSSAS), in accordance with IMO's PSSAS Guidelines A 24 / Res.982 Annex 1.2, have differences in their definitions, purposes, and limitation of activities. However, as confrontation between developing countries and developed countries arises over the definition of MPAS, Norway introduced that it has implemented MPAS without definition of MPAS and argued that more practical discussion would be needed for standards and processes of MPA designations rather than definition of MPAS. In this regard, the definition of MPAS is likely to become the key issue linked to the purpose of the BBNJ international instrument, the scope of MPAS and effects of MPAS.[42]

---

41 IUCN Resolution 17.38 of the IUCN General Assembly, 1988, reaffirmed in Resolution 19.46, 1994.

42 IISD, Summary of the third session of the preparatory committee on marine biodiversity beyond areas of national jurisdiction, 2017.4.10., p. 5. http://enb.iisd.org/download/pdf/enb25129e.pdf.

2 *The Criteria for Designation of MPAs*

One of the biggest challenges in establishing MPAs in ABNJ is based on what criteria it is possible to designate MPAs. Although the criteria of designation are of a scientific nature rather than a legal nature, a more precise analysis is needed in that the existing international regime, framework and organizations have different standards, and their legal effects of designation of MPAs and limits of activities are not the same.

There are a variety of relevant regimes for MPAs at present, and there are similar standards and different standards among those existing regimes. That is, MPAs can be designated in a number of different ways, each encompassing a number of different purposes. Examples include protection of the marine environment from pollution, as in the case of the IMO PSSAs, and protection of specific species, as in the case of the IWC whale sanctuaries, and fisheries closures.[43]

The common criteria that most regimes suggest in designating MPAs are uniqueness or rarity, diversity, naturalness, biological productivity, vulnerability, fragility, endangered or threatened species, and so on. On the other hand, since each regime has a characteristic that it focuses on in setting up MPAs, these characteristics provide different criteria for designating MPAs under specific regimes. These different criteria include migratory routes and dependency in the IMO's PSSAs; categories of human activities and protected areas in the case of the IUCN Guidelines; while the Antarctic Treaty emphasizes the historical value in its environmental protocol.

Given the diverse criteria for designation of MPAs in the existing international regimes, the designation criteria for MPAs in the UN BBNJ international instrument should be a practical standard that can be used for designation and follow-up, and should be universally compatible with existing international regimes.

3 *The Relations with Existing International Regimes*

Another key issue in future negotiations on MPAs in the UN BBNJ international instrument is its relations with existing international regimes. In order to designate MPAs in ABNJ based on the new international instrument, it is necessary to discuss the following issues: identification of sites; proposal of MPAs; consultation and proposal evaluation; and decision making on designation. In this case, it is necessary to identify relationships with existing international regimes in the process of consultation with neighboring coastal States and relevant stakeholders, and in the participation and empowerment of relevant

43 Daniela Diz Pereira Pinto, *Fisheries Management in Areas beyond National Jurisdiction—the impact of ecosystem based law-making*, Martinus Nijhoff Publisher, 2013, p. 166.

international, regional, and sectoral organizations. This is because the existing international regime has already designated and managed various protected areas for sustainable fishing, safe navigation, biodiversity, and habitat preservation.

Specifically, with regard to the relations with the existing international regime, it is key to maintain a balance between the "not to undermine" principle agreed upon by the international community in the process of debating the BBNJ international instrument, and the practical purposes of individual international regimes that have previously designated MPAs in ABNJ.[44] This is because the basic objective of the introduction of the BBNJ international instrument is to establish a comprehensive and integrated framework for achieving the conservation and sustainable use of marine biodiversity in ABNJ. However, this makes it inevitable that there will be conflict with the principle of "not undermine" other existing regimes because no specific discussions have taken place as to the meaning of "comprehensive and integrated framework" and "hybrid approach" including the meaning of "undermine".

Specifically, further discussions and legal and policy reviews are needed on the following issues: whether an area can be re-designated as an MPA based on a BBNJ international instrument even if it is already designated as a protected area in accordance with an existing regime; if possible, how to resolve the potential conflicts from difference in limits of activities and regulations among diverse existing regimes; and how to cooperate in the management of MPAs with existing international organizations after designation of MPAs.

For example, if IMO's MARPOL 73/28 Convention designates an area as a Special Area or PSSA where the emission of oil from a ship is prohibited, Parties to the Convention must obey it, but the instrument cannot impose obligations on non-Parties. In this case, if some or all of the waters are designated as MPAs under the BBNJ international instrument, what should be the limitations of the activities and how to resolve the conflicts of regulations? Those questions need to be approached legally and it is necessary to establish clear regulations in the process of negotiating texts on MPAs in the UN BBNJ international instrument.

4 *The Feasibility of MPAs*

The final challenge in designating MPAs in ABNJ is the feasibility of MPAs. In designating and managing MPAs in the ABNJ, there are a few issues to be considered. First, there are various categories of stakeholders involved in designating MPAs. In particular, it is not easy to make a decision on MPAs since it involves various stakeholders in the process of making proposals and decision-making. Second, to designate MPAs in ABNJ requires scientific evidence, but

44 UNGA RES 69/292, 2015.6.19.

it is a fact that relevant data information and research on ABNJ is insufficient. Third, to secure scientific data and evidence requires a lot of money and how to obtain such funds is the key. Finally, how and who manages the MPAs set in the ABNJ is an important issue when it comes to the effectiveness of MPAs.

Which organization, under what mandate, with what kind of methods and tools should manage and ensure that the MPAs in ABNJ can achieve its objectives properly? Given the existence of such realistic limitations, to designate MPAs in the ABNJ might lead to mass production of a "Paper Park" that exists only in documents, and may conflict with existing international regimes or organizations. Therefore, at the inter-governmental negotiation process of a UN BBNJ international instrument, which is expected to start in 2018, it will be desirable to proceed with a pilot MPAs project, involving scientists, based on information already held by relevant international organizations.[45]

Lastly, for the feasible MPAs in the ABNJ it is necessary to distinguish the legal gaps and the implementation gaps. The negotiation of a BBNJ international instrument should address the legal gaps, not the implementation gaps, since the implementation gaps are a question of political intention and willingness to protect the interest of the international community as a whole.[46]

V Conclusions

The 72nd session of the UNGA held on Christmas Eve (December 24, 2017) adopted the decision to convene the intergovernmental conference (IGC) to elaborate the texts of an international legally binding instrument under the UNCLOS on the conservation and sustainable use of marine biodiversity of ABNJ.[47] The IGC will hold four sessions of 10 working days each with the first meeting taking place in the second half of 2018, the second and third in 2019, and the fourth meeting in the first half of 2020, respectively. The first meeting will be held from September 4 to 17, 2018.

Given that from 2018, the United Nations will be working on the draft of an international legally binding instrument for BBNJ, many issues need to be further discussed before starting the draft process since there still exists divergence of opinions among states. These include the definition of MPAs, designation criteria and procedures, designated effects, post-designation

45 Su Jin Park, The Legal Framework and Relevant Issues on the Marine Protected Areas in ABNJ, the 41st Annual Conference of the Center for Oceans Law and Policy (Yogyakarta presentation material), 2017.5.17.

46 Rosemary Rayfuse, *supra* note 17, p. 189.

47 https://www.un.org/press/en/2017/ga11997.doc.htm.

management, decision making processes, cooperation and avoidance of conflicts with existing international regimes and organizations, financing, establishment of relevant organization, as well as capacity building and technology transfer.

In addition, another challenge before us is to make sure the BBNJ international instrument is harmonious with UNCLOS, UNFSA, RFMOs, and the IMO without any conflicts. We need to ensure that the establishment of MPAs is carried out in a way that manages the impact on existing activities in the ABNJ such as shipping, fisheries, and marine scientific research.

CHAPTER 9

Marine Genetic Resources beyond National Jurisdiction and Sustainable Development Goals: The Perspective of Developing Countries

A. Gusman Siswandi[*,**]

Abstract

Goal 14 of the Sustainable Development Goals on "Conserve and Sustainably Use the Oceans, Seas and Marine Resources" highlights the relationships between marine technology, marine biodiversity, and developing countries. In particular, Goal 14.a states: "Increase scientific knowledge, develop research capacity and transfer marine technology, taking into account the Intergovernmental Oceanographic Commission Criteria and Guidelines on the Transfer of Marine Technology, in order to improve ocean health and to enhance the contribution of marine biodiversity to the development of developing countries, in particular small island developing states and least developed countries". In this context, this article discusses whether the current international legal regime is adequate to encounter numerous challenges in achieving Goal 14, especially from the perspective of developing countries. It focuses on the utilization of marine genetic resources beyond national jurisdiction through marine scientific research activities in the field of marine biodiversity, which is also known as marine bioprospecting. This article argues that the current international legal regime concerning marine genetic resources beyond national jurisdiction is still heavily fragmented and inadequate to address the challenges in achieving Goal 14, especially in ensuring the contribution of marine biodiversity to the development of developing countries, small island developing states, and least developed countries. Thus, a more integrated

* Some parts of this article build on chapters from the author's Ph.D. Thesis, *Marine Bioprospecting: International Law, Indonesia, and Sustainable Development* (The Australian National University, 2013). The author would like to express his sincere gratitude to Professor Donald Rothwell of the Australian National University, Professor Matthew Rimmer of the Queensland University of Technology, Professor Liesbeth Lijnzaad of Maastricht University, and Emeritus Professor Alfred Soons of Utrecht University for their valuable comments and suggestions. All the remaining shortcomings are entirely mine. The author's PowerPoint presentation is available at http://www.virginia.edu/colp/pdf/yogya-siswandi.pdf

** S.H. (Unpad), LL.M. (London), Ph.D. (ANU), Lecturer at Universitas Padjadjaran Faculty of Law, Bandung, Indonesia. E-mail at ahmad.gusman@unpad.ac.id

 | DOI:10.1163/9789004366619_014

approach between the Sustainable Development Goals and relevant international instruments is critically needed.

1 Introduction

Numerous studies have revealed the impressive state of biological diversity in our oceans. This, for example, has been demonstrated by the Census of Marine Life, which was conducted from 2000 to 2010 and involved more than 540 expeditions as well as 2,700 scientists from more than 80 nations.[1] The staggering result of the Census has been described as follows:

> The Census encountered an anticipated riot of species, which are the currency of diversity. It upped the estimate of known marine species from about 230,000 to nearly 250,000. Among the millions of specimens collected in both familiar and seldom-explored waters, the Census found more than 6,000 potentially new species and completed formal descriptions of more than 1,200 of them.[2]

This astounding wealth of marine biodiversity has played an essential role in enhancing the quality of human life, for example, through the development of marine natural products.

It has been widely acknowledged that a large number of marine organisms have unique biological characteristics. Such organisms are equipped with appropriate mechanisms to survive by developing complex biological as well as chemical processes for defence, attack, and other still unknown purposes.[3] Obviously, marine organisms have become of increasing interest to scientists and thus have been researched extensively for pharmaceutical products. It has been recorded that over 14,000 new chemical entities have been identified from marine sources and at least 300 patents have been issued for marine natural products.[4]

---

1 J.H. Ausubel, D.T. Crist, and P.E. Waggoner, "First Census of Marine Life 2010: Highlights of a Decade of Discovery", available at <http://www.coml.org/highlights-2010/>.

2 *Ibid.* See also J. Montgomery and L. Carter, "Marine Science in the Past 25 Years: Main Findings and Trends", in: Vidas and Schei (eds.), *The World Ocean in Globalisation: Climate Change, Sustainable Fisheries, Biodiversity, Shipping, Regional Issues* (Leiden and Boston, Martinus Nijhoff Publishers, 2011), pp. 37–52.

3 F. de la Calle, "Marine Genetic Resources: A Source of New Drugs – The Experience of the Biotechnology Sector", 24 *International Journal of Marine and Coastal Law*, 2009, 209–220.

4 B. Hunt and A.C.J. Vincent, "Scale and Sustainability of Marine Bioprospecting for Pharmaceuticals", 35 *Ambio: A Journal of the Human Environment*, No. 2, 2006, 57. See also

The development of marine natural and pharmaceutical products involves the utilization of marine genetic resources through a particular activity known as "marine bioprospecting". This activity can be generally described as the process of searching for marine living resources from which valuable genetic material can be obtained and commercially exploited.[5] In recent years, the interests on marine bioprospecting have been significantly growing. A study in 2012 revealed that marine bioprospecting will increase in the years to come, providing new medicines for human illnesses as well as other innovative products for a number of industrial activities, such as biotechnology.[6] In addition, the profits gained from the utilization of marine genetic resources in marine biotechnology products have been particularly remarkable. In 2002, global sales of these products were estimated at about US$2.4 billion.[7] In the pharmaceutical industry, the annual profits from a sea sponge compound were reportedly between $50 million and $100 million, and cancer-fighting agents derived from marine organisms were estimated to be worth $1 billion in 2005.[8] Moreover, in general, the marine biotechnology industry itself is projected to reach US$4.6 billion by the year 2017.[9]

However, it has been widely observed that developed countries with their advanced biotechnology capacity would most likely have greater opportunities to reap the benefits from the utilization of biodiversity, including marine genetic resources. As one scholar has noted:

M.C. Leal *et al.*, "Trends in the Discovery of New Marine Natural Products from Invertebrates over the Last Two Decades – Where and What are We Bioprospecting?" 7 *PLoS ONE* e30580, No.1, 2012; P.M. Erwin, S. López-Legentil and P.W. Schuhmann, "The Pharmaceutical Value of Marine Biodiversity for Anti-cancer Drug Discovery", 70 *Ecological Economics*, No. 2, 2010, 445–451; D. Leary *et al.*, "Marine Genetic Resources: A Review of Scientific and Commercial Interest", 33 *Marine Policy*, No. 2, 2009, 183–194; and T.F. Molinski *et al.*, "Drug Development from Marine Natural Products", 8 *Nature Reviews Drug Discovery*, No. 1, 2009, 69–85.

5 Although there is no universally agreed definition, some scholars have attempted to formulate a detailed definition of bioprospecting. For instance, Farrier and Tucker defined bioprospecting as "the collection of small samples of biological material for screening in the search for commercially exploitable biologically active compounds or attributes such as genetic information". They also noted that the end-focus of bioprospecting is mostly on the design and development of pharmaceuticals as well as other types of commercial products sourced from biological resources. See especially D. Farrier and L. Tucker, "Access to Marine Bioresources: Hitching the Conservation Cart to the Bioprospecting Horse", 32 *Ocean Development and International Law*, 2001, 214.

6 Leal *et al.*, *supra* note 4, 1.

7 Leary *et al.*, *supra* note 4, 190.

8 *Ibid.* See also S. Arico, "Deep Sea: The Last Frontier", available at <http://www.unesco.org/new/en/natural-sciences/ioc-oceans/single-view-oceans/news/the_last_frontier/>.

9 "Growing International Interest in Oceanology to Benefit the Global Marine Biotechnology Market, According to New Report by Global Industry Analysts, Inc.", available at <http://www.prweb.com/releases/marine_biotechnology/marine_biomaterials/prweb9420116.htm>.

> Following the development of biotechnology, the value and significance of biogenetic resources has appeared increasingly prominent, playing a significant role in the spheres of world agriculture, food security, and the global economy. Nevertheless, there has been a differential approach between technologically advanced (most developed) countries and those (most developing) with rich genetic resources towards the issues of benefit sharing and technology transfer resulting from genetic resources, and who should have the entitlement of the intellectual property based on such resources.[10]

To some extent, this situation may lead to inequalities between developed countries on one hand and developing or least developed countries on the other. In this regard, international law has a pivotal role to resolve this issue by setting an international regime that could promote equalities among countries in the context of the utilization of genetic resources.

This situation has been exacerbated by some legal loopholes in the utilization of marine genetic resources, especially those beyond national jurisdiction. There are two main relevant international instruments in this regard, namely the United Nations Convention on the Law of the Sea (UNCLOS)[11] and the Convention on Biological Diversity (CBD).[12] These instruments basically deal with the utilization of marine living resources and genetic resources; however, they do not explicitly address marine bioprospecting activities and the utilization of marine genetic resources in areas beyond national jurisdiction, which include high seas and the deep sea bed (the Area). Should the utilization of marine genetic resources beyond national jurisdiction continue without a clear, well-defined legal regime, the benefits arising therefrom may only be enjoyed by states or private institutions that have more capacity and resources.

Meanwhile, in order to renew the global development commitments enumerated in the Millennium Development Goals, the international community has adopted the Sustainable Development Goals through a document titled "Transforming Our World: The 2030 Agenda for Sustainable Development".[13] This document comprises seventeen goals, including Goal 14, namely "Conserve

10 K. Ni, "The Incorporation of the CBD Mandate on Access and Benefit-sharing Into TRIPS Regime: An Appraisal of the Appeal of Developing Countries with Rich Genetic Resources", 1 *Asian Journal of WTO and International Health Law and Policy*, 2006, 434.

11 *United Nations Convention on the Law of the Sea*, opened for signature 10 December 1982, 1833 UNTS 3 (entered into force 16 November 1994).

12 *Convention on Biological Diversity*, opened for signature 5 June 1992, 1760 UNTS 79 (entered into force 29 December 1993).

13 *Transforming Our World: The 2030 Agenda for Sustainable Development*, UN Doc A/RES/70/1 (21 October 2015).

and Sustainably Use the Oceans, Seas and Marine Resources for Sustainable Development".[14] Under this Goal, paragraph 14.a particularly highlights the relationships between marine technology, marine biodiversity, and developing countries as follows:

> Increase scientific knowledge, develop research capacity and transfer marine technology, taking into account the Intergovernmental Oceanographic Commission Criteria and Guidelines on the Transfer of Marine Technology, in order to improve ocean health and to enhance the contribution of marine biodiversity to the development of developing countries, in particular small island developing states and least developed countries.[15]

This Goal is indeed of crucial importance to the specifically mentioned countries, all of which receive special attention throughout the agendas under the Sustainable Development Goals.[16] As far as the oceans and seas are concerned, these countries are generally exposed to numerous issues, such as marine pollution, degrading quality of the marine environment, and coastal poverty.[17]

In this regard, this article will further discuss the relationships between marine scientific research and technology, marine biodiversity, and developing countries as reflected under Goal 14.a, especially in the context of the utilization of marine genetic resources beyond national jurisdiction. For this purpose, this article will be divided into three parts. The first part will analyze relevant international legal instruments, particularly the provisions under UNCLOS and the CBD. The second part will further elaborate on the relationship between the Sustainable Development Goals and marine genetic resources. The third part will discuss the ongoing process on the development of an international legally binding instrument for marine biodiversity beyond areas of national

---

14 *Ibid.*

15 United Nations, "Goal 14: Conserve and Sustainably Use the Oceans, Seas and Marine Resources", available at <http://www.un.org/sustainabledevelopment/oceans/>.

16 *Transforming Our World*, *supra* note 13, para. 22. See also United Nations, "The Sustainable Development Goals Report 2016", available at <http://unstats.un.org/sdgs/report/2016/>.

17 See e.g., D. Griggs *et al.*, "Sustainable Development Goals for People and Planet", 495 *Nature*, No. 7741, 2013, 305–307; T.A. Stojanovic and C.J.Q. Farmer, "The Development of World Oceans and Coasts and Concepts of Sustainability", 42 *Marine Policy*, 2013, 157–165; and J. Campbell, E. Whittingham, and P. Townsley, "Responding to Coastal Poverty: Should We be Doing Things Differently or Doing Different Things?" in: Hoanh, Tuong, Gowing, and Hardy (eds.), *Environment and Livelihoods in Tropical Coastal Zones: Managing Agriculture-Fishery-Aquaculture Conflicts* (Wallingford, CABI International, 2006), pp. 274–292.

jurisdiction, focusing on the perspectives of developing countries with respect to marine genetic resources, and what should be the way forward.

2 The Current International Legal Regime

2.1 *UNCLOS*

2.1.1 Marine Scientific Research and Technology

The utilization of marine genetic resources through marine bioprospecting generally involves particular research activities as well as scientific methods. UNCLOS is therefore relevant in this context since it is the main international treaty in the field of the law of the sea that specifically regulates marine scientific research and technology transfer. It has been acknowledged as "a constitution for the oceans",[18] and it is also an instrument that reflects the principles of sustainable development as it embraces economic, social, and environmental aspects with respect to the utilization of the seas and oceans.[19]

In regard to marine scientific research, UNCLOS contains two relevant parts, namely Part XIII (Marine Scientific Research) and Part XIV (Development and Transfer of Marine Technology). The marine scientific research regime under UNCLOS is primarily established upon the regime of state sovereignty and jurisdiction at sea. In this regard, there are mainly seven maritime zones regulated under UNCLOS: the territorial sea; archipelagic waters; the contiguous zone; the Exclusive Economic Zone (EEZ); the continental shelf; the high seas; and the deep sea bed or the "Area" beyond the limits of national jurisdiction. Generally, according to UNCLOS, the coastal States have the right to regulate, authorize and conduct marine scientific research conducted within their jurisdiction, which includes the territorial sea, archipelagic waters, the EEZ and the continental shelf.[20]

In areas beyond national jurisdiction, including the high seas and the Area, UNCLOS basically applies freedom of marine scientific research. In this regard, article 257 of UNCLOS provides that all states, irrespective of their geographical location, and competent international organizations are entitled to conduct marine scientific research in the water column beyond the limits of the EEZ. Furthermore, under article 256, all states and competent international organizations have the right to conduct marine scientific research in the Area.

18 T. Koh, "A Constitution for the Oceans", available at <http://www.un.org/depts/los/convention_agreements/texts/koh_english.pdf>.

19 UNCLOS, Preamble.

20 UNCLOS, Arts. 21, 40, 56, 245, and 246.

In addition, article 143(1) provides that all marine scientific research within the Area in particular shall be carried out exclusively for peaceful purposes and for the benefit of mankind as a whole in accordance with Part XIII of UNCLOS.[21]

In conducting marine scientific research, all States and relevant stakeholders must adhere to the following four principles provided under article 240 of UNCLOS. First, marine scientific research must be conducted exclusively for peaceful purposes. Second, marine scientific research must be conducted with appropriate scientific methods and means compatible with UNCLOS. Third, marine scientific research must not unjustifiably interfere with other legitimate uses of the sea compatible with UNCLOS. Fourth, marine scientific research must be conducted in compliance with all relevant regulations adopted in conformity with UNCLOS, including those for the protection and preservation of the marine environment.

Furthermore, UNCLOS underscores the international cooperation in the field of marine scientific research, which is an important aspect in ensuring that all states could benefit from it. In this context, under articles 243 and 244, states and competent international organizations are obliged to cooperate in creating favourable conditions for the conduct of marine scientific research and, accordingly, to make available publication and dissemination of information and knowledge resulting from marine scientific research through appropriate channels.

In addition to marine scientific research, Part XIV of UNCLOS provides a number of provisions with respect to the promotion of the development and transfer of marine technology, as well as international cooperation in this field. In particular, UNCLOS also touches upon the issue of equality between states in enjoying the benefits resulting from the advancement of marine technology. Article 266 paragraphs 2 and 3 read:

> 2. States shall promote the development of the marine scientific and technological capacity of States which may need and request technical assistance in this field, particularly developing States, including land-locked and geographically disadvantaged States, with regard to the exploration, exploitation, conservation and management of marine resources, the protection and preservation of the marine environment, marine scientific research and other activities in the marine

21 The meaning of "the benefit of mankind as a whole" may be viewed in light of relevant provisions under Part XIII, particularly on the duty of researching States to publish the knowledge resulting from marine scientific research. See e.g., A.H.A. Soons, *Marine Scientific Research and the Law of the Sea* (The Hague, T.M.C. Asser Instituut, 1982), p. 229.

environment compatible with the Convention, with a view to accelerating the social and economic development of the developing States.
3. States shall endeavor to foster favorable economic and legal conditions for the transfer of marine technology for the benefit of all parties concerned on an equitable basis.

Although UNCLOS has provided the legal regime on marine scientific research and technology, there is still an ongoing debate whether these provisions have adequately addressed the issue of marine genetic resources.[22] The utilization of these resources through marine bioprospecting may embrace both the scientific research aspect and the commercial aspect; and in practice, it is often difficult to draw the line between these two. The 2005 Report of the United Nations Secretary-General on Oceans and the Law of the Sea noted:

> It is difficult to differentiate scientific research from commercial activities involving genetic resources, commonly referred to as bioprospecting. In most cases, genetic resources are collected and analyzed as part of scientific research projects, in the context of partnerships between scientific institutions and industry. It is only at a later stage that knowledge, information and useful materials extracted from such resources enter a commercial phase. The difference between scientific research and bioprospecting therefore seems to lie in the use of knowledge and results of such activities, rather than in the practical nature of the activities themselves.[23]

22 See e.g., M.W. Lodge, "Improving International Governance in the Deep Sea", 19 *International Journal of Marine and Coastal Law*, No. 3, 2004, 299; S. Arico and C. Salpin, *Bioprospecting of Genetic Resources in the Deep Seabed: Scientific, Legal and Policy Aspects* (Japan, United Nations University-Institute of Advanced Studies, 2005); K.M. Gjerde *et al.*, *Options for Addressing Regulatory and Governance Gaps in the International Regime for the Conservation and Sustainable Use of Marine Biodiversity in Areas beyond National Jurisdiction* (Gland, IUCN, 2008); A. Broggiato, "Marine Genetic Resources beyond National Jurisdiction – Coordination and Harmonization of Governance Regimes", 41 *Environmental Policy and Law*, 2011, 35–42; L. Glowka, "Evolving Perspectives on the International Seabed Area's Genetic Resources: Fifteen Years after the 'Deepest of Ironies'", in: Vidas (ed.), *Law, Technology and Science for Oceans in Globalisation: IUU Fishing, Oil Pollution, Bioprospecting, Outer Continental Shelf* (Leiden, Martinus Nijhoff Publishers, 2010), pp. 397–419; and D. Leary, "Moving the Marine Genetic Resources Debate Forward: Some Reflections", 27 *International Journal of Marine and Coastal Law*, No. 2, 2012, 435–448.

23 *Oceans and the Law of the Sea, Report of the Secretary-General, Addendum*, 60th session, Agenda Item 76 (a), UN Doc A/60/63/Add.1 (15 July 2005), para. 202.

Similarly, Rothwell and Stephens also observed that drawing a distinction between "pure" research on one hand and "applied" research on the other hand could be problematic at times. They stated that "there is considerable crossover between the two, with pure science often forming the basis for practical oceans management decisions, and for further, commercially-oriented, research".[24]

These uncertainties have triggered a fundamental question: could marine biodiversity research or marine bioprospecting activities be regarded as marine scientific research under UNCLOS? Some scholars have argued that marine scientific research pursued for economic gains and involving the exploitation of marine living resources would belong to the UNCLOS regime on resources utilization; thus, the provisions on Part XIII and Part XIV do not necessarily apply to this kind of research.[25]

Nevertheless, it should be noted that the development in marine biotechnology, particularly the utilization of marine genetic resources, had not taken place when UNCLOS was negotiated. Hence, there is a need to interpret the UNCLOS provisions under Part XIII and Part XIV in light of recent developments regarding the utilization of such resources. Within this context, it may be argued that since marine biodiversity research and marine bioprospecting involve scientific method and applications, they should be viewed as parts of marine scientific research. Matz, for instance, has noted:

> Because of the resemblance of the exploitation of genetic resources with scientific research, in respect to the UN Convention on the Law of the Sea the provisions on marine scientific research could be applicable. In general terms the expression "marine scientific research" is most often used to describe activities to expand scientific knowledge of the marine environment and its processes. ...The expansion of knowledge on marine genetic resources and activities as to their exploration, consequently, are to be considered marine scientific research.[26]

---

24 D.R. Rothwell and T. Stephens, *The International Law of the Sea* (Oxford, Hart Publishing, 2010), p. 321.

25 See e.g., P. Birnie, "Law of the Sea and Ocean Resources: Implications for Marine Scientific Research", 10 *International Journal of Marine and Coastal Law*, 1995, 242; and A. Soons, "Marine Scientific Research Provisions in the Convention on the Law of the Sea: Issues of Interpretation", in: Brown and Churchill (eds.), *The UN Convention on the Law of the Sea: Impact and Implementation* (Honolulu, The Law of the Sea Institute-William S. Richardson School of Law, 1989), p. 367.

26 N. Matz, "The Interaction between the Convention on Biological Diversity and the UN Convention on the Law of the Sea", in: Ehlers, Borgese, Wolfrum, and Hoszlig (eds.),

In addition, however, since marine bioprospecting also involves the extraction of marine genetic resources aiming at commercial gains, it would also be related to the resource utilization regime under UNCLOS. In this regard, the principles related to the preservation and protection of the marine environment, especially the principle of sustainable use of marine living resources, would be especially relevant and could also be applicable to marine bioprospecting activities.[27]

With respect to the institutional framework in the field of marine scientific research and technology, several UNCLOS provisions make a specific reference to "competent international organizations". For instance, Article 271 stipulates that:

> States, directly or through competent international organizations, shall promote the establishment of generally accepted guidelines, criteria and standards for the transfer of marine technology on a bilateral basis or within the framework of international organizations and other fora, taking into account, in particular, the interests and needs of developing States.[28]

In this context, the Intergovernmental Oceanographic Commission (IOC) was established in 1960 to promote international cooperation and to coordinate programmes related to marine science and technology.[29] It has also been recognized as the competent international organization in the fields of marine scientific research and technology transfer in accordance with UNCLOS.[30]

Within the ambit of Article 271, the IOC has adopted the "Criteria and Guidelines on the Transfer of Marine Technology" (IOC Guidelines). In general, this

---

Marine Issues: From a Scientific, Political, and Legal Perspective (Leiden, Martinus Nijhoff Publishers, 2002) p. 211. See also T. Scovazzi, "Is the UN Convention on the Law of the Sea the Legal Framework for All Activities in the Sea? The Case of Bioprospecting", in: Vidas (ed.), *supra* note 22, pp. 312–313.

27 A.G. Siswandi, *Marine Bioprospecting: International Law, Indonesia, and Sustainable Development* (Ph.D. Thesis, The Australian National University, 2013), pp. 81–82. See also H.N. Scheiber, "Economic Uses of the Oceans and the Impacts on Marine Environments: Past Trends and Challenges Ahead", in: Vidas and Schei (eds.), *supra* note 2, p. 95.

28 UNCLOS, Art. 271.

29 Intergovernmental Oceanographic Commission, "Resources, Meetings, Documents, People", available at <http://www.ioc-unesco.org/>. For general discussions about the IOC see P. Ehlers, "The Intergovernmental Oceanographic Commission: An International Organization for the Promotion of Marine Research", 15 *International Journal of Marine and Coastal Law*, 2000, 533–554.

30 *Ibid.*

document underscores the specific situation of developing countries by stipulating that in marine technology transfer, due regard should be given to:

> The needs and interests of developing States, particularly land-locked and geographically disadvantaged States as well as other developing States which have not been able to establish or develop their own capabilities in marine sciences, scientific research, observations of the oceans and coastal areas, and related technology, or to develop the infrastructure needed to achieve such ends.[31]

Nevertheless, this document only provides a relatively general framework for marine technology transfer and does not specifically address marine bioprospecting activities. This particular situation will be further discussed below.

2.1.2 Marine Genetic Resources beyond National Jurisdiction

As noted earlier, UNCLOS did not make a specific reference to marine genetic resources as well as their utilization. This has resulted in a fundamental issue as to whether UNCLOS could really be applied to activities pertaining to the utilization of marine genetic resources, especially those beyond areas of national jurisdiction. For instance, with respect to the Area, UNCLOS defines "resources" as "all solid, liquid or gaseous mineral resources *in situ* in the Area at or beneath the seabed, including polymetallic nodules";[32] thus it only covers non-living resources and does not include living or genetic resources.

With respect to the high seas regime, UNCLOS provides legal principles and standards for the utilization of living resources in the high seas. However, they are currently only applicable to fisheries. In this regard, UNCLOS stipulates that the high seas are open to all States and both coastal and land-locked States enjoy freedom of fishing, subject to the conditions related to conservation in Section 2.[33] Glowka noted however that it is possible to infer that the term "living resources" does not only cover fish and invertebrate communities, but also genetic resources.[34] Thus, it has been assumed that marine genetic resources would be freely accessible under the high seas legal regime.[35]

---

31 International Oceanographic Commission, "IOC Criteria and Guidelines on the Transfer of Marine Technology", available at <http://unesdoc.unesco.org/images/0013/001391/139193m.pdf>.

32 UNCLOS, Art. 133 (a).

33 UNCLOS, Art. 87.1 (e).

34 L. Glowka, "The Deepest of Ironies: Genetic Resources, Marine Scientific Research, and the Area", 12 *Ocean Yearbook*, 1996, 168.

35 *Ibid.* See also L. Glowka, "Genetic Resources, Marine Scientific Research and the International Seabed Area", 8 *Review of European Community and International Environmental*

The absence of detailed provisions concerning marine genetic resources beyond national jurisdiction would surely lead to legal uncertainties with respect to their utilization. Moreover, the application of the freedom of the high seas to the utilization of marine genetic resources in this area has been largely based on an assumption. Most importantly, legal uncertainties surrounding the utilization of marine genetic resources beyond national jurisdiction do not only concern the issue of the legal status of such resources and who has the right to enjoy the benefit arising from their utilization. There are also other issues at stake, especially the environmental ones. Without any clear standards and regulations, activities pertaining to the utilization of marine genetic resources may cause harm to the marine environment and could be detrimental to marine biodiversity conservation efforts.

Therefore, some clarifications on the legal and institutional issues in the context of marine genetic resources beyond national jurisdiction are urgently needed. With respect to the Area's genetic resources in particular, Glowka, for instance, has argued that:

> Creating a legal regime for the Area's genetic resources could have a number of positive aspects. It could secure the legal status of the Area's genetic resources, provide the basis for the fair and equitable utilization of the Area's genetic resources, and ensure conservation and sustainable use of the Area's organisms.[36]

However, he also observed that:

> If the international community decides to take up the issue, the positive aspects of the regime would have to be carefully weighed against possible negative aspects that, whether valid or not, certainly will be raised in the debate. The outcome of discussions will undoubtedly have to define a way to avoid hindering marine scientific research in the Area; creating unreasonable obstacles to commercial biotechnological development; unreasonably limiting commercial incentives, such as intellectual property rights, for work undertaken on the Area's genetic resources; and creating a negative political reaction analogous to the development of the Area's mining regime.[37]

Law, No. 1, 1999, 56. For more recent developments on this issue see L. Glowka, *supra* note 22, pp. 397–419.

36 Glowka, *supra* note 34, 169–170.

37 *Id.* at 70.

This observation has clearly signified a need for the international community to develop a specific legal regime for the utilization of marine genetic resources beyond national jurisdiction. The current development in this context will be further discussed in Parts 3 and 4 of this article.

2.2 *CBD*

The CBD is another international legal instrument that is relevant to marine bioprospecting activities and reflect the principles of sustainable development.[38] It has also been regarded as "the most successful biodiversity-related convention" to date.[39] According to article 1 of the CBD, the objectives of the Convention are threefold: the conservation of biological diversity; the sustainable use of its components; and the fair and equitable sharing of the benefits arising out of the utilization of genetic resources.

In accordance with article 15(1), the CBD recognizes that states have sovereign rights over their natural resources and that national governments have the authority to determine access to genetic resources. Within this context, article 15 further provides that the grant of access shall be on mutually agreed terms and subject to prior informed consent of the Contracting Party providing genetic resources. Following these principles, article 15 of the CBD requires each Party to take appropriate measures to ensure a fair and equitable sharing of benefits arising from the utilization of genetic resources with the provider country. Altogether, these provisions have established what is generally known as the access and benefit-sharing (ABS) mechanism in the utilization of genetic resources.

The principles regarding the ABS mechanism under the CBD have been reiterated in the 2010 Nagoya Protocol on Access to Genetic Resources and the Fair and Equitable Sharing of Benefits Arising from their Utilization to the Convention on Biological Diversity (Nagoya Protocol).[40] On matters relating to access, the Nagoya Protocol specifies measures that must be taken by the Parties to ensure that prior informed consent of the provider country and, where relevant,

38 M.C. Segger, "Sustainable Development in International Law", in: Bugge and Voight (eds.), *Sustainable Development in International and National Law* (Groningen, Europa Law Publishing, 2008), pp. 87–202.

39 V. Mauerhofer and F. Nyacuru, "Biodiversity, Migratory Species and Natural Heritage", in: Harris (ed.), *Routledge Handbook of Global Environmental Politics* (London and New York, Routledge, 2014), p. 483.

40 *Nagoya Protocol on Access to Genetic Resources and the Fair and Equitable Sharing of Benefits Arising from their Utilization to the Convention on Biological Diversity*, opened for signature 2 February 2011, UNEP/CBD/COP/DEC/X/1 of 29 October 2010 (entered into force 12 October 2014).

indigenous and local communities have been obtained. These include, among others, clear and transparent domestic access and benefit-sharing legislation; fair and non-arbitrary rules on accessing genetic resources; and the issuance of a permit as evidence of the decision to grant prior informed consent and of the establishment of mutually agreed terms.[41]

The ABS framework under the CBD and the Nagoya Protocol could provide some opportunities to ensure equality between the users and the providers of genetic resources. Article 5(4) of the Nagoya Protocol stipulates that shared benefits from the utilization of genetic resources "may include monetary and non-monetary benefits". In this regard, the Annex of the Protocol lists a number of examples of such benefits, including "collaboration, cooperation and contribution in scientific research and development programmes, particularly biotechnological research activities, where possible in the Party providing genetic resources".[42] As noted by Morgera and her colleagues, non-monetary benefit could also contribute to build the capacity of provider countries in utilizing their genetic resources.[43]

It could be argued that the marine scientific research and technology transfer regime under UNCLOS and the ABS provisions of the CBD and the Nagoya Protocol are unquestionably relevant to marine biodiversity research and bioprospecting activities. Nevertheless, there are still some legal problems and lacunae. On one hand, the CBD and the Nagoya Protocol, which regulate the utilization of genetic resources, do not contain particular provisions with respect to marine genetic resources and their scopes are limited to genetic resources within national jurisdiction. On the other hand, UNCLOS, does not explicitly address marine bioprospecting as a part of activities that would fall under the marine scientific research and technology transfer and it is also silent with respect to marine genetic resources beyond national jurisdiction.

41 Nagoya Protocol, Art. 6. For more detailed discussions on the Nagoya Protocol see e.g., C. Richerzhagen, "The Nagoya Protocol: Fragmentation or Consolidation", 3 *Resources*, No. 1, 2014, 135–151; M. Buck and C. Hamilton, "The Nagoya Protocol on Access to Genetic Resources and the Fair and Equitable Sharing of Benefits Arising from Their Utilization to the Convention on Biological Diversity", 20 *Review of European, Comparative and International Environmental Law*, No. 1, 2011, 47–61; A.G. Siswandi, "The Nagoya Protocol: Unfinished Business Remains Unfinished", in: Rimmer (ed.), *Indigenous Intellectual Property: A Handbook of Contemporary Research* (Cheltenham, Edward Elgar, 2015), pp. 334–364; and E. Morgera, E. Tsioumani, and M. Buck, *Unraveling the Nagoya Protocol: A Commentary on the Nagoya Protocol on Access and Benefit-sharing to the Convention on Biological Diversity* (Leiden, Brill, 2014).

42 Nagoya Protocol, Annex, Paragraph 2 (b).

43 E. Morgera, E. Tsioumani, and M. Buck, *supra* note 41, 133–134.

3 Sustainable Development Goals and Marine Genetic Resources

The utilization of marine genetic resources is relevant to Sustainable Development Goals since it is a particular activity that could contribute to conservation and sustainable use of marine biodiversity. In addition, this activity could represent the relationship between scientific research, technology, and sustainable development.

A growing attention with respect to the relationship between marine scientific research and sustainable development[44] has been reflected, for instance, in a document titled "A Blueprint for Ocean and Coastal Sustainability", prepared by multiple United Nations agencies preceding the United Nations Conference on Sustainable Development (Rio+20).[45] The document proposes four objectives regarding ocean and coastal sustainability; marine scientific research is explicitly stated under Objective 4, which states: "Actions supporting marine research, monitoring and evaluation, technology and capacity transfer as a means for improving knowledge, addressing emerging issues, developing capacities in support of sustainable use of the ocean".[46]

Nevertheless, this particularly important proposal was not adequately addressed in the result of the Rio+20 Conference, "The Future We Want".[47] Although this document does recognize "the important contribution of the scientific and technological community to sustainable development", and underlines the commitment of the Member States to take necessary actions to close the technological gap, particularly between developing and developed countries,[48] it only contains two paragraphs that explicitly address marine scientific research.[49] Furthermore, the section on Oceans and Seas only contains general and weak language with respect to sustainable development in the context of oceans.[50] Overall, the Rio+20 has been criticized as a forum that failed to deliver further measures to implement sustainable development.[51]

44 See e.g., H. Harden-Davies, "The Regulation of Marine Scientific Research: Addressing Challenges, Advancing Knowledge", in: Warner and Kaye (eds.), *Routledge Handbook of Maritime Regulation and Enforcement* (London and New York, Routledge, 2016), p. 214.

45 IOC/UNESCO, IMO, FAO, UNDP, *A Blueprint for Ocean and Coastal Sustainability* (Paris, IOC/UNESCO, 2011).

46 *Id.* at 7.

47 *The Future We Want*, UN Doc A/RES/66/288 (11 September 2012).

48 *Id.*, para. 48.

49 *Id.*, paras. 160 and 166.

50 See Siswandi, *supra* note 27, p. 493.

51 See e.g., A.D. Stasi, "The Normative Force of the Outcome Document 'The Future We Want'", in: Fitzmaurice, Maljean-Dubois, and Negri (eds.), *Environmental Protection and Sustainable Development from Rio to Rio+20* (Leiden and Boston, Brill Nijhoff, 2014), p. 25.

Despite these shortcomings, the adoption of "The Future We Want" document marked a new history in the evolution of sustainable development by introducing the Sustainable Development Goals, or post-2015 agenda, which serve to replace the Millennium Development Goals.[52] Following this initiative, a worldwide consultation was held to obtain input for the formulation of such goals through an intergovernmental process that also brought all stakeholders on board. It was conducted by involving over one million people, 88 national consultations, 11 thematic dialogues, and a global survey conducted through MY World.[53]

The Sustainable Development Goals were finally adopted through a document titled "Transforming Our World: The 2030 Agenda for Sustainable Development".[54] This document comprises seventeen goals, including Goal 14, namely "Conserve and Sustainably Use the Oceans, Seas and Marine Resources for Sustainable Development".[55] The Sustainable Development Goals are also accompanied by 169 associated targets, which are integrated and indivisible.[56] These goals and targets came into effect on 1 January 2016 and will serve as guidance for countries in making development-related decisions.[57]

Goal 14 has indeed some relevance to the utilization of marine genetic resources although it does not explicitly address such resources. As discussed earlier, the utilization of marine genetic resources generally involves the application of marine scientific research and technology through marine bioprospecting activities. In this context, the relationship between marine scientific research, marine biodiversity, and developing countries is clearly stated under Goal 14.a.

It is also important to note that during the discussions and negotiations prior to the adoption of the Sustainable Development Goals, some issues concerning the relationships between marine biodiversity and the utilization of genetic resources were brought forward by a number of countries, especially in the sessions of the United Nations General Assembly Open Working Group on Sustainable Development Goals held from 2013 to 2014.[58] Bolivia, for the Group

52 *The Future We Want, supra* note 47, paras. 245–251.

53 United Nations Development Group, "A Million Voices: The World We Want-A Sustainable Future with Dignity for All", available at <http://www.undp.org/content/undp/en/home/librarypage/mdg/a-million-voices--the-world-we-want.html>.

54 *Transforming Our World, supra* note 13.

55 *Ibid.*

56 *Id.*, para. 18.

57 *Id.*, para. 21.

58 The Open Working Group was established upon the decision adopted at the Rio+20 forum. See further P.S. Chasek *et al.*, "Getting to 2030: Negotiating the Post-2015 Sustainable

of 77 and China, for instance, noted that there is still a gap in the exploitation of the ocean's genetic resources.[59] Switzerland, France, and Germany, called for the improvement of access to genetic resources and the fair and equitable sharing of benefits as well as sustainable management of marine genetic resources.[60] Mexico and Peru emphasized the importance of the sustainable use and equitable sharing of benefits from biodiversity as a key to development.[61]

The Open Working Group sessions resulted in a zero draft of the Sustainable Development Goals, which proposed the elements of Sustainable Development Goals, including Goal 14.a. as it reads now.[62] It was followed up by the Intergovernmental Negotiations on the Post-2015 Development Agenda that was established during the UN General Assembly Special Event on the Millennium Development Goals.[63] However, there was not much further elaboration on marine scientific research and technology transfer, which made the proposed Goal 14.a almost left untouched since the adoption of the proposal from the Open Working Group.

Hence, it appears that there is still a lack of attention on marine scientific research in the elaboration of Goals 14 and 14.a of the Sustainable Development Goals. While Goal 14.a has mentioned explicitly the role of marine scientific research and technology, it does not elaborate in detail about how to achieve this role and how to implement the elements of sustainable development, especially in regard to marine biodiversity research and marine bioprospecting. This has also been reflected in the 2016 Sustainable Development Goals Report. The document does highlight explicitly three important areas under Goal 14: sustainable fisheries, marine protected areas, and marine pollution.[64]

Development Agenda", 25 *Review of European Community and International Environmental Law*, No. 1, 2016, 5–14. See also International Institute of Sustainable Development, "Summary of the Eighth Session of the UN General Assembly Open Working Group on Sustainable Development Goals: 3–7 February 2014", available at <http://www.iisd.ca/sdgs/owg8>; and International Institute of Sustainable Development, "Summary of the Tenth session of the UN General Assembly Open Working Group on Sustainable Development Goals: 31 March–4 April 2014", available at <http://www.iisd.ca/sdgs/owg10>.

59 International Institute of Sustainable Development, "Summary of the Eighth Session of the UN General Assembly Open Working Group on Sustainable Development Goals: 3–7 February 2014", available at <http://www.iisd.ca/sdgs/owg8>.

60 *Ibid.*

61 *Ibid.*

62 *Report of the Open Working Group of the General Assembly on Sustainable Development Goals*, UN Doc A/68/970 (12 August 2014).

63 See *Outcome Document of the Special Event to Follow Up Efforts Made Towards Achieving the Millennium Development Goals*, UN Doc A/RES/68/6 (28 January 2014).

64 United Nations, "The Sustainable Development Goals Report 2016", available at <http://unstats.un.org/sdgs/report/2016/>.

However, there is no particular reference to the role of marine scientific research in supporting the achievement of this particular goal.

The "Transforming Our World" document has stated that the Sustainable Development Goals will be followed up and reviewed using a set of global indicators.[65] In this regard, an Inter-Agency and Expert Group on Sustainable Development Goal Indicators was tasked to develop and propose such indicators.[66] As of June 2016, this Group has agreed to adopt 230 indicators.[67] However, the indicator on Goal 14.a still seems to be partial. Goal 14.a only has one indicator that states "proportion of total research budget allocated to research in the field of marine technology".[68]

Taking this current situation into account, it would be particularly challenging for marine scientific research to provide a significant contribution to the achievement of sustainable development. Therefore, the Sustainable Development Goals need to embrace the contemporary paradigm that puts scientific research as one of the key players in sustainable development. As one scholar has noted:

> There is an increased recognition that Science, Technology and Innovation (STI) can spur inclusive and sustainable development in multiple ways. For STI to be a driver for sustainable development it is important that Development Agendas are people-centered, creating an enabling environment for the power of STI to be a harness for development.[69]

Within this context, the indicator of Goal 14.a needs to be expanded and should not be merely about the budget allocated to marine research and technology. Rather, it should be equipped with more indicators that could demonstrate the role of marine scientific research in promoting the contribution of marine biodiversity to the economy of developing countries, least developed countries, and small island developing States.

65 *Transforming Our World*, *supra* note 13, para. 75.

66 *Ibid.*

67 *Final List of Proposed Sustainable Development Goals Indicators. Report of the Inter-Agency and Expert Group on Sustainable Development Goal Indicators*, UN Doc E/CN.3/2016/2/Rev.1 (19 February 2016), Annex IV. See also *Revised List of Global Sustainable Development Goal Indicators*, UN Doc E/CN.3/1017/2 <https://unstats.un.org/sdgs/indicators/Official%20Revised%20List%20of%20global%20SDG%20Indicators.pdf>.

68 *Id.* at 19.

69 L. Brito, "The Role of Science, Technology and Innovation Policies and Instruments for a Paradigm Shift towards Sustainable Development", in: Bolay *et al.* (eds.), *Technologies for Sustainable Development* (Switzerland, Springer International Publishing, 2014), p. 13.

In addition to Goal 14.a, Goal 14.c clearly reflects the importance of the conservation and sustainable use of oceans as well as their resources. It states:

> Enhance the conservation and sustainable use of oceans and their resources by implementing international law as reflected in the United Nations Convention on the Law of the Sea, which provides the legal framework for the conservation and sustainable use of oceans and their resources, as recalled in paragraph 158 of "The Future We Want".

It could be observed from this paragraph that UNCLOS has been regarded as the main international instrument that should be taken into account in achieving Goal 14.

Despite these positive relationships between marine scientific research and technology, marine biodiversity, and developing countries under Goal 14 of the Sustainable Development Goals, four questions at least remain unresolved. First, it is still not clear whether the terms "scientific knowledge", "research capacity", and "technology transfer" under Goal 14.a also include marine biodiversity research or marine bioprospecting activities. Second, Goal 14.a specifies the IOC Guidelines on marine technology transfer; however, it remains unclear whether this document is relevant to marine biodiversity research or marine bioprospecting activities.

Third, Goal 14.a aims "to enhance the contribution of marine biodiversity to the development of developing countries, in particular small island developing States and least developed countries", and it begs a question of whether sharing of benefits from the utilization of marine genetic resources could be included. Finally, Goal 14.c acknowledges UNCLOS as the primary instrument that should be implemented in the conservation and sustainable use of marine biodiversity; however, as discussed earlier, UNCLOS is still silent on marine genetic resources and marine bioprospecting, especially in areas beyond national jurisdiction.

4 The Perspective of Developing Countries

4.1 *The Development of an International Legally Binding Instrument*

It could be concluded from the above discussions that there is indeed a clear linkage between marine genetic resources, marine scientific research and technology, and sustainable development. Nevertheless, current international law has not yet provided a comprehensive instrument to address this linkage. In addition, the legal status of marine genetic resources beyond national jurisdiction remains unclear and developing countries are generally concerned about this situation as it may lead to their disadvantage.

In this regard, the United Nations has initiated negotiations on an international legally binding instrument (ILBI) under UNCLOS on the conservation and sustainable use of marine biodiversity of areas beyond national jurisdiction.[70] In particular, the negotiations address the following topics: marine genetic resources, including questions on the sharing of benefits; measures such as area-based management tools, including marine protected areas; environmental impact assessments; and capacity-building and the transfer of marine technology.[71] Prior to holding an intergovernmental conference to negotiate these issues, a Preparatory Committee was established to make substantive recommendations on the elements of a draft text of the international legally binding instrument.[72] The Committee was assigned to start its work in 2016 and to report to the United Nations General Assembly by the end of 2017.[73]

At the time of writing, the Committee has concluded four meetings, which were held between March 2016 and July 2017. Throughout these meetings, the developing countries have expressed their concerns with respect to the utilization of marine biodiversity, including marine genetic resources, beyond areas of national jurisdiction. For instance, the G77 and China proposed the following preambular element:

> Desiring by this new instrument to develop an effective regime of conservation and sustainable use of marine biological diversity of areas beyond national jurisdiction, including through a fair and equitable regime of access to and sharing of benefits of marine genetic resources. The principle of common heritage of humankind will contribute to the realization of the said objective as, through the application of this principle, the interests and needs of humankind as a whole, especially those of developing countries will be fairly addressed and taken care of.[74]

70 *Development of an International Legally Binding Instrument under the United Nations Convention on the Law of the Sea on the Conservation and Sustainable Use of Marine Biological Diversity of Areas Beyond National Jurisdiction*, UN Doc GA Res. 69/292 Agenda item 74 (*a*) (6 July 2015).

71 *Ibid.*

72 *Ibid.*

73 *Ibid.*

74 "Chair's Non-paper on Elements of a Draft Text of an International Legally-binding Instrument under the United Nations Convention on the Law of the Sea on the Conservation and Sustainable Use of Marine Biological Diversity of Areas beyond National Jurisdiction" (28 February 2017), available at <http://www.un.org/depts/los/biodiversity/prepcom_files/Chair_non_paper.pdf>. See also International Institute of Sustainable

This proposal clearly suggests the significance of legal issues arising from the utilization of marine genetic resources beyond national jurisdiction. The G77 and China's position is also supported by the Alliance of Small Island States (AOSIS) and the Caribbean Community (CARICOM).

Furthermore, G77 and China, supported by CARICOM and Jamaica, also proposed the following guiding principles with respect to the utilization of marine genetic resources beyond national jurisdiction:[75]

- The principle of common heritage of mankind must underpin the new regime governing marine genetic resources of areas beyond national jurisdiction;
- No claim or exercise of sovereignty or sovereign rights;
- Exclusively for peaceful purposes;
- Equitable sharing of benefits; and
- Governed by an international regime.

On more technical issues, especially with respect to access to marine genetic resources beyond national jurisdiction and sharing of the benefits arising from their utilization, some recommendations that reflect the interests of developing countries include:[76]

- Developing the models based on UNCLOS, CBD/Nagoya, International Treaty on Plant Genetic Resources for Food and Agriculture (ITPGRFA),[77] and the Antarctic Treaty System;[78]
- Taking into account the needs and interests regarding marine scientific research as well as the development opportunities of the developing countries, including future generations;
- Due consideration accorded to Small Island Developing States (SIDS); and
- Transfer of technology and strengthening the research capabilities of developing countries.

Development, "Summary of the Third Session of the Preparatory Committee on Marine Biodiversity Beyond Areas of National Jurisdiction 27 March–7 April 2017" available at <http://enb.iisd.org/download/pdf/enb25129e.pdf>.

75 *Ibid.*

76 *Ibid.*

77 *International Treaty on Plant Genetic Resources for Food and Agriculture*, opened for signature 3 November 2001, 2400 UNTS (entered into force 29 June 2004).

78 The Antarctic Treaty System includes: *The Antarctic Treaty*, opened for signature 1 December 1959, 402 UNTS 71 (entered into force 23 June 1961); *Protocol on Environmental Protection to the Antarctic Treaty*, opened for signature 4 October 1991, 30 ILM (1991) (entered into force 14 January 1998); *Convention for the Conservation of Antarctic Seals*, opened for signature 1 June 1972, 1080 UNTS 175 (entered into force 11 March 1978); *Convention on the Conservation of Antarctic Marine Living Resources*, opened for signature 1 August 1980, 1329 UNTS 47 (entered into force 7 April 1982).

Through its fourth meeting, the Preparatory Committee delivered its recommendations to the General Assembly containing the elements that could be considered in drafting the text of the ILBI. With respect to marine genetic resources, the proposed elements include:[79]

- Scope;
- Access and benefit-sharing, including objectives, principles and approaches, and benefit-sharing modalities; and
- Intellectual property rights.

In addition, the recommendations include proposed elements on capacity-building and transfer of marine technology, which are considerably relevant to developing countries in terms of the utilization of marine genetic resources. Under the paragraph on "Types of and Modalities for Capacity-building and Transfer of Marine Technology", the recommendations note:

> The text would elaborate on forms of cooperation and assistance in relation to marine genetic resources, including questions on the sharing of benefits, measures such as area-based management tools, including marine protected areas, and environmental impact assessments.[80]

Despite the fact that these recommendations have represented the interests of developing countries, especially with respect to marine genetic resources beyond national jurisdiction, some critical issues remain unresolved and still need further discussions. For instance, it has been suggested that:

> With regard to marine genetic resources, including the question of the sharing of benefits, further discussions are required on whether the instrument should regulate access to marine genetic resources; the nature of these resources; what benefits should be shared; whether to address intellectual property rights; and whether to provide for the monitoring of the utilization of marine genetic resources of areas beyond national jurisdiction.[81]

---

79 *Report of the Preparatory Committee Established by General Assembly Resolution 69/292: Development of an Internationally Legally Binding Instrument under the United Nations Convention on the Law of the Sea on the Conservation and Sustainable Use of Marine Biological Diversity of Areas beyond National Jurisdiction* (*Advance, Unedited Version*), available at <http://www.un.org/depts/los/biodiversity/prepcom_files/Procedural_report_of_BBNJ_PrepCom.pdf>.

80 *Ibid.*

81 *Ibid.*

These outstanding issues would surely pose some challenges in developing a specific international legal framework for the utilization of marine genetic resources beyond national jurisdiction as well as in ensuring its contributions to the achievement of Sustainable Development Goals, particularly Goal 14.

4.2 *The Way Forward*

UNCLOS, the CBD and the Nagoya Protocol are currently the main international instruments relevant to the marine scientific research and technology transfer, particularly in relation to marine biodiversity research and marine bioprospecting. Their provisions could serve as the international legal framework that could support the Sustainable Development Goals on oceans. However, these instruments are still rather fragmented and have resulted in a number of legal lacunae that may be counterproductive, especially to the implementation of Goal 14.a. This article has presented several issues that may arise from the utilization of marine genetic resources beyond national jurisdiction, and, arguably, they are the most complex legal hurdles that must be addressed by developing countries.

In this regard, the development of the ILBI under UNCLOS for marine biodiversity beyond national jurisdiction is undeniably crucial. UNCLOS should reflect an integrated approach in bringing together all the elements of sustainable development in the context of marine biodiversity, since it has already acknowledged that "the problems of ocean space are closely interrelated and need to be considered as a whole".[82] In fact, "integration" is also one of the most important principles of international law relating to sustainable development.[83] In the previous section, this article has described the role that marine biodiversity research could play in sustainable development. However, the lack of clarity on the legal aspects of marine biodiversity research, especially marine bioprospecting, may impede such an important role.

Although the current provisions on marine scientific research and technology transfer under UNCLOS have been complemented by the guidelines on the implementation of marine scientific research and technology transfer,[84]

---

82 UNCLOS, Preamble. See also R. Barnes, "The Law of the Sea Convention and the Integrated Regulation of the Oceans", 27 *International Journal of Marine and Coastal Law*, 2012, 859–866.

83 International Law Association New Delhi Conference (2002), *Legal Aspects of Sustainable Development*, available at <https://ila.vettoreweb.com/Storage/Download.aspx?DbStorageId=1200&StorageFileGuid=f646a782-aa62-486c-b615-8590ffa7e2db>.

84 United Nations Division for Ocean Affairs and the Law of the Sea Office of Legal Affairs, *Marine Scientific Research: A Revised Guide to the Implementation of the Relevant Provisions of the United Nations Convention on the Law of the Sea* (New York, United Nations, 2010).

they are mainly procedural and might not have a substantial effect on the regulation of marine bioprospecting activities, especially those undertaken beyond national jurisdiction. The IOC Guidelines, however, have provided a framework that is notably relevant to the implementation of Goal 14.a, particularly in advocating marine scientific research and technology transfer that could enhance the contribution of marine biodiversity to the development of developing countries. Therefore, there is a need for the IOC to play a more active role in the implementation of these guidelines in light of the Sustainable Development Goals.[85]

The ABS mechanism under the CBD and the Nagoya Protocol could be utilized as a proper instrument to create equality between the users and the providers of genetic resources. Through this mechanism, the stakeholders in genetic resource utilization could secure their interests and ensure that such utilization would contribute to the conservation of biological diversity. However, when it comes to the utilization of marine genetic resources beyond national jurisdiction, the CBD and the Nagoya Protocol provisions are difficult to implement since their scopes are limited within national jurisdiction. Consequently, the ABS mechanism so far could only be applied to the utilization of marine genetic resources within national jurisdiction.

Nonetheless, taking into account all of the environmental threats to marine biodiversity, it is essential that the ABS mechanism in marine bioprospecting activities contribute to tackling such threats by creating incentives for users and providers of marine genetic resources to conserve marine biodiversity. This has actually been the ultimate objective of the CBD. However, it has been observed that the three objectives of the CBD are still not appropriately linked. Dutfield, for instance, has noted that "the text of the CBD is unbalanced, in the sense that so much of it deals with matters that are not directly related to conservation of biological diversity".[86]

In this context, numerous scholars have proposed arguments that support the ABS mechanism as a means to create an incentive for biodiversity conservation and, ultimately, for achieving the goals of sustainable development. For example, ten Kate and Laird argued that:

85 For further discussions on the role of the IOC, especially with respect to biodiversity beyond national jurisdiction, see H. Harden-Davies, "Marine Science and Technology Transfer: Can the Intergovernmental Oceanographic Commission Advance Governance of Biodiversity beyond National Jurisdiction?" 74 *Marine Policy*, 2016, 260–267.

86 G. Dutfield, "Sharing the Benefits of Biodiversity: Is There a Role for the Patent System?" 5 *Journal of World Intellectual Property*, No. 6, 2002, 900.

> Sustainable economic activities based on biodiversity can also serve to boost sustainable development by providing an incentive for conservation and an alternative to more destructive income-generating schemes. …The provision of value-added derivatives of genetic resources, such as extracts for screening, or of pre-bred materials for crop development, or the processing of material into finished products for local, regional or international markets also provides jobs.[87]

As also suggested by Richerzhagen:

> ABS does not take effect only in one dimension. Besides its impact on conservation and benefit-sharing, ABS can have a large-scale impact on provider countries' economic and social situation. It has the potential to alleviate poverty and support economic development. For example, ABS can assist countries in their development goals if it attracts foreign direct investment, drives the establishment of new industries, and supports the improvement of research potential. Therefore, ABS needs to be seen in a much broader sense of sustainable development and cannot be reduced to only environmental objectives.[88]

These views demonstrate that there is a clear linkage between the ABS mechanism, conservation of biological diversity and sustainable development. In this light, the ABS regulation should not be merely placed in the context of exercising a sovereign right over natural resources and exploring the wealth from the utilization of biodiversity. Rather, it should be placed within a broader context, which is the conservation and sustainable use of biodiversity, and this view should also be extended to the utilization of marine genetic resources beyond national jurisdiction.

Lastly, the elaboration and indicators of the Sustainable Development Goals should be further developed as they need to articulate stronger commitments in implementing the principles of international law relating to sustainable development.[89] This is particularly true with respect to Goal 14.a as, currently, there is only one indicator to measure this particularly important goal. In addition, the experience of countries regarding the implementation of

87 K. ten Kate and S.A. Laird, "Biodiversity and Business: Coming to Terms with the 'Grand Bargain'", 76 *International Affairs*, No. 2, 2000, 262.

88 C. Richerzhagen, "Effective Governance of Access and Benefit-sharing under the Convention on Biological Diversity", 20 *Biodiversity and Conservation*, No. 10, 2011, 2259.

89 See also T. Pogge and M. Sengupta, "The Sustainable Development Goals (SDGs) as Drafted: Nice Idea, Poor Execution", 24 *Washington International Law Journal Association*, 2015, 571–587.

Goal 14 is still limited, particularly the developing countries, least developed countries, and small island developing States.[90]

In this regard, a high-level United Nations Conference to Support the Implementation of Sustainable Development Goal 14 was held in June 2017 and adopted a resolution titled "Our Ocean, Our Future: Call for Action".[91] The resolution reiterates the commitment in achieving Goal 14. In particular, it stresses the importance of marine scientific research as a part of conservation and sustainable use of the oceans, seas, and marine resources. It also confirms a commitment to:

> [D]edicate greater resources to marine scientific research, such as interdisciplinary research and sustained ocean and coastal observation, as well as the collection and sharing of data and knowledge, including traditional knowledge, in order to increase our knowledge of the ocean, to better understand the relationship between climate and the health productivity of the ocean, to strengthen the development of coordinated early warning systems on extreme weather events and phenomena, and to promote decision-making based on the best available science, to encourage scientific and technological innovation, as well as to enhance the contribution of marine biodiversity to the development of developing countries, in particular small island developing States and least developed countries.[92]

This paragraph again demonstrates that marine scientific research, marine biodiversity, and sustainable development are closely interlinked. Nevertheless, the existing void in international legal instruments, especially with respect to the utilization of marine genetic resources beyond national jurisdiction, may pose considerable challenges to such positive relationships. The resolution on "Our Ocean, Our Future: Call for Action" as mentioned above still needs to be complemented by concrete measures to ensure the achievement of Goal 14 through the utilization of marine genetic resources beyond national jurisdiction by providing benefits to developing countries as well as to the conservation and sustainable use of marine biodiversity.

---

90 The European Union is an example of a few countries that have been studied in this context. See W. Rickels, *et al.*, "Indicators for Monitoring Sustainable Development Goals: An Application to Oceanic Development in the European Union", 4 *Earth's Future*, 2016, 252–267.

91 *Our Ocean, Our Future: Call for Action*, UN Doc A/RES/71/312, Seventy-first session, Agenda Items 19 and 73 (*a*) (14 July 2017).

92 *Id.*, para. 13 (*f*).

5 Conclusion

The Sustainable Development Goals could be regarded as a new chapter that brings together the international community in promoting and implementing sustainable development in the context of oceans. It is also an important instrument that brings together marine scientific research, technology transfer, marine biodiversity, and explicitly recognizes their crucial roles in sustainable development. The existing situation, however, reflects that these elements are still heavily fragmented and more concerted efforts are still needed, especially in ensuring their contribution to the development of developing countries, small island developing States, and least developed countries.

This article has attempted to present the inadequacy of the current international legal regime with respect to marine genetic resources beyond national jurisdiction. There is still a need to clarify the scope of marine scientific research under UNCLOS and to amend the IOC Guidelines so that these documents can be applied to marine bioprospecting activities. The CBD and Nagoya Protocol can assist in this regard but only in cases where the utilization of marine genetic resources is conducted within national jurisdiction.

Ultimately, the implementation of the Sustainable Development Goals, including Goal 14, can only work if they are supported by stronger legal implementations provided under relevant international instruments. Otherwise, these goals will remain utopian. International law has a significant role to play in bringing together the elements of sustainable development in the context of marine scientific research and technology transfer, especially in the field of marine biodiversity. As Jordan and O'Riordan have noted, strong sustainability can only be supported by binding policy integration and strong international agreements.[93]

Bibliography

Books and Articles

Arico, S. and C. Salpin, *Bioprospecting of Genetic Resources in the Deep Seabed: Scientific, Legal and Policy Aspects*, Japan: United Nations University-Institute of Advanced Studies, 2005.

Barnes, R., "The Law of the Sea Convention and the Integrated Regulation of the Oceans", *International Journal of Marine and Coastal Law* 27 (2012), 859–866.

93 A. Jordan and T. O'Riordan, "Sustainable Development: The Political and Institutional Challenge", in: Kirkby, O'Keefe, and Timberlake (eds.), *The Earthscan Reader in Sustainable Development* (London, Earthscan, 1995), pp. 287–289.

Birnie, P., "Law of the Sea and Ocean Resources: Implications for Marine Scientific Research", *International Journal of Marine and Coastal Law* 10 (1995), 242.

Brito, L., "The Role of Science, Technology and Innovation Policies and Instruments for a Paradigm Shift towards Sustainable Development", in: Bolay et al. (eds.), *Technologies for Sustainable Development*, Switzerland: Springer International Publishing, 2014.

Broggiato, A., "Marine Genetic Resources beyond National Jurisdiction – Coordination and Harmonization of Governance Regimes", *Environmental Policy and Law* 41 (2011), 35–42.

Buck, M. and C. Hamilton, "The Nagoya Protocol on Access to Genetic Resources and the Fair and Equitable Sharing of Benefits Arising from Their Utilization to the Convention on Biological Diversity", *Review of European, Comparative and International Environmental Law* 20, No. 1 (2011), 47–61.

Calle, F. de la, "Marine Genetic Resources: A Source of New Drugs – The Experience of the Biotechnology Sector", *International Journal of Marine and Coastal Law* 24 (2009), 209–220.

Campbell, J., E. Whittingham, and P. Townsley, "Responding to Coastal Poverty: Should We be Doing Things Differently or Doing Different Things?" in: Tuong Hoanh, Gowing, and Hardy (eds.), *Environment and Livelihoods in Tropical Coastal Zones: Managing Agriculture-Fishery-Aquaculture Conflicts*, Wallingford: CABI International, 2006.

Chasek, P.S. et al., "Getting to 2030: Negotiating the Post-2015 Sustainable Development Agenda", *Review of European Community and International Environmental Law* 25, No. 1 (2016), 5–14.

Dutfield, G., "Sharing the Benefits of Biodiversity: Is There a Role for the Patent System?" *Journal of World Intellectual Property* 5, No. 6 (2002), 900.

Ehlers, P., "The Intergovernmental Oceanographic Commission: An International Organization for the Promotion of Marine Research", *International Journal of Marine and Coastal Law* 15 (2000), 533–554.

Erwin, P.M., S. López-Legentil and P.W. Schuhmann, "The Pharmaceutical Value of Marine Biodiversity for Anti-cancer Drug Discovery", *Ecological Economics* 70, No. 2 (2010), 445–451.

Farrier, D. and L. Tucker, "Access to Marine Bioresources: Hitching the Conservation Cart to the Bioprospecting Horse", *Ocean Development and International Law* 32 (2001), 214.

Gjerde, K.M. *et al.*, *Options for Addressing Regulatory and Governance Gaps in the International Regime for the Conservation and Sustainable Use of Marine Biodiversity in Areas beyond National Jurisdiction*, Gland: IUCN, 2008.

Glowka, L., "Evolving Perspectives on the International Seabed Area's Genetic Resources: Fifteen Years after the 'Deepest of Ironies'", in: Vidas (ed.), *Law, Technology and Science for Oceans in Globalisation: IUU Fishing, Oil Pollution, Bioprospecting, Outer Continental Shelf*, Leiden: Martinus Nijhoff Publishers, 2010.

Glowka, L., "Genetic Resources, Marine Scientific Research and the International Seabed Area", *Review of European Community and International Environmental Law* 8, No. 1 (1999), 56.

Glowka, L., "The Deepest of Ironies: Genetic Resources, Marine Scientific Research, and the Area", *Ocean Yearbook* 12 (1996), 168.

Griggs, D. et al., "Sustainable Development Goals for People and Planet", *Nature* 495, No. 7741 (2013), 305–307.

Harden-Davies, H., "Marine Science and Technology Transfer: Can the Intergovernmental Oceanographic Commission Advance Governance of Biodiversity beyond National Jurisdiction?" *Marine Policy* 74 (2016a), 260–267.

Harden-Davies, H., "The Regulation of Marine Scientific Research: Addressing Challenges, Advancing Knowledge", in: Warner and Kaye (eds.), *Routledge Handbook of Maritime Regulation and Enforcement*, London and New York: Routledge, 2016b.

Hunt, B. and A.C.J. Vincent, "Scale and Sustainability of Marine Bioprospecting for Pharmaceuticals", *Ambio: A Journal of the Human Environment* 35, No. 2 (2006), 57.

IOC/UNESCO, IMO, FAO, UNDP, *A Blueprint for Ocean and Coastal Sustainability*, Paris: IOC/UNESCO, 2011.

Jordan, A. and T. O'Riordan, "Sustainable Development: The Political and Institutional Challenge", in: Kirkby, O'Keefe, and Timberlake (eds.), *The Earthscan Reader in Sustainable Development*, London: Earthscan, 1995.

Kate, K. ten and S.A. Laird, "Biodiversity and Business: Coming to Terms with the 'Grand Bargain'", *International Affairs* 76, No. 2 (2000), 262.

Leal, M.C. *et al.*, "Trends in the Discovery of New Marine Natural Products from Invertebrates over the Last Two Decades – Where and What are We Bioprospecting?" *PLoS ONE* e30580 7, No.1 (2012).

Leary, D. *et al.*, "Marine Genetic Resources: A Review of Scientific and Commercial Interest", *Marine Policy* 33, No. 2 (2009), 183–194.

Leary, D., "Moving the Marine Genetic Resources Debate Forward: Some Reflections", *International Journal of Marine and Coastal Law* 27, No. 2 (2012), 435–448.

Lodge, M.W., "Improving International Governance in the Deep Sea", *International Journal of Marine and Coastal Law* 19, No. 3 (2004), 299.

Matz, N., "The Interaction between the Convention on Biological Diversity and the UN Convention on the Law of the Sea", in: Ehlers, Borgese, Wolfrum, and Hoszlig (eds.), *Marine Issues: From a Scientific, Political, and Legal Perspective*, Leiden: Martinus Nijhoff Publishers, 2002.

Mauerhofer, V. and F. Nyacuru, "Biodiversity, Migratory Species and Natural Heritage", in: Harris (ed.), *Routledge Handbook of Global Environmental Politics*, London and New York: Routledge, 2014.

Molinski, T.F. et al., "Drug Development from Marine Natural Products", *Nature Reviews Drug Discovery* 8, No. 1 (2009), 69–85.

Montgomery, J. and L. Carter, “Marine Science in the Past 25 Years: Main Findings and Trends”, in: Vidas and Schei (eds.), *The World Ocean in Globalisation: Climate Change, Sustainable Fisheries, Biodiversity, Shipping, Regional Issues*, Leiden and Boston: Martinus Nijhoff Publishers, 2011.

Morgera, E., E. Tsioumani, and M. Buck, *Unraveling the Nagoya Protocol: A Commentary on the Nagoya Protocol on Access and Benefit-sharing to the Convention on Biological Diversity*, Leiden: Brill, 2014.

Ni, K., “The Incorporation of the CBD Mandate on Access and Benefit-sharing Into TRIPS Regime: An Appraisal of the Appeal of Developing Countries with Rich Genetic Resources”, *Asian Journal of WTO and International Health Law and Policy* 1 (2006), 434.

Pogge, T. and M. Sengupta, “The Sustainable Development Goals (SDGs) as Drafted: Nice Idea, Poor Execution”, *Washington International Law Journal Association* 24 (2015), 571–587.

Richerzhagen, C., “Effective Governance of Access and Benefit-sharing under the Convention on Biological Diversity”, *Biodiversity and Conservation* 20, No. 10 (2011), 2259.

Richerzhagen, C., “The Nagoya Protocol: Fragmentation or Consolidation”, *Resources* 3, No. 1 (2014), 135–151.

Rickels, W., et al., “Indicators for Monitoring Sustainable Development Goals: An Application to Oceanic Development in the European Union”, *Earth's Future* 4 (2016), 252–267.

Rothwell, D.R. and T. Stephens, *The International Law of the Sea*, Oxford: Hart Publishing, 2010.

Scheiber, H.N., “Economic Uses of the Oceans and the Impacts on Marine Environments: Past Trends and Challenges Ahead”, in: Vidas and Schei (eds.), *The World Ocean in Globalisation: Climate Change, Sustainable Fisheries, Biodiversity, Shipping, Regional Issues*, Leiden and Boston: Martinus Nijhoff Publishers, 2011.

Scovazzi, T., “Is the UN Convention on the Law of the Sea the Legal Framework for All Activities in the Sea? The Case of Bioprospecting”, in: Vidas (ed.), *Law, Technology and Science for Oceans in Globalisation: IUU Fishing, Oil Pollution, Bioprospecting, Outer Continental Shelf*, Leiden: Martinus Nijhoff Publishers, 2010.

Segger, M.C., “Sustainable Development in International Law”, in: Bugge and Voight (eds.), *Sustainable Development in International and National Law*, Groningen: Europa Law Publishing, 2008.

Siswandi, A.G., “The Nagoya Protocol: Unfinished Business Remains Unfinished”, in: Rimmer (ed.), *Indigenous Intellectual Property: A Handbook of Contemporary Research*, Cheltenham: Edward Elgar, 2015.

Siswandi, A.G., *Marine Bioprospecting: International Law, Indonesia, and Sustainable Development*, Ph.D. Thesis, The Australian National University, 2013.

Soons, A., "Marine Scientific Research Provisions in the Convention on the Law of the Sea: Issues of Interpretation", in: Brown and Churchill (eds.), *The UN Convention on the Law of the Sea: Impact and Implementation*, Honolulu: The Law of the Sea Institute-William S. Richardson School of Law, 1989.

Soons, A.H.A., *Marine Scientific Research and the Law of the Sea*, The Hague: T.M.C. Asser Instituut, 1982.

Stasi, A.D., "The Normative Force of the Outcome Document 'The Future We Want'", in: Maljean-Dubois Fitzmaurice, and Negri (eds.), *Environmental Protection and Sustainable Development from Rio to Rio+20*, Leiden and Boston: Brill Nijhoff, 2014.

Stojanovic, T.A. and C.J.Q. Farmer, "The Development of World Oceans and Coasts and Concepts of Sustainability", *Marine Policy* 42 (2013), 157–165.

International Documents

Convention for the Conservation of Antarctic Seals, opened for signature 1 June 1972, 1080 UNTS 175 (entered into force 11 March 1978).

Convention on Biological Diversity, opened for signature 5 June 1992, 1760 UNTS 79 (entered into force 29 December 1993).

Convention on the Conservation of Antarctic Marine Living Resources, opened for signature 1 August 1980, 1329 UNTS 47 (entered into force 7 April 1982).

Development of an International Legally Binding Instrument under the United Nations Convention on the Law of the Sea on the Conservation and Sustainable Use of Marine Biological Diversity of Areas Beyond National Jurisdiction, UN Doc GA Res. 69/292 Agenda item 74 (*a*) (6 July 2015).

Final List of Proposed Sustainable Development Goals Indicators. Report of the Inter-Agency and Expert Group on Sustainable Development Goal Indicators, UN Doc E/CN.3/2016/2/Rev.1 (19 February 2016), Annex IV.

International Treaty on Plant Genetic Resources for Food and Agriculture, opened for signature 3 November 2001, 2400 UNTS (entered into force 29 June 2004).

Nagoya Protocol on Access to Genetic Resources and the Fair and Equitable Sharing of Benefits Arising from their Utilization to the Convention on Biological Diversity, opened for signature 2 February 2011, UNEP/CBD/COP/DEC/X/1 of 29 October 2010 (entered into force 12 October 2014).

Oceans and the Law of the Sea, Report of the Secretary-General, Addendum, 60th session, Agenda Item 76 (a), UN Doc A/60/63/Add.1 (15 July 2005).

Our Ocean, Our Future: Call for Action, UN Doc A/RES/71/312, Seventy-first session, Agenda Items 19 and 73 (*a*) (14 July 2017).

Outcome Document of the Special Event to Follow Up Efforts Made Towards Achieving the Millennium Development Goals, UN Doc A/RES/68/6 (28 January 2014).

Protocol on Environmental Protection to the Antarctic Treaty, opened for signature 4 October 1991, 30 ILM (1991) (entered into force 14 January 1998).

Report of the Open Working Group of the General Assembly on Sustainable Development Goals, UN Doc A/68/970 (12 August 2014).

The Antarctic Treaty, opened for signature 1 December 1959, 402 UNTS 71 (entered into force 23 June 1961).

The Future We Want, UN Doc A/RES/66/288 (11 September 2012).

Transforming Our World: The 2030 Agenda for Sustainable Development, UN Doc A/RES/70/1 (21 October 2015).

United Nations Convention on the Law of the Sea, opened for signature 10 December 1982, 1833 UNTS 3 (entered into force 16 November 1994).

United Nations Division for Ocean Affairs and the Law of the Sea Office of Legal Affairs, *Marine Scientific Research: A Revised Guide to the Implementation of the Relevant Provisions of the United Nations Convention on the Law of the Sea*, New York: United Nations, 2010.

Internet Materials

"Chair's Non-paper on Elements of a Draft Text of an International Legally-binding Instrument under the United Nations Convention on the Law of the Sea on the Conservation and Sustainable Use of Marine Biological Diversity of Areas beyond National Jurisdiction" (28 February 2017), available at <http://www.un.org/depts/los/biodiversity/prepcom_files/Chair_non_paper.pdf>.

"Growing International Interest in Oceanology to Benefit the Global Marine Biotechnology Market, According to New Report by Global Industry Analysts, Inc.", available at <http://www.prweb.com/releases/marine_biotechnology/marine_biomaterials/prweb9420116.htm>.

Arico, S., "Deep Sea: The Last Frontier", available at <http://www.unesco.org/new/en/natural-sciences/ioc-oceans/single-view-oceans/news/the_last_frontier/>.

Ausubel, J.H., D.T. Crist, and P.E. Waggoner, "First Census of Marine Life 2010: Highlights of a Decade of Discovery", available at <http://www.coml.org/Highlights-2010>.

Intergovernmental Oceanographic Commission, "Resources, Meetings, Documents, People", available at <http://www.ioc-unesco.org/>.

International Institute of Sustainable Development, "Summary of the Eighth Session of the UN General Assembly Open Working Group on Sustainable Development Goals: 3–7 February 2014", available at <http://www.iisd.ca/sdgs/owg8>.

International Institute of Sustainable Development, "Summary of the Tenth session of the UN General Assembly Open Working Group on Sustainable Development Goals: 31 March–4 April 2014", available at <http://www.iisd.ca/sdgs/owg10>.

International Institute of Sustainable Development, "Summary of the Eighth Session of the UN General Assembly Open Working Group on Sustainable Development Goals: 3–7 February 2014", available at <http://www.iisd.ca/sdgs/owg8>.

International Institute of Sustainable Development, "Summary of the Third Session of the Preparatory Committee on Marine Biodiversity Beyond Areas of National

Jurisdiction 27 March–7 April 2017" available at <http://enb.iisd.org/download/pdf/enb25129e.pdf>.

International Law Association New Delhi Conference (2002), *Legal Aspects of Sustainable Development*, available at <https://ila.vettoreweb.com/Storage/Download.aspx?DbStorageId=1200&StorageFileGuid=f646a782-aa62-486c-b615-8590ffa7e2db>.

International Oceanographic Commission, "IOC Criteria and Guidelines on the Transfer of Marine Technology", available at <http://unesdoc.unesco.org/images/0013/001391/139193m.pdf>.

Koh, T., "A Constitution for the Oceans", available at <http://www.un.org/depts/los/convention_agreements/texts/koh_english.pdf>.

Report of the Preparatory Committee Established by General Assembly Resolution 69/292: Development of an Internationally Legally Binding Instrument under the United Nations Convention on the Law of the Sea on the Conservation and Sustainable Use of Marine Biological Diversity of Areas beyond National Jurisdiction (Advance, Unedited Version), available at <http://www.un.org/depts/los/biodiversity/prepcom_files/Procedural_report_of_BBNJ_PrepCom.pdf>

Revised List of Global Sustainable Development Goal Indicators, UN Doc E/CN.3/1017/2 <https://unstats.un.org/sdgs/indicators/Official%20Revised%20List%20of%20global%20SDG%20indicators.pdf>.

United Nations Development Group, "A Million Voices: The World We Want-A Sustainable Future with Dignity for All", available at <http://www.undp.org/content/undp/en/home/librarypage/mdg/a-million-voices--the-world-we-want.html>.

United Nations, "Goal 14: Conserve and Sustainably Use the Oceans, Seas and Marine Resources", available at <http://www.un.org/sustainabledevelopment/oceans/>.

United Nations, "The Sustainable Development Goals Report 2016", available at <http://unstats.un.org/sdgs/report/2016/>.

United Nations, "The Sustainable Development Goals Report 2016", available at <http://unstats.un.org/sdgs/report/2016/>.

PART 3

Status of Deep Seabed Minerals

CHAPTER 10

Status of Deep Seabed Minerals: Introductory Remarks

*Alfonso Ascencio-Herrera**

On behalf of the Secretary-General of the International Seabed Authority ("the Authority"), Mr. Michael Lodge, I wish to express our deep gratitude to the organizers of the conference: the Coordinating Ministry for Maritime Affairs of the Republic of Indonesia and the Center for Oceans Law and Policy,** for the kind invitation to the Authority to moderate this panel and for acknowledging the importance of the deep seabed mineral resources in the context of Sustainable Development Goal 14.[1]

The presentations focus on the mineral resources of the international seabed area, known as the "Area"; i.e. the seabed and ocean floor and subsoil thereof, beyond the limits of national jurisdiction.[2] According to Part XI of the 1982 Convention on the Law of the Sea, the use of the Area shall be exclusively for peaceful purposes,[3] and, significantly, it also provides that activities in the Area are to be carried out for the benefit of mankind as a whole, taking into particular consideration the interests and needs of developing countries,[4] which include LDCs, SIDS and land-locked States. The International Seabed Authority is an autonomous organization through which its members shall organize and control the mineral exploration and exploitation activities in the deep ocean floor beyond national jurisdiction.[5] The Authority is a 23-year-old organization and is one of the three institutions created by the Convention. The other two entities are the International Tribunal for the Law of the Sea

* Legal Counsel and Deputy to the Secretary-General of the International Seabed Authority.

** My gratitude also goes to Ms. Michelle Bond and Ms. Katarina Hovden for their research.

1 UNGA Res. 70/1 Transforming our world: the 2030 Agenda for Sustainable Development (21 October 2015) UN Doc A/Res/70/1, p. 14.

2 See Article 1(1) of the United Nations Convention on the Law of the Sea (adopted 10 December 1982, entered into force 16 November 1994) 1833 UNTS 3 ("the Convention" or "LOSC").

3 See LOSC, Article 141.

4 See LOSC, Article 140; and Michael W. Lodge, "The Deep Seabed" in: Rothwell, Donald R, Oude Elferink, Alex G., Scott, Karen N. and Stephens, Tim (eds.), *Oxford Handbook of the Law of the Sea,* Oxford University Press, 2015, p. 230.

5 See LOSC, Article 157(1); and Michael W Lodge, op. cit. at p. 231.

 | DOI:10.1163/9789004366619_015

and the Commission on the Limits of the Continental Shelf.[6] At present, the Authority has 168 members, which also include the European Union.[7] Moreover, as of today, the Authority has approved 29 plans of work for exploration covering more than 1.3 million square kilometres[8] of the seabed. Contractors include States, State entities and private corporations sponsored both by developed and developing States.[9]

The mission of the International Seabed Authority is a unique one, yet limited in scope. Its mandate is circumscribed to administer the use of the mineral resources of the Area,[10] while at the same time, adopting appropriate regulations to ensure the effective protection of the marine environment from harmful effects that may arise from exploration and exploitation activities in the Area.[11] The Authority also has the role to promote and encourage the conduct of marine scientific research in the Area, and coordinate and disseminate the results of such research, when available.[12] Even in the context of that confined mandate, the combination of those aspects, together with the aspiration of the Convention on the Law of the Sea to promote the economic and social advancement of all peoples of the world,[13] reflects the basic three-pillared architecture of the notion of Sustainable Development.[14]

As a consequence, the Authority provides a useful platform for countries wishing to develop a blue economy by promoting economic growth, environmental sustainability, social inclusion and the improvement of livelihoods.

Indeed, the Authority has an important role to play in the constellation of Agenda 2030. Its mandate has the potential of contributing towards the fulfillment of the various SDG14 associated targets, in particular: 14.2, 14.5, 14.7, 14(a) and 14(c).

With regard to target 14.2,[15] the Authority's task relating to the marine environment is to set the regulatory conditions under which deep sea mining

6 See LOSC, Annex II and Annex VI.

7 Member States <www.isa.org.jm/member-states> accessed 30 September 2017.

8 See <www.isa.org.jm/sites/default/files/documents/EN/SG-Stats/iho-2017.pdf>.

9 Deep Seabed Minerals Contractors (International Seabed Authority) <www.isa.org.jm/deep-seabed-minerals-contractors> Accessed 30 September 2017.

10 See LOSC, Article 157(1).

11 See LOSC, Article 145.

12 See LOSC Article 143(2); and Lodge, Michael W., "Implementation of the Common Heritage of Mankind" in: Scheiber, Harry N. and Paik, Jin-Hyun (eds.), *Regions, Institutions and Law of the Sea*, Leiden/Boston: Martinus Nijhoff Publishers, 2013, pp. 140–141.

13 LOSC, Preamble.

14 See *supra* note 1, at p. 3.

15 "By 2020, sustainably manage and protect marine and coastal ecosystems to avoid significant adverse impacts, including by strengthening their resilience, and take action for

can proceed without causing serious harm to the marine environment.[16] The Exploration Regulations attempt to strike a balance between a precautionary approach to activities in the Area and an incremental approach to regulation, with an emphasis on gathering sufficient data to establish an environmental baseline during the exploration phase, in order to assess the range of potential environmental impacts.[17] It is obvious that the environmental considerations will need to be far more stringent during the exploitation phase.

On target 14.5,[18] it is worth recalling that in 2012 the Authority took an important step with the formal adoption of the "first regional scale environmental management plan for the deep seabed".[19] That year, the Authority's Council adopted the Environmental Management Plan for the *Clarion-Clipperton* Zone.[20] Perhaps, one of the most striking features of the Plan is that it creates a network of nine Areas of Particular Environmental Interest, where mining activities are currently not allowed. This network represents a significant portion of the size of the *Clarion-Clipperton* Zone. In this regional context, the Environmental Management Plan contributes in a significant way to the implementation of that target.

The role and programs of the Authority are also relevant for targets 14.7[21] and 14(a).[22] In this regard, the Authority has the possibility to contribute to the sustainable growth of ocean economies of developing States, in the following three aspects:

their restoration in order to achieve healthy and productive oceans". See *supra* note 1, at p. 23.

16 Lodge, Michael, "Protecting the Marine Environment of the Deep Seabed" in: *Research Handbook on International Marine Environmental Law*, Rayfuse, Rosemary (ed.), Cheltenham, UK, Edward Elgar, 2015, p. 157.

17 Ibid., at p. 158.

18 "By 2020, conserve at least 10 per cent of coastal and marine areas, consistent with national and international law and based on the best available scientific information. See *supra* note 1, at p. 24.

19 Lodge, *supra* note 16, at p. 168.

20 Decision of the Council relating to an environmental management plan for the Clarion-Clipperton Zone (26 July 2012) ISBA/18/C/22; Lodge, *supra* note 16, pp. 165–168.

21 "By 2030, increase the economic benefits to small island developing States and least developed countries from the sustainable use of marine resources, including through sustainable management of fisheries, aquaculture and tourism". See *supra* note 1, at p. 24.

22 "Increase scientific knowledge, develop research capacity and transfer marine technology, taking into account the Intergovernmental Oceanographic Commission Criteria and Guidelines on the Transfer of Marine Technology, in order to improve ocean health and to enhance the contribution of marine biodiversity to the development of developing countries, in particular small island developing States and least developed countries". See *supra* note 1, at p. 24.

First, promoting the participation of developing States in deep seabed mining activities[23] and sharing the economic benefits on the basis of equity.[24] Currently, a number of Small Island Developing States are sponsoring entities that are exploring polymetallic nodules in the Area,[25]

Second, promoting and encouraging the conduct of marine scientific research in the Area and by disseminating the results from such research,[26] and

Third, promoting capacity-building and training opportunities,[27] in cooperation with State entities and industrial contractors carrying out deep seabed exploration, and through the Authority's Endowment Fund for Marine Scientific Research in the Area.[28]

In relation to Target 14(c),[29] the Authority has adopted regulations, governing exploration for all three types of mineral resources—nodules,[30] sulphides,[31] and cobalt crusts[32]—and is currently developing draft exploitation regulations, a part of which will be the regulation of environmental aspects. Together, these instruments will form a complete mining code.[33]

Lastly, it is worth highlighting that the Authority took an active role during the UN Ocean Conference in support of the implementation of SDG 14, which was held from 5 to 9 June in New York City, in the following three ways:

---

23 See LOSC, Article 148.

24 See LOSC, Article 140(2); Lodge, Michael W., "Implementation of the Common Heritage of Mankind" *supra* note 12, at pp. 137–140.

25 Kiribati, Tonga, Nauru and Cuba (through the Interoceanmetal Joint Organization); "Deep Seabed Minerals Contractors" (International Seabed Authority) <www.isa.org.jm/deep-seabed-minerals-contractors> accessed 30 September 2017.

26 See LOSC, Article 143(2); Lodge, *supra* note 24, pp. 140–141.

27 See LOSC, Annex III, Article 15.

28 "Endowment Fund" (International Seabed Authority) <www.isa.org.jm/contractors/endowment-fund> accessed 30 September 2017.

29 "Enhance the conservation and sustainable use of oceans and their resources by implementing international law as reflected in UNCLOS, which provides the legal framework for the conservation and sustainable use of oceans and their resources, as recalled in paragraph 158 of 'The Future We Want'". See *supra* note 1, at p. 24.

30 The Regulations on Prospecting and Exploration for Polymetallic Nodules in the Area (approved 13 July 2000, amended 25 July 2013 and 24 July 2014) ISBA/6/A/18, ISBA/19/A/9, ISBA/19/A/12, ISBA/20/A/9.

31 The Regulations on Prospecting and Exploration for Polymetallic Sulphides in the Area (approved 7 May 2010, amended 25 July 2013 and 24 July 2014) ISBA/16/A/12/Rev.1, ISBA/19/A/12, ISBA/20/A/10.

32 The Regulations on Prospecting and Exploration for Cobalt-rich Ferromanganese Crusts in the Area (approved 27 July 2012, amended 25 July 2013) ISBA/18/A/11, ISBA/19/A/12.

33 Lodge, *supra* note 4, p. 241.

(a) First, organizing a side event with the Governments of Tonga and Mexico, as well as with the United Nations Department for Economic and Social Affairs ("UN-DESA"); and participating as a speaker at various side events, including one on increasing the participation of land-locked countries in the implementation of SDG 14;

(b) Second, participating as one of the panelists in Partnership Dialogue 7 on enhancing the conservation and sustainable use of oceans and their resources, by implementing international law as reflected by the Convention; and,

(c) Third, registering a set of seven voluntary commitments,[34] mainly:

 (i) Enhance the role of women in marine scientific research through targeted capacity-building;

 (ii) Encourage dissemination of research results through the Authority's Secretary-General Award for Deep-Sea Research Excellence;

 (iii) Improve the assessment of essential ecological functions of the deep sea oceans through long-term underwater oceanographic observatories in the Area; and

 (iv) Enhancing deep sea marine biodiversity assessment through the creation of online taxonomic atlases linked to deep sea mining activities in the Area.

 In partnership with UN-DESA:

 (v) Abyssal initiative for Blue Growth: Advance SDG 14 and the quest for a Blue Economy through the promotion of socio-economic benefits for developing countries, including SIDS, and increasing scientific knowledge and research capacity: perspectives on deep seabed mining.

 In cooperation with the African Minerals Development Centre (AMDC):

 (vi) Foster cooperation to promote the sustainable development of Africa's deep seabed resources in support of Africa's Blue Economy; and

 (vii) Map the Blue Economy of Africa to support decision-making, investment and governance of activities undertaken on the continental shelf and in adjacent international seabed areas.

34 "ISA Registers Seven Voluntary Commitments to Support the Implementation of SDG14 at the 2017 Ocean Conference" (International Seabed Authority, 5 June 2017), <www.isa.org.jm/news/isa-registers-seven-voluntary-commitments-support-implementation-sdg14-2017-ocean-conference> accessed 4 July 2017.

CHAPTER 11

Efforts to Enhance Deep Seabed Activities and Korean Law on Exploration for and Exploitation of Resources in the Deep Seabed Area

Seokwoo Lee and Hee Cheol Yang***

Abstract

The enactment of Korean domestic law, in addition to accommodating for the domestic measures required by UNCLOS, is understood as a measure of Korean nationals to actively comply with the international and domestic standard of activities in the Area and to ensure sincere implementation of such activities. Naturally, since the Korean Law on Exploration for and Exploitation of Resources in the Deep Seabed Area (hereinafter referred to as "Korean Law on Deep Seabed Area") itself is an integration of UNCLOS and the ISA rules into the Korean law, with an aim to administer Korean nationals' activities in the Area and to strengthen the international performance of it, this Korean Law on Deep Seabed Area will not be prominent except for the procedural contents related to notification-authorization-issuance of warranty.

However, it is worth noting that the Authority's management capacity during the later exploitation phase, as was the case with exploration phase, needs to be improved. Efficient management of activities in the Area has significant limits in its function and effectiveness notwithstanding fulfillment of the sponsoring States' and contractors' duties. As such, it is meaningful that the Korean Law on Deep Seabed Area re-angled its focus from exploitation to management of activities in the Area.

* Seokwoo Lee is Professor of International Law, Inha University Law School, Korea (2003-present) and Chairman of the Foundation for the Development of International Law in Asia (DILA) (2012-present). He is the Co-Editor-in-Chief of *Asia-Pacific Journal of Ocean Law and Policy* (APJOLP) and *Asian Yearbook of International Law* (Asian YBIL), the Executive Editor of *The Korean Journal of International and Comparative Law* (KJICL), the Co-Series Editor of the book series entitled *Maritime Cooperation in East Asia*, and the Associate Editor of *Brill Research Perspectives in the Law of the Sea*, all published by Brill. He has authored more than 80 publications in English in addition to his more than 80 publications in Korean. His representative recent book publication in English is: *The Making of International Law in Korea: From Colony to Asian Power* (co-authored, Brill/Nijhoff (2016)). He holds a D. Phil. (Oxford), LL.M.s (NYU, Minnesota, and Korea University), and LL.B. (Korea University).

** Hee Cheol Yang is the Director, Ocean Policy Institute, Korea Institute of Ocean Science and Technology (KIOST), Korea.

 | DOI:10.1163/9789004366619_016

I United Nations Convention on the Law of the Sea and the International Seabed Authority in General

According to Article 157(1) of the United Nations Convention on the Law of the Sea (UNCLOS or Convention), the International Seabed Authority (the ISA or Authority) is an "autonomous international organization through which state parties to the Convention organize and control activities in the Area, particularly with a view to administering the related mineral resources." Article 145 articulates that the fundamental competences of the Authority are to "regulate exploration and mining in the deep seabed and to ensure that the marine environment is protected from any harmful effects, which may arise during exploration and mining activities." The limits of the Authority are expressed in Articles 87, 112, and 147. One of the crucial roles of the Authority is its organizational function, through which state parties to the Convention are to "organize and control the activities in the Area, particularly with a view to administering the resources of the Area." Therefore, analogous to other issues pertaining to the Area, large industrialized states and smaller developing states are largely divided as regards to how much power is divested in the Authority.

The Authority comes into play because the mineral resources retrieved from the Area are common heritage of mankind (see Art. 136–137), and no one party (state or individual investor) has exclusive rights of exploitation to the same. In fact, it is for this very reason that the Authority is necessary—to safeguard investors that intend to carry out long-term development of the resources in the Area, by giving them exclusive rights to the resources of those areas for the duration of their contracts or licenses with the Authority.

However, conflicts between states regarding the Authority dealt in Part XI of the Convention were not resolved until the Agreement Relating to the Implementation of Part XI of the UNCLOS of 10 December 1982 was enacted in 1994. The 1994 Agreement in essence altered the decision-making process so that individual state interests override that of mankind's. This was primarily due to the protest of large-margin states, or in the opinion of the United States, "unfairly and unnecessarily granted a disproportionate voice to developing countries that have little or no investment in seabed mining operations." Further, it changed the relationship of the internal organs of the Authority, namely the Assembly, Council and Secretariat. Finally, it guaranteed that the state with the largest GDP to always have a seat on the Council, allowing policies of developed states to dominate mankind.

As can be observed from the overview of issues regarding the provisions of the Convention, this paper will provide a modest proposal to come to amicable solutions on the responsibilities of states *versus* international bodies in regulating activities within "common areas" of the ocean, coupled with growing environmental concerns that at times oppose states' interests.

II Korea's Legislative Direction for Exploitation of Resources in the Deep Seabed and Ocean Floor

In anticipation of proceeding to the phase of exploitation of resources of the deep seabed and ocean floor (the Area), procurement of the sponsoring States' profit structure, threat to the environment caused by resource exploitation and the like, conflict structure was raised for the purposes of international alignment of interests between the international community and sponsoring States. As such, the international community is taking measures to minimize damage to the marine environment caused by activities in the Area, for instance, by instructing that the best environmental practices apply.

While the 2011 International Tribunal for the Law of the Sea (ITLOS) advisory opinion, which demands each State to take domestic measures, is not legally binding, there is a fair chance that it will be raised during the Authority's site consideration process as a measure to procure implementation, an important factor for the sponsoring State. As such, the domestic measures could operate as a means to encourage active participation for common heritage of mankind, because it requires its nationals' activities in the Area to comply with international duties and sponsors implementation in the areas of environment, technology, finances and so forth.

Negative perceptions towards exploitation of resources of the Area and the insistence on revenue return at the exploitation phase both stems from the difference in how the common heritage of mankind is used. Korea, as a pioneer investor, has already enhanced its understanding of the Area by conducting exploration in the Area over twenty years and thereby applied the best environmental practices while attempting to minimize hazardous elements associated at the exploitation phase.

The enactment of Korean domestic law, in addition to accommodating for the domestic measures required by UNCLOS, is understood as a measure of Korean nationals actively complying with the international and domestic standard of activities in the Area and to ensure sincere implementation of such activities. Naturally, since the Korean Law on Exploration for and Exploitation of Resources in the Deep Seabed Area (hereinafter referred to as the Korean Law on Deep Seabed Area) itself is an integration of UNCLOS and the International Seabed Authority rules into the Korean law with an aim to administer Korean nationals' activities in the Area and to strengthen the international performance of it, this Korean Law on Deep Seabed Area will not be prominent except for the procedural contents related to notification-authorization-issuance of warranty.

However, it is worth noting that the Authority's management capacity during the later exploitation phase, as was the case with exploration phase, needs to be improved. Efficient management of activities in the Area has significant limits in its function and effectiveness, notwithstanding fulfillment of the sponsoring States' and contractors' duties. As such, it is meaningful that the Korean Law on Deep Seabed Area re-angled its focus from exploitation to management of activities in the Area.

The international community's approach to resources of the deep seabed and ocean floor or Area used to be that resources of the Area were purely under the purview of the Authority. But the other approach in the Area outside the limits of national jurisdiction, however, is anticipated to focus on the 200 nautical miles of the Area newly incorporated into the continental shelf of the coastal States in accordance with Article 76 of UNCLOS. Accordingly, Korea's legal or political response must also mirror this dichotomy in its approach and analysis. Under Article 76 of the Convention, exploitation of the resources in the areas expanded as continental shelf of the coastal State and exploitation of resources in the Area beyond national jurisdiction differ in their legal character in that they are each governed by Article 82 of the Convention, Part XI of the Convention and 1994 Implementing Agreement respectively. However, both share the commonality that the distribution of interests occurs in accordance with the Authority's discussion and that both are in the deep sea area with distinct features of the development of resources on land and in the water, and as such, we may approach or respond to both in much the same way.

Korea is in urgent need of implementing a legislative structure that incorporates institutional support for the advancement of Koreans and resources of the Area beyond the national jurisdiction. Precedents of early legislation on the resources of the Area include Unites States' Deep Seabed Hard Mineral Resources Act and Germany's Act of Interim Regulation of Deep Seabed Mining[1] in 1980. The United Kingdom's Deep Sea Mining Temporary Provisional Act and France's Law on the Exploration and Exploitation of Mineral Resources of the Deep Seabed were enacted in 1981. Japan's Law on Interim Measures for Deep Seabed Mining and Russia's Edict on Provisional Measures to Regulate the Activity of Soviet Enterprises Relating to the Exploration and Exploitation

1 After becoming a signatory of the Convention in 1994, Germany enacted legislation with an overall character on the marine in 1995, including one section dedicated to regulation of the exploitation of the Area. Korea Ocean Research & Development Institute, Report on Development of Deep Sea Mineral Resources (CRPM32100-1782-5) (Ansan: Ministry of Land, Transport and Maritime Affairs, 2006), p. 60.

of Mineral Resources of Seabed Areas Beyond the Limits of the Continental Shelf were enacted in 1982. Italy's Law on the Exploration and Exploitation of the Mineral Resources of the Deep Seabed was enacted in 1985. However, such legislation was applied merely as a transitional measure before the Convention took effect and does not include environmental changes related to exploitation of the Area.

The Authority's demand for each state's legislation or implementation of administrative measures related to the Area[2] became official through the 2011 Seabed Disputes Chamber's advisory opinion.[3] Considering that the Seabed Disputes Chamber had made recommendations on the specific details of what should be incorporated into domestic law depending on different circumstances, a more meaningful approach to practical legislation on the Area would be the enactment of legislation after 2011. For instance, according to the Seabed Disputes Chamber, the sponsoring State's obligations under the treaty and related documents is divided into an obligation to secure compliance and direct obligation. The Seabed Disputes Chamber suggests as examples of direct obligations:

1. An obligation to assist the Authority by taking all measures necessary to ensure compliance under Article 153(4) of the Convention;
2. An obligation to apply a preventative approach;
3. An obligation to best environmental management measures;
4. An obligation to take measures to ensure the provision of compliance in the event where the Authority issues an emergency order for protection of the marine environment;
5. An obligation to specify means of achieving compliance; and
6. An obligation to conduct assessment of the impact on the marine environment and the like.

Additionally, the Seabed Disputes Chamber specifies that the sponsoring State's legislation should include provisions in addition to the Convention and the Exploitation Regulation on mineral resources of the Area, including the contracting party's financial and technological capabilities, conditions for the issuance of certificate and sanctions for the contracting party's failure to fulfill its obligations. With respect to the domestic measures, while recognizing discrepancies among different legal systems, it is advised that the contracting party be closely supervised and also that an implementing agency for adjustment of activities with the Authority be established.

2 ISBA/18/C/8, Add.1.

3 ITLOS Case No.17.

According to the progress of legislation reported to the Authority, approximately 30 different States have reported on the status of enactment of the related legislation. Instances of actual enactment of such legislation include: Cook Island's Seabed Minerals Act 2009; Czech Republic's Exploration for and Exploitation of Mineral Resources from the Seabed beyond Limits of National Jurisdiction (Act No. 158/2000 of 18 May 2000); Germany's Seabed Mining Act of 6 June 1995 (the Act), amended by article 74 of the Act of 8 December 2010 (Federal Law Gazette I, p. 1864); Guyana's Maritime Zones Act 2010 (Act No. 18 of 2010); United Kingdom's Deep Sea Mining (Temporary Provisions) Act 1981 and enactment of its legislation (Deep Sea Mining Act 2014); Singapore's Deep Seabed Mining Act 2015(No. 6 of 2015); China's Law of the CRP on Exploration for and Exploitation of Resources in the Deep Seabed Area 2016; Tonga's Seabed Minerals Act 2014 (2014); Fiji's International Seabed Mineral Management Decree 2013 (2013); Federated States of Micronesia's Seabed Resources Act of 2014 (2014); Tuvalu's Tuvalu Seabed Minerals Act 2014 (2014) and so forth.

Each legislation is differently composed depending on each respective State's legislative structure and some legislation qualifies only as a legislation for simple exploitation on marine mineral resources instead of the official legislation on the Area as advised by the Authority. Some legislation targets marine mineral resources both within and outside the national jurisdiction and others display a highly similar set of provisions and contents. For instance, Fiji and Nauru's enacted official legislation on the Area beyond national jurisdiction containing 57 and 56 provisions respectively, which stated that the Seabed Minerals Authority be established to domestically administer the related activities.[4] The similarity of both state's legislation is attributable to the EU's model law of Pacific-African Caribbean Pacific (P-ACP) region[5] which operated as their basis. On the other hand, Tuvalu and Tonga differs from this model law in that they enacted legislation which encompasses marine mineral resources both within and outside the national jurisdiction, and each consist of 129 and 121 provisions, respectively. These two states, in particular, include diverse financial burdens such as application fees, sponsorship payments, taxes, and loyalty, among them.[6]

States with relatively simpler legislative systems include China, Singapore and Korea, which is currently in the process of initiating enactment of the

4 Article 6, Fiji International Seabed Mineral Management Decree 2013; Article 7, Nauru International Seabed Minerals Act 2015.

5 P-ACP group includes Cook Islands, Federated States of Micronesia, Fiji, Kiribati, Marshall Islands, Nauru, Niue, Palau, Papua New Guinea, Samoa, Solomon Islands, Timor Leste, Tonga, Tuvalu and Vanuatu.

6 Article 93–94, Tonga Seabed Minerals Act; Article 100, Tuvalu Seabed Minerals Act.

legislation. Each of these states' law on the Area is composed of 29 provisions (China), 25 provisions (Singapore), 28 provisions (Korea), and the Korean and the Chinese legislation incorporates step-by-step rules on securing the mines from the Authority. Distinctive features of the Korean legislation include "prospecting, exploration or exploitation" and the law regulates procedures for notification in the case of prospecting, and the procedures for obtaining a permit in the case of exploration or exploitation. Singapore and China's legislation include only the exploration or exploitation of the resources within its permit system.[7] A noteworthy feature of the Chinese legislation is that it facilitates the overall management of information by the nation through the support of construction and operation of public infrastructure related to the Area, exchange and cooperation of science and technology related to the Area, and sharing of achievements,[8] dissemination of deep sea science,[9] and submission of copies of all materials and samples.[10] Likewise, the Korean Law on Deep Seabed Area includes regulation of development of science technology for support of the activities in the Area and provision of information, dissemination of international cooperation achievements. Notwithstanding this, the legislation does not directly require submission of information related to the activities in the Area.[11] However, both the Korean and Singaporean legislation include regulations to impose conditions,[12] and thereby could potentially conduct holistic management policies by, for instance, requiring submission of related information as a prerequisite of the permit.

In addition, another distinctive feature of the Korean Law on Deep Seabed Area is that unlike other legislation that mandates the imposition of measures of marine environment protection as an obligation of the contracting party, it simultaneously regulates the obligations of a state and obligation of persons engaged in the activities in the Area.[13] In particular, the Korean Law on Deep Seabed Area requires as part of the obligations of the state, the duty to take measures to protect the marine environment and to implement an emergency

7 Article 8, Law of the CRP on Exploration for and Exploitation of Resources in the Deep Seabed Area 2016; Article 6, Singapore Deep Seabed Mining Act 2015.

8 Article 16, Law of the CRP on Exploration for and Exploitation of Resources in the Deep Seabed Area 2016.

9 Article 17, Law of the CRP on Exploration for and Exploitation of Resources in the Deep Seabed Area 2016.

10 Article 18, Law of the CRP on Exploration for and Exploitation of Resources in the Deep Seabed Area 2016.

11 Article 16, Korean Law on Deep Seabed Area.

12 Article 11, Korean Law on Deep Seabed Area; Article 10, Singapore Deep Seabed Mining Act 2015.

13 Article 19, Korean Law on Deep Seabed Area.

action plan to be triggered in case of serious infringement on the marine environment. This requires taking state measures which may be necessitated by the activities in the Area over and above simple supervision and inspection.

III The Korea Law on Exploration for and Exploitation of Resources in the Deep Seabed Area

III.1 *Need for Legislation*

There is an absence of domestic regulation for implementation of international treaties related to activities in the Area. Resources of the Area are collectively managed by the international community in accordance with UNCLOS, and international law related to the Area strongly requires the enactment of pertinent regulations. Resources of the Area refers to all mineral resources *in situ* in the Area and subsoil thereof, beyond the limits of national jurisdiction and as of 2017, approximately 30 States have officially enacted their respective legislations.

The international community is moving towards commercialization of resources of the Area for the first time, and there is a defect in the domestic legislation as to the regulation of Korea's Area. Korea, which has secured its mining lot in the Area, is in urgent need of enacting a domestic law pertinent to the Authority and ITLOS's implementation of duty to manage and fulfill state liability exemption prerequisites.

There is a need to procure ground rules on implementation of domestic measures for immunity of States that sponsor activities in the Area (hereinafter the "sponsoring States"). UNCLOS Article 139, Article 153(4), Annex III Article 4(4), ISA Decision of the Council (ISBA/17/C/20, para.3), Decision of the Council (ISBA/18/C/8, Add.1) demands a provision in the domestic law that secures activities in the Area.

ITLOS was established in 1995 in Hamburg, Germany in accordance with Annex VI of the UNCLOS. In 2011, ITLOS recommended in its advisory opinion (ITLOS Case No.17) that the sponsoring States put forward domestic legislation in order to be exempt from related state liability. In its 2011 advisory opinion, ITLOS ruled on the sponsoring States' indirect and direct obligations, and emphasized that specific measures be legislated into the sponsoring State's domestic law.

The ground rules for stable exploitation of resources of the Area from the enactment of the Korean Law on Deep Seabed Area are as follows:

1. The first rule is stable exploitation of resources of the Area in accordance with nationals' support of activities in the Area and enactment of a law on exemption from state liability in accordance with

the duty to comply with international law. It responds with a paradigm change by international community's active management of the Area's resources, which follows the "first-come, first-served basis," and reinforcement of state party's management obligation; and

2. The second rule is that Korea needs to cater to the demand to internationally conform by renewing its manganese nodule exploration contract (expired in 2016) for five additional years, and prepare to proceed into the commercial (exploitation) phase within the next five years etc.

Reflecting this development, there is progress on introducing the law on exploitation of mineral resources of the Area. Domestically, the exploitation of resources of the Area has been under the purview of the Ministry of Trade, Industry and Energy. Notwithstanding this, there is no legislation with the focus on exploitation of mineral resources of the Area, the commercial development period of which was uncertain. The Ministry of Ocean and Fisheries' approach in regards to the exploitation of resources of the Area rests in the pre-commercial stage, with the focus on "investigation—securing the mining area—selection of priority mining area" and there is insufficient recognition of the need of a legislation on exploitation.

However, recently the Authority has been giving more weight to the establishment of domestic law as a precondition for the granting of a mining block. This could operate as an obstacle to the further securing of the mining area and the entrance into the exploitation phase. When Brazil applied for the exploitation of manganese nodules, states such as the Netherlands raised the issue of the absence of domestic legislation related to the Area.

Within the next five years, Korea, as a contracting State of the Area, is in urgent need of a legal basis as grounds to support the policy, promotion of marine enterprise, and leading technological development for the transition into the commercial phase.

Enactment of legislation related to the Area ensures exploitation of the resources of the Area as the leading features of the Ministry of Ocean and Fisheries and could contribute towards stabilization of strategic mineral supply, which now heavily relies on foreign sources. Mining of the Area, which necessitates cutting-edge marine development technology and tremendous investment costs, requires institutional and financial support measures. It needs the basis for promotion of the industry in the Area by stimulating participation of private enterprises.

In 2012, the Ministry of Ocean and Fisheries (formerly known as the Ministry of Land, Transport and Maritime Affairs) initiated the domestic legislation in accordance with the Authority's trend of expanding activities in the Area, and in 2013 devised its first draft (Marine Mineral Resources Development Act). However, the then Ministry of Land, Transport and Maritime Affairs was designed to include "domestic and foreign" spatially in the direction of the 2013 draft, and was designed to include "other domestic minerals of the domestic waters."

III.2 *Legislative Direction and Organizational Structure*

III.2.1 Legislative Direction

Enactment of new legislation for the authorization process of conducting exploration and exploitation of resources of the Area accepts international law related to activities in the Area. Thus, the new legislation administers Korean activities in the Area, and clarifies the international obligation for the guarantee of implementation of obligations.

The legislation is comprised of basic plans of work for exploitation of resources of the Area, authorization process for the people engaged in the Enterprise, compliance with international norm and protection of marine environment etc., including basic plans of work for activities in the Area (Section 2), authorization of activities in the Area (Section 3), supplementary rules (Section 5). This legislation also ensures international competitiveness for activities in the Area by regulating support of activities in the Area, establishment of the Enterprise (Section 4), encouragement of joint ventures (Section 3) and promotion of activities in the Area.

III.2.2 Structure and Contents

This legislation is composed of six Sections and 28 provisions:

> Section 1 (General Provisions) points out the purpose, definitions, relationship with international conventions, and responsibility and liability. Section 2 (Basic Plans of Work for Activities in the Area) is about formulation of basic plans of work for exploration and exploitation of the Area, and the establishment of a review committee for activities in the Area. Section 3 (Authorization of Activities in the Area) touches upon issues such as notification of prospecting, issuance of the Enterprise permit, encouragement of joint ventures, issuance of certificate etc., revocation of the Enterprise permit, termination of support, rights and obligations of the persons engaged in the Enterprise, and transfer of rights of persons engaged in the Enterprise etc. Section 4 (Promotion of Activities in the Area) is about assistance with activities in the Area, establishment of the Enterprise, and delegation of rights, etc. Section 5 (Supplementary Rules) has provisions on protection and preservation of marine environment,

protection and preservation of rare or fragile ecosystems, emergency measures; inspection and supervision, corrective orders, and charges. Finally, Section 6 (Penalties) mentions penalties, joint penal provision, and administrative fines.

III.2.3 Main Contents of the Korean Law on Deep Seabed Area

The main contents of the Korean Law on Deep Seabed Area are as follows:

First, application and scope of administration includes activities in the Area beyond the national jurisdiction (Korean Law on Deep Seabed Area, Articles 1–4). The legislation provides for regulation on implementation of international obligations and domestic procedures for the people of the Republic of Korea engaged in activities in the Area (prospecting, exploration, exploitation) administered by the Authority. "Resources of the Area" refers to all mineral resources *in situ* in the Area and subsoil thereof, beyond the limits of national jurisdiction, and is defined by UNCLOS as common heritage of mankind and all rights in the resources of the Area are vested in mankind as a whole, on whose behalf the Authority shall act.

Second, the legislation provides a formulation of basic plans of work for exploration and exploitation of the Area, Establishment of a review committee for activities in the Area (Korean Law on Deep Seabed Area, Articles 6–7). It formulates basic plans of work for systematic and effective management of the exploitation of resources of the Area, and refers matters related to the development and alteration of basic plans to the review committee for activities in the Area.

The legislation provides for submission of contents such as the basic purpose of activities in the Area, direction and strategies of activities in the Area, promotion and support of the Enterprise, scientific and technological development for activities in the Area, international cooperation related to activities in the Area, securing resource stability in cases of emergency, and collection and distribution of information needed for activities in the Area. The legislation provides for establishment of a review committee comprised of related departments to review matters related to authorization, modification, revocation, and termination of the Enterprise. The Chairperson of the committee shall act as a deputy minister of the Ministry of Oceans and Fisheries, and the committee shall consist of no more than 20 members.

Third, the legislation outlines notification, authorization, support and revocation (Korean Law on Deep Seabed Area, Articles 8–9 and Articles 11–13). The legislation provides that the people of the Republic of Korea who seek to undertake prospecting are categorized into notification (prospecting), and permit (exploration, exploitation) depending on the level of activities and must submit related plans and applications. The legislation's approach in prospecting, exploration, or exploitation is the same as that of the Authority's Exploitation Regulation with the approach of notification and authorization.[14] The legislation requires persons seeking to undertake prospecting to report to the minister of Ministry of Oceans and Fisheries and must first undergo the notification process.

The legislation requires persons who seek to conduct an Enterprise to obtain a permit from the minister of the Ministry of Oceans and Fisheries (10 years) and to apply separately for the exploration permit and for the exploitation permit. The following documents must be submitted: proof of financial, investment and technical capabilities for the Enterprise; assessment of the impact of exploration or exploitation on the marine environment; and an emergency action plan to be triggered in case of serious infringement on the marine environment.

The legislation strengthens implementation of international law and measures to administer Korean compliance by stipulating that the persons engaged in the Enterprise shall not proceed with the exploration or exploitation until having become a contracting person through an exploration or exploitation contract with the Authority. The legislation provides that the applications for the Enterprise are to be reviewed in the order of submission and where deemed necessary, a joint venture of two or more persons to the Enterprise is recommended with a view to strengthen international competitiveness. The Authority's Exploitation Regulation also reviews applications in the order they were submitted, and while this principle has been domestically adopted, the possibility of strengthening national coordination has been introduced.

The legislation also provides that in the case of issuing a permit for the Enterprise, the minister of the Ministry of Oceans and Fisheries must issue a certificate that is necessary for persons engaged in the Enterprise to enter into a contract with the Authority. The permit for the Enterprise is a domestic procedure for issuance of a certificate necessary for persons engaged in the Enterprise to enter into a contract with the Authority and as such, the government assesses the ability to implement international obligations during the procedure of issuing a permit.

14 The Authority's Exploitation Regulation requires persons seeking to undertake prospecting to notify the related plans to the Secretary General of the Authority, and to refrain from carrying out the prospecting until confirmation that such notification has been filed.

In addition to revocation of the permit, the legislation provided for termination of support in certain cases, such as obtaining the permit through deceit or other illegal means. In such cases, the minister of the Ministry of Oceans and Fisheries shall submit a written submission to the Authority, and the termination takes effect six months after the Secretary General of the Authority's receipt of the submission in order to protect the rights of the sponsor.

Fourth, the legislation describes the rights, obligations, and transfer of rights for persons engaged in the Enterprise (Korean Law on Deep Seabed Area, Articles 14–15). The legislation provides that the persons engaged in the Enterprise have the exclusive exploration or exploitation rights only to certain resources within the territory of the contract signed with the Authority and urges the fulfillment of contractual obligations. The legislation clarifies the rights of persons engaged in the Enterprise and urges implementation of contractual duties with the Authority and provides for compliance with domestic regulations related to safety.

The legislation also provides for the assignment or assumption of rights and obligations to be made only to the people of the Republic of Korea, and to follow procedures for modification of the permit. The Enterprise hinges a state's international responsibility and as such, deems assignment of its rights as a critical modification in the permit.

Fifth, the legislation encourages joint ventures (Korean Law on Deep Seabed Area, Article 10). The legislation provides for encouragement of joint ventures of more than two persons for the purpose of strengthening international competitiveness and promoting efficiency, as the Enterprise is accompanied by a large-scale budget and technology development. The legislation provides for adjustments to resolve an overlap where there is more than one person who intends to undertake the Enterprise in the same area for the same resources of the Area prescribed by the Authority.

Sixth, the legislation promotes activities in the Area (Korean Law on Deep Seabed Area, Articles 16–18). The legislation ensures support for the promotion of activities in the Area such as provision of relevant information, training of professional personnel, development of science and technology equipment, and international cooperation. It includes expansion of support and promotion, such as support of industries related to activities in the Area, support by training of professional personnel and support of research institutes related to activities in the Area. The legislation further provides for establishment of the professional Enterprise group for implementation and support of the activities in the Area and provides that the group may carry out the government's mandate related to the activities in the Area.

IV Response to Development of Extended Continental Shelf beyond 200 Nautical Miles of the Coastal States under Article 76 of the Convention

Given the status of technological developments related to exploitation of the Area, distribution of coastal States' interests pursuant to Article 82 of the Convention does not appear to have been executed in the very foreseeable future.[15] However, considering the adoption of relevant rules for exploration of manganese nodules in the past and the discussion process of hydrothermal exploration regulations, specific discussions in the international community about Article 82 of the Convention need to begin at this point. This is due to the conflict of intergovernmental interests being stronger in the case of Article 82 of the Convention because it discusses the distribution of profits from the development of the continental shelf that was incorporated into the sovereign jurisdiction of one country, whereas the discussion of the Area is one that involves the common heritage of mankind.

With the gradual stabilization of the spatial parameters of the Area, creation of an order for the purposes of legal interpretation and implementation of development of non-living resources in the continental shelf beyond 200 nautical miles of the coastal States became a new task for the international law of the sea. The Convention is working on a device to return the resources of the Area, which is the common heritage of mankind susceptible to interference by expansion of the coastal States' continental shelf, to the international community, and Article 82 of the Convention is part of such an effort. Accordingly, Article 82 of the Convention is considered to be one of the most controversially compromised and one of the most innovative products of the Convention. However, before a realistic power of execution can be achieved, the exploitation of resources and the associated profit-sharing in the continental shelf beyond 200 nautical miles of the coastal States, in which a state has its sovereign rights, remains a problem to be solved. Ultimately, with respect to international law, while the method of reaching an appropriate compromise between competing states at the time of signing UNCLOS would be followed, an important issue remains in that process. This is to decide to what extent the views of coastal States with the expanded continental shelf will be accepted.

To begin with, "all products produced in the mining area in the period beginning five years after the start of production" or "evaluation on and value of products" and relationship with obligations, run the risk of freezing exploitation of the resources in a state's continental shelf beyond 200 nautical miles

15 Chinese ISA homepage: http://china-isa.jm.chineseembassy.org/chn/xwdt/t539030.htm.

of the coastal States. This may create new conflicts in the process of regular in-kind donations and cash payments. For instance, the issue of how to realistically achieve realization of the investment loss in the case where development is delayed due to a *force majeure* event at the development stage of coastal States, or in the case of resuming a previously delayed work for a certain reason, may lead to a long debate between the international community and coastal States. With respect to the *force majeure* factor at the phase of exploration or exploitation, the Regulation on Exploitation of Manganese Nodules in the Area instructs that

> The Contractor shall not be liable for an unavoidable delay or failure to perform [any of its obligations/the Exploitation Activities] under this Contract due to *force majeure*, provided the Contractor has taken all reasonable steps to overcome the delay or obstacle to performance … has failed to complete the preparations necessary to proceed in to the exploitation phase or is under the economic circumstances where the Contractor cannot proceed into the development phase, the Contract may apply for an extension of the exploration plan for no more than five years, under the recommendations of the committee and with the permission of the board of directors.

Such *force majeure* clause related to the Area may also operate as a flexible negotiating factor inducing a compromise between parties in the case of the expanded continental shelf of the coastal State that falls within the parameters of Article 82 of the Convention. However, notwithstanding the need to adequately reflect environmental conditions for promoting development of the coastal States, it is imperative that there be measures for the distribution of development benefits to the international community and for the acceleration of commercial development.

With respect to a fair method of distribution, it is clear that the discussions on the distribution method and standard of development of non-living resources in the continental shelf beyond 200 nautical miles of the coastal States will be preceded with much controversy and complexity compared to that of the Area. While it is necessary to recognize the difference in stance on distribution by targeting the continental shelf that is included in the national jurisdiction, it is anticipated that a certain standard will be provided by the regulation on exploitation of mineral resources of the Area.

From February 11–13, 2009, the Authority and Chatham House of the United Kingdom jointly conducted a seminar in London on the issue of financial or in-kind payment associated with the development of non-living resources in the continental shelf beyond 200 nautical miles of the coastal States, which is regulated by Article 82 of the Convention. During the seminar it was predicted that given the recent application progress for the continental shelf of coastal States and the related exploitation of resources, the receipt of payment will be conducted under Article 82 of the Convention in 2015 at the earliest.[16] Admittedly, the build-up of issues were also highlighted with respect to the jurisdiction and procedural problems between coastal States and the Authority when it comes to the issue of profit-sharing under Article 82 of the Convention. Besides the national implementation of Article 82 of the Convention, it is clear that the legal relationship surrounding mutual distribution of benefit and rights will be complicated if domestic interests are involved. Moreover, for the obligations imposed on states by international law, the relevant state must establish domestic legal regulations as to whether the state will shoulder the payment of relevant costs or whether such burden will be shifted to the producers.

Indeed, Article 82 of the Convention is not directly connected to Korea's understanding with respect to jurisdiction. However, expansion of coastal States' continental shelf by the Commission on the Limits of the Continental Shelf means that resources of the Area (minerals, energy, etc.) that were previously under the control of the Authority will be reduced and as such, ultimately signifies a serious narrowing down of room for economic advancement. That is, regardless of whether or not Korea becomes the direct profit sharing target under Article 82 of the Convention, there must be a formulation of theoretical and policy directions related to Article 82 of the Convention. This determination is bound to arise as a new agenda for Korea's profits of the Area under the international maritime order. This is because at the heart of such discussions lies the issue of Korea's interest and its participation. This involves the issues of exploitation of resources of the Area connected to the coastal State's continental shelf, profit-sharing, and the major mineral resources and all other non-living resources as is currently discussed by the Authority.

V Conclusion

"The Area and its resources" are defined by UNCLOS as common heritage of mankind. Related norms of the Convention and the Authority require all

16 During the meeting, there were discussions of payment of costs or payment in kind associated with the exploitation of the resource, the amount and time of such payments, the direct powers and dispute settlement mechanisms of the Authority in the relevant area for the purposes of enforcing the specific problems of the aforementioned clauses, and the results of the seminar were to be referred by the Conference of the Parties or the Authority. Chinese ISA homepage, http://china-isa.jm.chineseembassy.org/chn/xwdt/t539030.htm.

state parties to take domestic legal and administrative measures necessary for support and implementation of the activities in the Area. According to Article 1(1.3) of UNCLOS, "activities in the Area" means all activities of exploration for, and exploitation of, the resources of the Area.

To reflect the limits in the current law and the need for new legislation related to the Area, as a State Party to the Convention and the Authority, this legislation seeks to regulate Korean nationals' activities in the Area to be conducted in accordance with international law. As a sponsoring State, the legislation seeks to enforce legal responsibilities and obligations prescribed by international law and international judicial bodies, and other law enforcement agencies as part of domestic law.[17] It provides necessary procedures such as support of Korean nationals' activities in the Area (permit etc.) and provides for international cooperation and compliance with international norms, and thereby promote Enterprises in the Area in order to secure a stable supply of resources.

Key factors to consider are sponsorship obligations of Korean nationals who engage in the Enterprise beyond the limits of national jurisdiction, measures to prevent environmental pollution, the support of related industries, domestic accommodation of international norms, long-term promotion of enterprise in the Area, ensuring implementation of international norms by the persons engaged in the enterprise, and the sharing of information about the Area and cooperation.

---

17 UNCLOS Article 139, ITLOS advisory opinion, Decision of the ISA Council (ISBA/18/C/8, ISBA/18/C/8/Add.1).

CHAPTER 12

The Role of Science for Environmental Impact Evaluation Resulting from Ocean Mining

*Tomohiko Fukushima**

Abstract

As expectations rise for the development of deep-sea mining, there is a greater demand for strictness in environmental impact assessments (EIAs). However, no standardized methods for evaluating the impact of mining on deep-sea ecosystems have been established because unlike risk assessments for human health, EIA does not have an absolute evaluation focal point. Therefore, it is essential to review assessment methods from both social science and natural science fields. Recently, as scientists have started to actively participate in the policymaking process, and policymakers have started to learn about scientific progress, the gap between the two groups has been getting narrower. On the one hand, the primary objective of science is the pursuit of truth. To that end, raw data is gathered, which then becomes information, from which eventually principles and laws are deduced. However, that path is a long and winding road. Deduced information diverges, and the laws are repeatedly revised; however, this is the essence of science. On the other hand, resource development is an economic activity. The goals of developers converge on the pursuit of profit. Although they may be pursuing profit, in places like the deep sea, where knowledge about the ecosystem is limited, developers must rely on the expertise of scientists to conduct EIAs. However, while developers seek convergent conclusions, scientists often indicate results that are divergent. Therefore, sometimes scientists end up hindering the progress of development, policymaking, and legislation. However, the author believes that science's biggest contribution is in making the process of decision-making on ocean resource development more deliberate. We come to know our ignorance when diverse scientific findings are put forth and, each time, past knowledge is revised. Through that

* The author works for JAMSTEC (Japan Agency for Marine-Earth Science and Technology). His field of expertise is deep-sea biology and deep-sea environment studies. He received a doctorate of science from Kyoto University in 2004. The title of his thesis is "Ecological Characteristics of Deep-Sea Benthic Organisms in Relation to Manganese Nodules Development Practices". The author's PowerPoint presentation is available at http://www.virginia.edu/colp/pdf/yogya-fukushima.pdf.

 | DOI:10.1163/9789004366619_017

process, we can study the advantages and disadvantages of ocean development from various perspectives.

1 The Current Situation of the Development of Deep-Sea Mineral Resources

In Areas Beyond National Jurisdiction (ABNJs) as well as maritime boundaries under national jurisdiction, expectations are rising for seabed resource development. For the former, this is obvious when looking at the changes in the number of applications for exploration area[1] with the International Seabed Authority (ISA). The increase is particularly striking for private enterprises since their involvement began in 2014 (Fig. 12.1).

As for maritime boundaries under the national jurisdiction of coastal nations, Nautilus Minerals was the first to acquire a license for exploration in Papua New Guinea (PNG) in 1997. Since then, licenses for exploration have continued to be issued by Pacific island states. In 2010, Diamond Fields International acquired a license for development in the Red Sea, and following that Nautilus Minerals acquired one in PNG in 2011[2] (Fig. 12.2).

In any event, the situation surrounding seabed mineral resources is developing rapidly. Naturally, the establishment of technology and systems to conduct undersea EIAs is becoming an urgent matter.

2 The Current Situation of Environmental Impact Assessment (EIA)

With rising expectations for seabed resource development, there have been increased activities related to EIA initiatives. The ISA held workshops on the taxonomy and identification of benthic organisms in 2013, 2014, and 2015 with the goal of developing a suitable EIA.[3] In these workshops, a large number of taxonomists were invited from the ISA to present the latest information on the

1 ISA has issued mining regulations for each minerals (polymetallic nodules; ISBA/19/C/17, polymetallic sulfides; ISBA/16/A/12/Rev.1, and cobalt-rich ferromanganese crusts; ISBA/18/A/11). Countries or organizations that make exploration contracts in advance with ISA and acquire a mining area that can exclusively explore.

2 The acquisition status of the exploration area is detailed in the following article. T. Fukushima and M. Nishijima "Taxonomic problems in environmental impact assessment (EIA) linked to Ocean Mining and possibility of new technology developments", in R. Sharma (ed.), *Deep-Sea Mining* (Gewerbestrasse, Springer, 2017), pp. 465–482.

3 The workshops were held in Berlin (Germany), Uljin (Korea) and Ghent (Belgium) respectively. Contents of each WS are introduced in ISA's HP. https://www.isa.org.jm/workshops (Accessed 15 Sep. 2017).

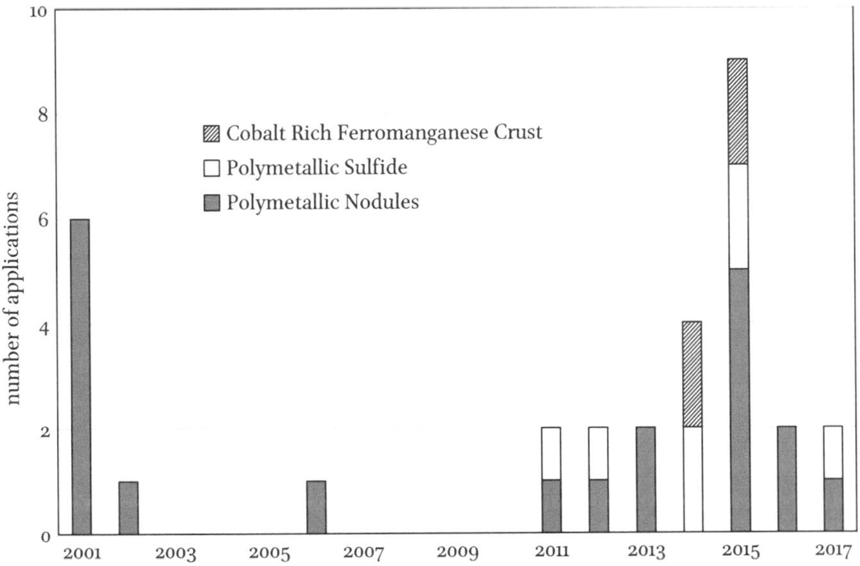

FIGURE 12.1 Changes in the application number of exploration area to International Seabed Authority.

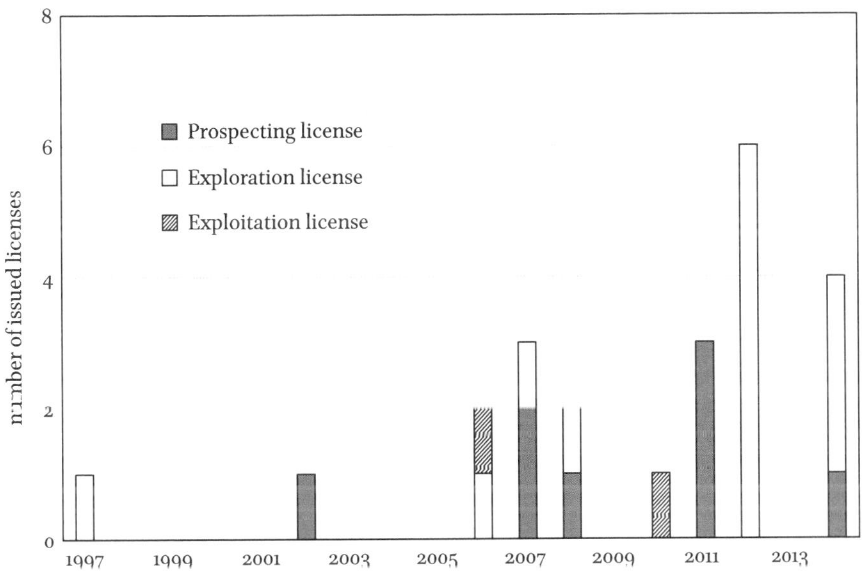

FIGURE 12.2 The number of licenses permitted in the jurisdiction areas of Coastal States. Data was referred from Fukushima and Nishijima, "Taxonomic problems in environmental impact assessment (EIA) linked to Ocean Mining and possibility of new technology developments" in R. Sharma (ed.), *Deep-Sea Mining* (2017) pp. 465–482.

classification of megafauna, macrofauna, and meiofauna. In addition to these technical efforts, the ISA held workshops on writing environmental regulations in 2016 and 2017, in which EIA mechanisms were discussed. Here as well, experts in oceanography are taking leading roles in the discussion alongside legal scholars.

Besides the ISA, scientists in European countries are driving the discussion of environmental research on seabed resource development. MIDAS (Managing Impacts of Deep-Sea Resource Exploitation)[4] is a project in which natural scientists, social scientists, and legal experts from 11 European countries and 32 agencies work alongside industry and NGOs to study ways to reduce the environmental impact accompanying seabed resource development. JPI Oceans (Joint Programming Initiative Healthy and Productive Seas and Oceans)[5] has 25 research organizations from 11 European countries involved in studying the ecological aspects of deep-sea mining and is proposing environmental management plans based on the results. INDEEP (International Network for Scientific Investigation of Deep-Sea Ecosystems)[6] is a project that focuses on narrowing the knowledge gap between science and society, and considers sustainable environmental management plans for the sea floor. It has 500 participants from 38 countries. Furthermore, DOSI (Deep Ocean Stewardship Initiative)[7] aims for integrated efforts between science, technology, policy, law, and economics to promote ecosystem management for the sea floor that will host resource development in the future (Table 12.1).

In Japan, DORD (Deep Ocean Resources Development Co., Ltd.)[8] and JOGMEC (Japan Oil, Gas and Metals National Corporation)[9] conduct activities aimed at developing deep-sea mineral resources, but for EIAs there is an initiative to develop a council of external scientists. Furthermore, JAMSTEC (Japan Agency for Marine-Earth Science and Technology)[10] and NIES (National Institute of Environmental Studies)[11] are working on developing the technology needed for environmentally responsible resource exploitation and management (Table 12.2).

4 MIDAS; http://www.eu-midas.net/ (Accessed 15 Sep. 2017).

5 JPI Oceans; http://www.jpi-oceans.eu/ (Accessed 15 Sep. 2017).

6 INDEEP; http://www.indeep-project.org/ (Accessed 15 Sep. 2017).

7 DOSI; http://dosi-project.org/ (Accessed 15 Sep. 2017).

8 DORD; http://www.dord.co.jp/english/business/index.html (Accessed 15 Sep. 2017).

9 Narita, T. *et al.*, "Summary of Environmental Impact Assessment for Mining Seafloor Massive Sulfides in Japan", *J. Shipping and Ocean Engineering* 5 (2015), 103–114.

10 SIP, Cross-ministerial Strategic Innovation Promotion Program, Next-generation technology for ocean resources exploration (Zipangu in the Ocean program), https://www.jamstec.go.jp/sip/images/pamphlet_e.pdf. (Accessed 15 Sep. 2017).

11 National Institute for Environmental Studies (NIES), http://www.nies.go.jp/index-e.html (Accessed 15 Sep. 2017).

TABLE 12.1 List of international collaborative projects.

Project	Supported by	Members	Terms
Blue Mining	EP7	19 research organisations from across Europe (11 countries)	2014–2018
MIDAS	EP7	32 organisations from across Europe (11 countries)	2013–2015
JPI Oceans	each member country	20 member countries	2011–
INDEEP	Total foundation	more than 500 members from 38 countries	2011–2016
DOSI	INDEEP etc.	185 members in mailing lists	2013–
EcoDEEP	each organization	Ifremer and Jamstec	2015–

EP7: European Commission's Framework 7 programme

TABLE 12.2 Technical protocols for EIA of deep-sea impact assessment issued by the JAMSTEC and the NIES.

NO	Title	Publication
SIP Protocol NO.1	Application of environmental metagenomic analyses for environmental impact assessment	JAMSTEC
NO.2	Genetic connectivity survey manuals	JAMSTEC
NO.3	A rapid method to analyze meiofaunal assemblages using an Imaging Flow Cytometer	JAMSTEC
NO.4	Acquisition of long-term monitoring images near the deep seafloor by Edokko Mark 1	JAMSTEC
NO.5	Microstructure measurements around deep sea floor-Direct measurements of the deep sea turbulence flow	JAMSTEC
NO.6	Onboard bioassay for seawater quality monitoring using a delayed fluorescence of microalgae	NIES

As evident from the above examples, with its insufficient scientific knowledge base compared to coastal zones, the presence of scientists is essential for seabed mining in the deep sea. However, the primary goal of science is not policymaking or making regulations. Therefore, scientific knowledge and the presence of scientists do not always function effectively in the policy- and regulation-making processes. The next section will therefore discuss the characteristics and primary role of science.

3 The Inherent Role of Science

The objective of science is to learn the truth that penetrates the natural world. However, it is a long and winding road to get there. To approach the truth, one must gather experimental facts one by one and deduce the laws that relate to those facts. In this way, theory is developed from known information, and the foundation of science is the inductive approach, which conjectures new conclusions based on these theories.

To deduce laws, eliminating "*ad hoc* assumptions"[12] and arriving at explanations that are as simple as possible, is considered beautiful. This is the principle of parsimony. The deductive method, in contrast to the inductive method, adds information to previously established theory to derive conclusions. If the assumed theory is correct, the correct conclusion will always be deduced because the method of retrieving information is implicitly contained within the assumption. In other words, the volume of information does not increase.

However, the volume of information can increase when one uses the inductive method. But unlike deduction, the conclusions may not always be correct because even if a conjecture is based on 99 observed facts, the 100th case may not always go according to the conjecture. In other words, the results provided by science carry the latent risk of revision.

As this is thought of as a matter certainly, there is no hesitation to repudiate prior conclusions. For example, the use and disuse theory[13] advanced by Lamarck was repudiated by Darwin's theory of evolution, and as a rebuttal to

12 *Ad hoc* assumption: this is one of the supplementary hypotheses to deny rebuttal of certain theory. It is added later.

13 Use and disuse theory: Organisms can inherit the characteristics acquired over their lifetime to descendants. The theory means that it will be the driving force for evolution.

Darwin's statement that "nature does not leap",[14] Stephen Jay Gould and Niles Eldredge advanced the theory of "punctuated equilibrium".[15]

In this way, the world of science is characterized by an emphasis on new discoveries (different from the past) and ingenuity (discovering unexplored and unknown things). Therefore, scientists are keenly aware of the issue or chronological priority and aim to publish their findings in academic journals before others do so. For example, Darwin, who had been hesitating to publish his theory of evolution, read a letter from Alfred Russel Wallace and decided that he should hurry up and publish. Even further in the past, there is the famous example of the competition between Newton and Liebniz over the discovery of differential and integral calculus. Furthermore, proof with regard to science is the passive endorsement of being unable to refute a hypothesis concerning the target phenomenon, and is no more than the best explanation for the time being. This too is the context for facilitating the revision of past discoveries.

Based on these characteristics of science, as EIAs must inevitably be associated with seabed resource development, information divergence and the repeated revisions of scientific knowledge are gaining attention.

4 Scientific Results Diverge

Scientific results diverge. Even so, the circumstances differ depending on the position of the person receiving the information. In other words, even though the results are consistent from a scientific perspective, they diverge with regard to EIAs.

In the 1990s, four mining simulation tests were conducted to demonstrate the effects of manganese nodule mining on the environment. The tests were conducted by the United States, Japan, the IOM,[16] and India (Table 12.3).

14 "Nature does not leap" were words Charles Darwin mentioned in the sense that evolution occurs by accumulation of small genetic variations. Charles Darwin, *Origin of Species*, 1859, p. 194.

15 Punctuated equilibrium: the theory that the evolution of living things does not progress gradually, but it repeats the period of sudden evolution and stagnation of change.

16 IOM (Interoceanmetal) is an international organization established in 1987 to encourage the development of deep-sea mineral resources. Present Member States are: Bulgaria, Cuba, Czech Republic, Poland, the Russian Federation, and Slovakia.

TABLE 12.3 Summary of benthic disturbance in the BIE-type experiment (Fukushima, 2004,[a] revised).

Experiment name	BIE II	JET	IOM	INDEX
performed by	USA	Japan	Joint organization	India
start from	1993	1994	1995	1997
experiment site	Central Pacific	Central Pacific	Central Pacific	Indian Ocean
impact area	150×3000m	200×2000m	120–230m×2500m	3000×200m

a Fukushima, T., Ecological characteristics of deep-sea benthic organisms in relation to manganese nodules development practices. Thesis Kyoto Univ., 2004.

There were differences of degree, but all the test results showed that unnatural disturbances from the mining simulations temporarily reduced the population of benthic organisms (Table 12.4).

Regarding this phenomenon, as the results were related to several phenomena that started with the unnatural disturbances, it was possible to give a scientific explanation that the food environment for benthic organisms had worsened. In other words, the results suggested that mining adversely impacted the community of benthic organisms. On the contrary, in the test conducted by Japan, while it was only for a limited classification group, separate results showed that species diversity increased due to the mining simulation (Fig. 12.3). If we interpret these results as a scientific outcome, it proves the theory of mid-scale disturbances on the sea floor. However, if we consider the number of appearances or level of diversity in terms of EIA, it turns out that mining has both negative and positive diverging effects on the deep-sea environment. How these truths should be evaluated will be left to the judgment of stakeholders, but they are bound to complicate that judgment.

5 Scientific Results Are Repeatedly Revised

While what follows does not concern benthic organisms in deep sea floor mineral resources, it demonstrates another characteristic of science: the repeated revision of results.

Benthic organisms that are generally sensitive to environmental pollution are sometimes used as sensors for discovering the state of pollution. These are called pollution indicator species and are often used in environmental assessments. The bristle worm *Praprionospio pinnata*, located along the coasts of

TABLE 12.4 Comparisons of abundance of benthic organisms (meiofauna) between their original levels (control levels) and levels after benthic disturbance, as seen in monitoring results (Fukushima, 2004(18), revised; original data from[a,b,c,d,e]).

Experiment	Organisms	1st Monitoring	2nd Monitoring	3rd Monitoring
BIE II	Nematoda	higher (NS)	lower (*)	–
	Harpacticoida	NS	NS	–
JET	Nematoda	lower (**)	lower (NS)	higher (NS)
	Harpacticoida	lower (NS)	higher (NS)	higher (NS)
IOM	total	lower (NS)	lower (*)	higher (NS)
	Nematoda	lower (NS)	lower (*)	higher (NS)
	Harpacticoida (ab)	lower (NS)	lower (*)	higher (NS)
INDEX	total abundance	lower (<50%)	lower (<41%)	lower (<13%)
	NO of Group	unchanged	decrease	decrease

** $p<0.01$, * $p<0.05$ significant difference in statistical analysis, NS; not significant

a Ozturgut, E., Trueblood, D.D. and Lawless, J. "An overview of the United States' Benthic Impact Experiment", *Proceedings of the International Symposium on Environmental Studies for Deep Sea Mining*, MMAJ, Tokyo (1997), 23–32.

b Shirayama, Y. and Fukushima, T. "Responses of meiobenthos community to the rapid resedimentation". *Proceedings of the International Symposium on Environmental Impact of Deep Sea Mining*, MMAJ, Tokyo (1997), 187–196.

c Radziejewska, T., Rokincka-Praxmajer, J. and Stoyanova, V. "IOM BIE revisited: meiobenthos at the IOM BIE site 5 years after the experimental disturbance". *Proceedings of the 4th ISOPE Ocean Mining Symposium*, Szczecin, Poland (2001), 63–68.

d Radziejewska, T., Szamalek, K. and Kotlinski, R. "Marine environment in the IOM Area (Clarion-Clipperton Region, Subtropical Pacific): current knowledge and future needs", *Proceedings of the 5th ISOPE Ocean Mining Symposium, isope,* Tsukuba (2003), 188–193.

e Sharma, R. *et al.* "Monitoring effects of simulated disturbance at INDEX site: current status and future activities", Proceedings of the 5th ISOPE Ocean Mining Symposium, ISOPE, Tsukuba (2003), 208–215.

Japan, is known as an excellent species in areas of the ocean where pollution has progressed, and has been used as an environmental pollution sensor. However, at some point, the taxonomic system was changed and *Praprionospio pinnata* was divided into four different species (Fig. 12.4).[17] If all four of those species were equipped with the ecological feature of being suitable pollution indicators,

17 *Praprionospio pinnata* was reconstituted into four species, *P. patiens, P. cordifolia, P. coora and P. oceanensis.* Of these, *P. patiens* can be a polluted indicator species, but *P. coora* and *P. oceanensis* are rather distributed in poor nutrient waters.

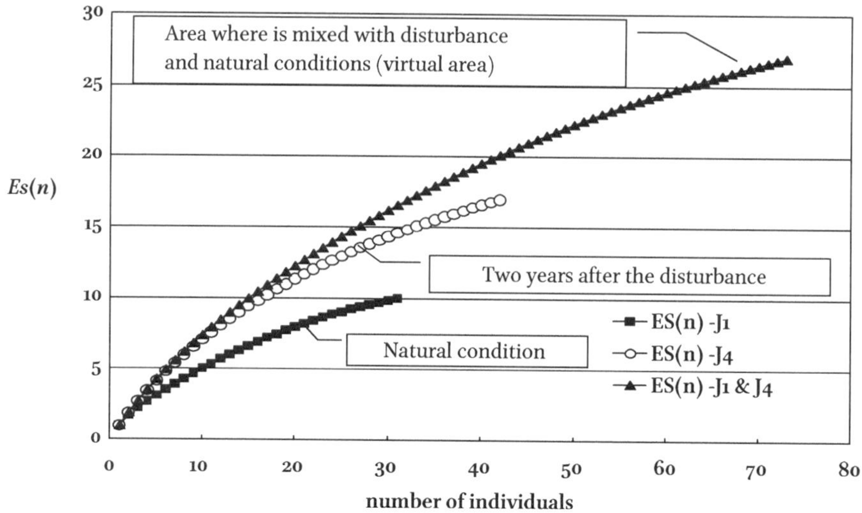

Es (*n*) is one of the index to express species diversity. It can be calculated by the following equation

$$Es\,(n) = \sum_{i=1}^{S}\left(1 - \frac{{}_{N-Ni}C_{n}}{{}_{N}C_{n}}\right) \quad (n \leqq N)$$

S; total number of species, N: total number of individuals, Ni; Number of individuals of the (i) species

FIGURE 12.3 Comparisons species diversity (*Es* (*n*)) between the natural condition and after the artificial disturbed condition. (Fukushima, T., Ecological characteristics of deep-sea benthic organisms in relation to manganese nodules development practices. Thesis Kyoto Univ., 2004)

previous environmental assessment results would not have had to change. However, the results showed that of the four species, only one could become a sensor for environmental pollution and the other three could not. Two in particular were actually found to inhabit openly oligotrophic sea areas. We can no longer go back and verify previous results, but we now cannot refute the possibility that the results of past environmental assessments using these indicator species were incorrect. Following this logic a step further, ocean management plans may have been considered based on the results of those environmental assessments, and may have affected the policies that came afterward.

Fortunately, as the taxonomic and ecological expertise regarding deep benthic organisms was insufficient, none of them were widely approved as indicator species. However, in the future, when development is advancing and

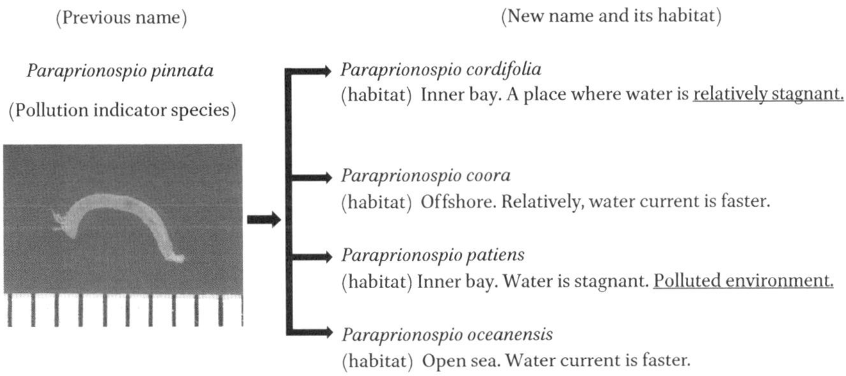

FIGURE 12.4 Name change of species which was used as pollution indicator species.
PHOTO: AUTHOR

rigorous EIAs and certain indicator species are needed, issues like this one may be unavoidable.

6 The Contributions of Science

As I have stated earlier, the goal of science is to understand the principles that penetrate the natural world. However, it is a winding road to get there. Therefore, in the decision-making process, science cannot suggest only convenient information. Scientists maintain a spirit of criticism and do not hesitate to revise past theories and laws. Therefore, even if something is suggested once, and even if those results had major significance for decision-making, they may be retracted. Therefore, scientists may hinder the progress of development, policymaking, and legislation. In this way, scientists can potentially confuse developers, policymakers, and lawmakers. However, that is the true shape of science, and people who demand results that go along with certain policies are disqualified as scientists. Conversely, precisely because science makes new, unexpected discoveries and ascertains the past, it makes us aware of our own ignorance and humbles us in the face of nature.

This also applies to deep sea floor mineral resource development. The deep sea has started to become more familiar owing to the advances of marine science. Just as we are learning about the unique ecosystems of the deep sea, it has become clear that the deep sea is a place of great interest for other industries.[18]

18 The deep sea is expected to be an attractive place for various industries including genetic resources. Fukushima, T., and Okamatsu, A. "Current Issues in Seafloor Massive Sulfide Mining Development", *Proc. 20th Int. Offshore and Polar Eng. Conf.* (2010), 181–185.

In this context, when arguments arise that pit preservation against utilization, or one form of utilization against another, it is difficult to make adjustments from a unidirectional point of view. Thus, the objective data and information provided by science are important. This, in itself, is what I consider science's biggest contribution to seabed mineral resource development.

Acknowledgements

This paper was produced for the proceedings of the 41st Annual Conference of the Center for Oceans Law and Policy held in May 2017. The conference was convened by the Ministry of Maritime Affairs of Republic of Indonesia and the Center for Oceans Law and Policy, University of Virginia School of Law. The International Seabed Authority (ISA), which took charge of one of the session in the conference, has offered the opportunity to the author to participate the conference. As for the contents of presentation, Mr. Ascencio-Herrera who is deputy to the Secretary General of ISA and is the chair of the session, and Dr. Shirayama who is Executive Director of the JAMSTEC supplied valuable suggestions. The author was financially supported by a project "next-generation technology for ocean resources exploration", which is one of "Cross-ministrial strategic innovation promotion program (SIP)". The present author would like to express his gratitude to everyone concerned this symposium. Finally, the author thanks Crimson Interactive Pvt. Ltd. (Ulatus) (www.ulatus.jp) for their assistance in manuscript translation and editing.

CHAPTER 13

The International Seabed Authority's Publication of Revised Draft Exploitation Regulations: On the Right Track but Significant Work Remains to Be Done

*Steven Potter**

Abstract

In early August 2017, the ISA published a revised draft of the exploitation regulations designed to govern the exploitation of nodules, sulphides and crusts on and under the seabed beyond national jurisdictions (the 'Area'). This paper touches on the current status of exploration activities in the Area and the regulatory framework currently in place governing exploration activities in the Area but is principally focused on the process currently being undertaken by the ISA in developing and finalizing the exploitation regime and the extent to which stakeholder concerns have been addressed in the revised draft exploitation regulations.

1 Background and Introduction

August 2017 saw the publication by the International Seabed Authority (ISA) of its second working draft exploitation regulations (Revised Draft Exploitation Regulations) designed to regulate exploitation by Contractors of polymetallic nodules, polymetallic sulphides and ferromanganese crusts[1] on and under the seabed in the 'Area', that part of the seabed beyond national jurisdiction.

The revised draft is another positive step forward by the ISA on the road to establishing a workable regulatory regime governing deep seabed mining activities in the Area. However, a significant amount of work remains to be

* Steven Potter, Partner, Pinsent Masons. The author's PowerPoint presentation is available at http://www.virginia.edu/colp/pdf/yogya-potter.pdf.

1 The author expects that once the working draft is near finalization, there will be separate (but very similar) regulations for each of these three resources (as is currently the case with the exploration regime).

 | DOI:10.1163/9789004366619_018

carried out to finalise the regulatory regime and provide a workable regulatory framework that incentivises Contactors to commit significant investment and resources to develop exploitation projects in the Area. These efforts must also address concerns of State Parties to the United Nations Convention of the Law of the Sea (the Convention), NGOs and other environmental groups of the potential negative environmental impact of deep seabed mining.

This paper will briefly examine:
- the current status of exploration activities in the Area;
- the current regulatory framework governing exploration activities in the Area;
- the process for developing and finalising the exploitation regulatory regime; and
- the August 2017's draft exploitation regulations and the extent to which this revised draft addresses key concerns of Contractors.

2 Status of Exploration Activities and Exploration Regulatory Regime

2.1 *Current Exploration Activities*

As of May 2017, the ISA had entered into 27 15-year exploration contracts with exploration Contractors.[2] These Contractors include State Parties to the Convention, State-owned entities and privately owned companies.[3] Seventeen of the contracts are for exploration of polymetallic nodules in the Clarion-Clipperton Fracture Zone ('CCZ') and Central Indian Ocean Basin. There are six contracts for exploration of polymetallic sulphides in the South-West Indian Ridge, South-East Indian Ridge, Central Indian Ridge and the Mid-Atlantic Range and four contracts for exploration for cobalt-rich crusts in the Pacific and South Atlantic Oceans.[4] Sponsoring States include Belgium, Brazil, China, France, Germany, Japan, Russia and the UK.

At the ISA's latest annual sessions in Jamaica in August 2017, an application by the Government of Poland for an exploration contract for polymetallic sulphides in the Mid-Atlantic Ridge was approved. This, combined with the

2 ISA, Report of the Secretary-General, "*Status of contracts for exploration and related matters*"; ISBA/23/C/7, paragraph 2.

3 The Convention also establishes the Enterprise as an organ of the ISA to engage in exploration and exploitation but the Enterprise has not yet been "operationalized". This will likely need to occur once the exploitation regime is in place and is another significant piece of work on the ISA's "to do" list.

4 *Supra* note 3, at Annex 1.

approval last year of South Korea's application for a cobalt rich crusts exploration contract East of the Northern Mariana Islands in the Pacific Ocean (which in August 2017 had yet to be signed), will bring the number of exploration contracts to 29. As interest in seabed mining grows and the ISA continues to make progress developing the regulatory pathway to exploitation, there is now a very real prospect of deep seabed mining taking place in the Area within the next decade.

2.2 *Current Exploration Regulatory Regime*

The exploration regulatory regime governing Contractors carrying out exploration activities in the Area has been in place for some time. In addition to Part XI, Annex III of the Convention, and the Agreement relating to the Implementation of Part XI of the United Nations Convention on the Law of the Sea of 10 December 1982 (the '*1994 Agreement*'), there are currently three principal regulations in place. In 2000, regulations on prospecting and exploration of polymetallic nodules[5] were first approved with substantially similar regulations for prospecting and exploration of polymetallic sulphides[6] and cobalt-rich crusts[7] approved in 2010 and 2012 respectively. These regulations are supplemented currently by four recommendations[8] of the ISA's Legal and Technical Commission (LTC), which exploration Contractors must "observe to the extent reasonably practicable".[9]

While prospecting (which does not grant the relevant entity any exclusive rights) only requires a notification to be made to the ISA,[10] each of these three regulations provides for a standard form exploration contract to be entered into by the successful applicant, hence the moniker 'Contractor'. These exploration contracts have an initial term of 15 years and are the principal legal

5 Regulations on Prospecting and Exploration for Polymetallic Nodules in the Area, ISBA/6/A/18. These have been amended a number of times, including most recently in 2014, see ISBA/20/A/9.

6 Regulations on Prospecting and Exploration for Polymetallic Sulphides in the Area, ISBA/16/A/12/Rev.1. These have been amended a number of times, including most recently in 2014, see ISBA/20/A/10.

7 Regulations on Prospecting and Exploration for Cobalt rich Ferromanganese Crusts in the Area, ISBA/18/A/11.

8 LTC Recommendations for the guidance of contractors: (1) on the content, format and structure of annual reports; (2) for the reporting of actual and direct exploration expenditure; (3) for the assessment of the possible environmental impacts arising from exploration for marine mammals in the Area; (4) and Sponsoring States relating to training programmes under plans of work for exploration; see https://www.isa.org.jm/mining-code/Recommendations.

9 *Supra* note 6, Annex IV (Standard Clauses for Exploration Contracts), Section 13.2(c).

10 *Supra* note 6, Regulation 3.

mechanism through which legally enforceable rights and obligations are established between each sponsored Contractor and the ISA. Importantly, each exploration contract makes it clear that the sponsored Contractor must comply with the Convention and the rules, regulations and decisions of the ISA. This contractual mechanism, along with the relevant sponsoring State's regulatory regime, is key because the Convention only creates rights and obligations amongst Convention State Parties (and the ISA).

The ISA is not the sole regulating authority for exploration and exploitation activities in the Area. The Convention provides that each sponsoring State also has significant obligations and, as those familiar with this area of law will know, these obligations were considered in some detail by the Seabed Disputes Chamber in its advisory opinion on responsibility and liability of Sponsoring States (the Advisory Opinion).[11] An important element of each sponsoring State's due diligence obligation to ensure compliance with the contract, the Convention, 1994 Agreement and the rules, regulations and procedures of the ISA is to take reasonably appropriate measures in its legal system which must consist of the implementation of national laws, regulations and administrative measures. As we move towards exploitation, we are now seeing a recent (and somewhat belated) increase in the number of sponsoring States implementing principal legislation governing exploration and exploitation activities of their sponsored Contractor(s) in the Area, including Singapore[12] in 2015 and China[13] and France[14] in 2016. The author also understands that South Korea is now actively considering introducing legislation in the near future.

Sponsoring State legislation is a crucial part of the regulatory regime, particularly in respect of compliance and enforcement related matters. While the Advisory Opinion states that it is an obligation for Sponsoring States to have this legislation in place, it expressly states that having this in place at the time of contract award is not a "condition precedent" to the ISA's award of the contract.[15] Given the increased risks arising from exploitation compared to

11 Responsibilities and Obligations of States Sponsoring Persons and Entities with respect to Activities in the Area, Seabed Disputes Chamber of the International Tribunal for the Law of the Sea, Advisory Opinion of 1 February 2011.

12 Deep Seabed Mining Act 2015.

13 Law of the People's Republic of China on Exploration for and Exploitation of Resources in the Deep Seabed Area, effective 1 May 2016. This law was introduced more than 15 years after China sponsored COMRA's first exploration contract in 2001. Observers should also note that there are no express provisions in this new law providing for decisions of the Seabed Disputes Chamber to be enforceable in China—cf. Singapore's Deep Seabed Mining Act, Sections 18 and 19.

14 Ordonnance no. 2016-1687 du 8 decembre 2016 relative aux espaces maritimes relevant de la souverainete ou de la juridiction Republique Francaise.

15 *Supra* note 12, paragraph 219.

exploration (which principally involves resource identification and environmental baseline studies, and, in the case of nodules, does not involve drilling), there is a strong argument for the LTC to insist that it will not recommend approval of an application for an exploitation contract to any applicant whose sponsoring State has not yet introduced adequate legislation. Indeed, there are still a number of sponsoring States that, despite sponsoring one or more contractors for more than 15 years, have still not introduced the necessary primary legislation.

Notably, a new draft regulation 91 (*DR 91*) has been included in the Revised Draft Exploitation Regulations. DR 91 is proposed to apply to Sponsoring States and requires them to "take all necessary measures to secure compliance by Contractors whom they have sponsored with:

- Part XI of the Convention, the 1994 Agreement and the rules, regulations and procedures of the Authority and the terms and conditions of the exploitation contract"; and
- a number of specific regulations including, for example, the new inspection related regulations (DRs 85, 86 and 87) and ISA compliance notices.

This appears to add little to what already comprises the sponsoring States' due diligence obligation and the author views this DR as more akin to a placeholder for discussion as DR 91 is further developed prior to finalization. What is needed is a clear understanding of how overlapping ISA and sponsoring State responsibilities will work together in practice. For example, which authority will take the lead in supervision, compliance and enforcement matters? This outstanding issue does appear to be recognised by the ISA and the author understands that a matrix of responsibility is likely to be published by the ISA for discussion prior to finalisation of the exploitation regime. This matrix will also include flag State responsibilities.[16]

3 Process for Developing and Finalizing the Exploitation Regulatory Regime

3.1 *2014 through 2015—Initial Consultation Process*

Although calls began in 2010 for work to start developing the exploitation regulatory regime, the process only began in March 2014 with the publication by the ISA Secretariat of a survey for stakeholders to respond to.[17] This was

16 See ISA Technical Study: No. 17. *'Towards an Environmental Management Strategy for the Area'*, Section 6.1.3.

17 See Developing a Regulatory Framework for Mineral Exploitation in the Area, February 2014 available at https://www.isa.org.jm/sites/default/files/isa-ssurvey.pdf.

followed in 2015 by the publication of two consultations papers: (1) *'Developing a Regulatory Framework for Mineral Exploitation in the Area'*;[18] and (2) *'Discussion Paper on the Development and Implementation of a Payment Mechanism in the Area for consideration by Members of the [ISA] and all stakeholders'*.[19] The former contained a framework for the exploitation regime which closely followed the structure of the current exploration regime. Both consultation papers were subject to stakeholder review and comment and are available for review on the ISA's website. Of particular note is the level of engagement by stakeholder groups, particularly NGOs and State Parties to the Convention. Of the 48 submissions received by the ISA, 19 were from NGOs, nine were from sponsored Contractors with only nine out of 168 State Parties to the Convention submitting comments to the ISA. The relatively limited engagement by State Parties in the consultation process is an issue of which the ISA Secretariat is keenly aware of and is seeking to address as it was also an issue in the stakeholder consultation exercise recently carried out by the ISA Secretariat in respect of the Revised Draft Exploitation Regulations.

3.2 *July 2016—Publication of Initial Draft Exploitation Regulations*

July 2016 saw the publication by the ISA of a detailed first working draft of the exploitation regulations (*'Initial Draft Exploitation Regulations'*).[20] At that time, it was envisaged that separate environmental regulations and ISA inspectorate regulations would be prepared and therefore the Initial Draft Exploitation Regulations did not address these important areas of regulation. The Initial Draft Exploitation Regulations comprised eleven parts covering the following key areas:

- Applications for approval of exploitation contracts (Part II & Annex 1);
- Terms of Exploitation Contract (Part III, Part IV & Annex VII);
- Financial Terms (Part V);
- Information gathering and handling (Part VI);
- Enforcement and Penalties (Part IX);
- Settlement of Disputes (Part X); and
- Review of the Authority's Regulations (Part XI).

18 See Developing a Regulatory Framework for Mineral Exploitation in the Area, March 2015 available at https://www.isa.org.jm/files/documents/EN//Survey/Report-2015.pdf.

19 See A Discussion Paper on the Development and Implementation of a Payment Mechanism in the Area, March 2015 available at https://www.isa.org.jm/files/documents/EN//WorkingPapers/DiscussionPaper-FinMech.pdf.

20 See Working draft of Exploitation Regulations and Standard Contract Terms available at https://www.isa.org.jm/files/documents/EN/Regs/DraftExpl/Draft_ExplReg_SCT.pdf.

We will consider some of the key concerns noted by Contractors in Sections 3.5.1 to 3.5.3 of this paper and the extent to which they have been addressed in the Revised Draft Exploitation Regulations, which were published in August 2017. One important point to note from the initial draft is the clear difference in regulatory approach contemplated compared to the exploration regime. For various reasons, including budgetary limitations, the ISA's approach under the exploration regime has been to apply "a light touch". For example, with one exception, every sponsored application made to the ISA for an exploration contract has been approved and, despite several Contractors doing very little work under their 15-year exploration contracts, no compliance notices have ever been issued by the ISA.[21] This contrasts with the approach proposed by the ISA in the Initial Draft Exploitation Regulations. Indeed, one valid criticism raised by a number of contractors in respect of the initial draft was that the proposed regime provided the ISA with too much control over commercial and operational matters and that these matters would be better left to be managed by the Contractor.

Two examples of this were the proposed development and production related obligations imposed on Contractors in the Initial Draft Exploitation Regulations.

> *Development*
> The approach to development was unusually onerous for sponsored Contractors including an obligation to comply in all material respects with certain minimum expenditure commitments in respect of the development of the mining project that were to be set out in the sponsored Contractor's mining plan.[22] While minimum expenditure commitments are common in many State's mining and oil and gas regulatory regimes, they are typically only imposed during the exploration phase and not during the development and mining phases.

21 See ISA, Summary Report of the Chair of the Legal and Technical Commission on the work of the Legal and Technical Commission during the eighteenth session of ISA, ISBA/18/C/20 ("In general terms, exploration proceeded slowly during the reporting period; the Commission expresses concern that one contractor has not carried out any exploration activities during the first 10-year period under the contract as it enters the final 5-year phase. The Commission finds this deeply troubling and invites the contractor to look into this matter accordingly; Some contractors report no work at all under exploration, mining and metallurgy related activity….").

22 Initial Draft Exploitation Regulations, Annex VI, Section 10.3.

Production

The draft also proposed imposing unduly restrictive production related obligations on sponsored Contractors, including compliance with specified production rates rather than leave this to be determined by the Contractor based on operating and market conditions.[23] Expecting a sponsored Contractor to maintain production at a specific level during adverse market conditions would require them to continue to mine when uneconomic to do so. Furthermore, while temporary reductions in production due to market conditions would be permitted, this required the consent of the ISA Council (which currently only meets once a year) and the payment of a yet to be determined fee.

The Initial Draft Exploitation Regulations were open to stakeholder submissions which were submitted in November 2016. These submissions are discussed below at Section 3.4.

3.3 *January 2017—Publication of Draft Environmental Regulations*

A detailed '*Discussion Paper on the development and drafting of Regulations on Exploitation of Mineral Resources in the Area*' was published by the ISA Secretariat in January 2017.[24] This paper contained detailed draft regulations (with a number of placeholders) governing environmental matters (Draft Environmental Regulations). The draft is split into 16 parts covering the following key matters:

- Guiding principles, Restrictions and Prohibitions (Part II);
- Initial Environmental Assessment (e.g. baseline report; scoping report) (Part III);
- Preparation of Environmental Plans (Part V);
- Public Consultation (Part VI);
- Compensatory Measures (including a proposed environmental liability trust fund) (Part X); and
- Compliance, Supervision, Enforcement and Penalties (Part XIV).

The Draft Environmental Regulations also marked an initial attempt to populate the required contents of a number of exploitation related environmental plans that a sponsored Contractor will need to prepare for ISA approval, including an environmental scoping report, environmental impact statement, environmental management and monitoring plan, closure plan and emergency response and contingency plan.

---

23 *Id.* Annex VI, Sections 11, 12 and 13.

24 See Developing a Regulatory Framework for Mineral Exploitation in the Area available at https://www.isa.org.jm/files/documents/EN/Regs/DraftExpl/DP-EnvRegsDraft25117.pdf.

As this draft was published after the November 2016 deadline for stakeholder comments, it has not been the subject of a formal stakeholder consultation process in the same way as the Initial Draft Exploitation Regulations. However, the author notes that most NGO submissions and a number of State Party submissions (e.g., Germany and Australia) focused mainly on environmental issues notwithstanding that the Initial Draft Exploitation Regulations only contained minimal references to the environment as these were to be dealt with in separate environmental regulations. For those interested in a range of views on the Draft Environmental Regulations, it is worth reviewing ISA Technical Study No. 17 that was produced following the March 2017 workshop in Berlin that focused principally on discussing the Draft Environmental Regulations.[25] A number of significant issues considered at the workshop included:

- whether to combine the exploitation regulations, environment regulations and inspectorate regulations into one final document (which now looks to be the approach that will be taken by the ISA);
- the development of acceptable environmental impact criteria for the application process for use as evaluation criteria by the LTC;
- that the division of responsibilities and tasks among Sponsoring States, the ISA and its organs required clarification to allow for effective supervision and enforcement of Contractor activities in the Area;
- the possibility of establishing a new organ or section within the ISA with responsibility for environmental related matters (this echoes the views of a number of Environmental NGOs whom, as part of their submissions on the Initial Draft Exploitation Regulations, recommended the establishment of a separate commission within the ISA with responsibility for environmental matters);
- that the definition of Interested Persons (relevant in the context of public consultation) was too narrow as it was limited to directly affected persons 'in the opinion of the [ISA];
- that the content, role and function of Environmental Impact Assessment should be clearly specified in the regulations; and
- that adaptive management was crucial in order to proceed in situations where there is uncertainty that cannot be resolved before development, as well as to ensure a precautionary reaction to unanticipated effects.

Clearly, both the Draft Environmental Regulations and ISA Technical Report No. 17 are important pieces of work in the ISA's building block approach to developing the exploitation regime but both documents have since been

25 See Towards an ISA Environmental Management Strategy for the Area, ISA Technical Study: No. 17 available at https://www.isa.org.jm/sites/default/files/files/documents/berlinrep-web.pdf.

superseded by the second draft of the Exploitation Regulations published in August 2017 (Revised Draft Exploitation Regulations).[26] This revised draft integrates exploitation, environmental and inspection related regulations into a more concise and streamlined standalone document. Given that the operative provisions (i.e., the regulations) in the Revised Draft Exploitation Regulations only run to approximately 50 pages of text compared to a combined 150 pages of text for the operative provisions in Initial Draft Exploitation Regulations and the Draft Environmental Regulations, the author expects that additional environmental provisions will be added to the next draft (e.g., to add in any agreed environmental incentive mechanisms such as the proposed seabed sustainability fund and the environmental liability trust fund). The author also expects a number of technical and administrative matters relating to environmental issues to be dealt with by way of LTC Recommendations for the Guidance of Contractors (which must be "observed" by Contractors) that will also need to be developed and finalised by the ISA before any application for an exploitation contract is made by a Contractor.

3.4 *Stakeholder Submissions—Initial Draft Exploitation Regulations*

The ISA received 43 submissions from stakeholders on the Initial Draft Exploration Regulations. Of those submissions, ten were from Contractors with a repeat of the limited participation by States with only eight State submissions (including a submission by the United States, which is not an ISA member State). More engagement by ISA member States in the development of the exploitation regulatory regime and other activities and meetings of the ISA going forward will be crucial and the ISA Secretariat is currently seeking to improve engagement with ISA Member States generally by "reorganising the meeting schedule of various organs of the [ISA], better communication with permanent missions and member States and increased outreach and communications activities".[27]

In terms of transparency, most stakeholders consented expressly to the publication of their submissions on the ISA's website.[28] Disappointingly, three

26 See Draft Regulations on Exploitation of Mineral Resources in the Area, ISBA/23/LTC/CRP.3, 8 August 2017 available at https://www.isa.org.jm/files/documents/EN/Regs/DraftExpl/ISBA23-LTC-CRP3-Rev.pdf.

27 ISA, Report of the Secretary General, "*Comments by the Secretary-General on the recommendations contained in the final report of the periodic review of the International Seabed Authority pursuant to article 154 of the United Nations Convention on the Law of the Sea.*" ISBA/23/A/5/Rev.1, Paragraph 4(a).

28 Contributions to the working draft exploitation regulations available at http://www.isa.org.jm/files/documents/EN/Regs/DraftExpl/Comments/Comments_Listing.pdf.

sponsored exploration Contractors,[29] the International Maritime Organisation, Korea Institute of Ocean Science and Technology and the United States did not expressly consent to the publication of their comments. For submissions on the Revised Draft Exploration Regulations that are due in December 2017,[30] the ISA has specified that submissions will be published unless the stakeholder expressly states that it is not to be published. Given the importance of transparency in the development of the exploitation regime, the author would suggest that the ISA takes a further step and refuses to take account of any submissions that are not available for public review.

For those interested in reviewing sponsored exploration Contractor comments on the Initial Draft Exploitation Regulations, the author considers that the most comprehensive submissions were made by Naura Ocean Resources Inc (NORI), Global Sea Mineral Resources NV (GSR), UK Seabed Resources Ltd (UKSRL) and Ocean Mineral Singapore (OMS), all of whom are privately owned Contractors. This is not surprising given that it is privately owned Contractors that have taken a leading role in working with the ISA in moving towards exploitation. One of many examples of this is the work carried out by UKSRL and GSR in developing a financial model to assist the ISA in modelling the impact of different *ad valorem* royalty rates on proposed polymetallic nodules mining projects in the CCZ. This model is further discussed below in the context of the review of the new payment provisions contained in the Revised Draft Exploitation Regulations.

We next consider the recently published Revised Exploitation Regulations, including the extent to which it addresses key issues raised by sponsored exploration Contractors in their submissions on the Initial Draft Exploitation Regulations.

3.5 *August 2017—Publication of Revised Draft Exploitation Regulations*[31]

The revised draft was published on the ISA's website in August 2017. As was the case with the Initial Draft Exploitation Regulations, the revised draft has not been signed off by the LTC and the ISA Secretariat expressly states that "no attempt has been made to reflect the views of the [LTC] on the draft, which

29 China Ocean Minerals Research and Development Organisation; Marawa Research and Exploration Ltd; and China Minmetals Corporation.

30 The ISA has requested that submissions are received from stakeholders by 17 November 2017 and no later than 31 December 2017.

31 Draft Regulations on Exploitation of Mineral Resources in the Area, ISBA/23/LTC/CRP.3, 8 August 2017 available at https://www.isa.org.jm/files/documents/EN/Regs/DraftExpl/ISBA23-LTC-CRP3-Rev.pdf.

remains a work in progress".[32] We do know however that the ISA Secretariat prepared the revised draft taking into account, amongst other things, the LTC's response to stakeholder comments on the Initial Draft Exploitation Regulations so it is not unreasonable to assume that changes made to the draft to take account of stakeholder comments have been made based on discussions between the ISA Secretariat's drafting team and the LTC.

As mentioned earlier, the revised draft includes environmental regulations and detailed inspection related regulations and it therefore appears that what is now proposed is for exploitation, environmental and inspection related regulations to be integrated into a single set of regulations. This is to try and reduce the risk of duplication and inconsistency arising between different sets of regulations (particularly between the Environmental Regulations and Exploitation Regulations), a concern raised by a number of stakeholders including the United Kingdom[33] and UKSRL[34] and that was also discussed at the Berlin workshop in March 2017.

The revised draft comprises twelve parts and includes a number of newly populated annexes and appendices. Key parts and appendices include:

- Applications for approval of exploitation contracts (Part II);
- Exploitation Contracts (Part III);
- Environmental Matters (Part IV);[35]
- Contractor Obligations (Part V);[36]

32 ISA, Note by the Secretariat, "*Draft Regulations on exploitation of mineral resources in the Area*", ISBA/23/C/12, paragraph 7.

33 Submission of the United Kingdom Government in Response to the ISA July 2016 Report on Developing a Regulatory Framework for Mineral Exploitation in the Area, paragraph 6 available at https://www.isa.org.jm/files/documents/EN/Regs/DraftExpl/Comments/UK.pdf.

34 UK Seabed Resources Submission in Response to the International Seabed Authority's Report on Developing a Regulatory Framework for Mineral Exploitation in the Area, paragraph 3 available at https://www.isa.org.jm/files/documents/EN/Regs/DraftExpl/Comments/UKSR.pdf.

35 This part is new but reflects a number of matters covered in the Draft Environmental Regulations. It includes general principles (e.g., protection and conservation of the Marine Environment, Best Environmental Practice, the precautionary approach and an ecosystem approach), and regulations covering the environmental scoping report, environmental impact statement, environmental management and monitoring plan and closure plan, ongoing obligations in respect of the marine environment and environmental performance.

36 This part is also new and results from the ISA Secretariat drafting team's decision to move a large number of provisions from the standard form exploitation contract into the regulations. This is discussed in Section 3.5.3 of this paper in the context of regulatory stability. Note also the new regulation (DR 31) that requires that all mining vessels and collectors

- Review and modification of a Plan of Work for Exploitation (Part VI);
- Financial Terms (Part VII);[37]
- Information Gathering and Handling (Part VIII);[38]
- Administrative Fees (Part X); and
- Inspections (Part XI).[39]

This paper does not provide a regulation-by-regulation review of the changes made in the Revised Draft Exploitation Regulations but it is clear from a review of the revised draft that the Secretariat's drafting team have considered and sought to address key legitimate concerns raised by many stakeholders. Examples include the removal of the onerous minimum development expenditure requirement, the removal of the Contractor's requirement to seek ISA Council approval for a reduction in production rates and the extension of the Sponsoring State's termination notice from six months to twelve months. The paper does however consider a more detailed review of changes to the provisions dealing with contract duration, financial terms and regulatory stability.

3.5.1 Contract Duration

The Initial Draft Regulations tentatively proposed an initial contract duration of no greater than 20 years or such shorter period as applicable to the expected economic life of the mining activities.[40] It also proposed that the Contractor may renew the exploitation contract for periods of not more than 10 years if certain conditions are satisfied, including that sufficient commercial quantities of the resource are available for recovery during the period of the relevant extension and that the Contractor has complied in all respects with the Convention, the 1994 Agreement, the Contract, the Exploitation Regulations and other rules, regulations and procedures of the ISA.

This tentative 20 year period was considered to be too short for those Contractors that did comment on this issue, particularly given that it is likely to

be fitted with electronic monitoring systems to record the date, time and location of all mining activities.

37 This part has been significantly revised and now includes a proposed ad valorem royalty related formula (but with the royalty rate yet to be included).

38 This includes a revised definition of Confidential Information. There is still a significant gap between Contractor and NGO views on the approach to be taken in relation to confidentiality. Contractors generally consider that the definition of Confidential Information is too narrow, whilst many NGOs consider it to be too broad.

39 This part has been significantly revised to include detailed and broad ISA Inspector powers. While this was to be expected, perhaps the drafting team has overreached slightly with the proposed inspector power to test mining equipment to destruction. See DR 86(e).

40 *Supra* note 23, Part III, DR 14(1).

take some Contractors up to five years from entry into the exploitation contract to achieve first commercial production as they will not begin to commit significant capital expenditure (and perhaps even finalise their funding arrangements) to develop the mining project (including building, testing and commissioning bespoke mining vessels, deep seabed recovery equipment and onshore processing facilities) until the exploitation contract is awarded.[41] A number of Contractors proposed an initial contract period of up to 30 years including UKSRL and NORI with OMS proposing 30 years from the commencement of commercial production. Tonga Offshore Mining Ltd proposed an initial term of up to 40 years with extensions lasting for a similar period.

The Revised Draft Exploitation Regulations have addressed the concern of those Contractors requesting 30 years and now provide for an initial maximum contract period of 30 years.[42] The revised draft also makes clear that the contract terms applicable to any extension are the standard terms in force at the time that the ISA approves the renewal.[43]

One additional issue that will need to be factored into each Contractor's assessment of the acceptability of this change is the revised process for ISA approval of certain key Contractor plans, notably a revised Environmental Management and Monitoring Plan and a revised Closure Plan. The revised draft provides for initial versions of these plans to be approved as part of the exploitation application process,[44] however, significantly, revised versions of these two plans must be prepared and approved by the ISA no later than twelve months prior to the proposed commencement of production.[45] It would therefore not be a surprise to see Contractors seeking approval for these revised plans relatively quickly after contract award as they are unlikely to commit significant project development capital expenditure until these revised plans have been approved but will not want to wait too long post contract award to commence development activities given the likely length of the development and ramp-up period.

3.5.2 Financial Terms of Exploitation Contract

Broadly reflecting the discussions from the first payment workshop in San Diego in May 2016, Part V of the Initial Draft Exploitation Regulations provided

41 OMS submission, pg. 6 available at https://www.isa.org.jm/files/documents/EN/Regs/DraftExpl/Comments/OMS.pdf.
42 Revised Draft Exploitation Regulations, DR 13(1).
43 *Id.* DR 13(5).
44 *Id.* DR 4(3)(g).
45 *Id.* DR 29(1)(c) and (d).

for two principal annual fees[46] to be paid by a Contractor as well as tentatively proposing an *ad valorum* royalty payable on sales value or the deemed sales value of the mineral ore sold or removed without sale from the contract mining area. In addition to the two principal annual fees, other proposed fees included application related fees for an exploitation contract, exploitation contract renewal, transfer of an interest in an exploitation contract, the use of an exploitation contract as security and material change to a plan of work. A placeholder was also included referring to the future inclusion of potential environmental fees, including environmental liability trust fund contributions.

The Initial Draft Exploitation Regulations also suggested that the 'system' of payments adopted under Part V would be subject to a separate period of stability (presumably longer than the five year period of stability proposed before the Council may review and apply amendments to the regulations), before it was reviewed and subject to change by the ISA Council.[47]

With the exception of NORI, a number of Contractors did not provide significant comments on Part V of the initial draft, presumably taking the same approach as UKSRL who noted that it was actively participating in the payment related workshops and that future workshops would provide significant guidance for further development of the payment provisions.[48] Some did note however that a period of stability would be required not just in respect of the system but also to protect against increases in royalty rates, in order to incentivise early movers. This is also consistent with discussions at the three payment regime workshops that what is being principally considered to incentivise early movers is an initial light *ad valorem* royalty before moving to a full *ad valorem* royalty with the flexibility of allowing the ISA to impose a different system (e.g., profit sharing) further down the track (towards the latter part of the commercial production period during the initial (now 30-year) contract period).

Two payment regime related workshops have occurred since stakeholder submissions were made in November 2016 and prior to the publication of the Revised Draft Exploitation Regulations and it's worth noting the key outcomes of these workshops.

---

46 An annual contract administration fee (the amount is not specified and is intended to reflect the ISA's costs of administering the contract) payable from contract signature and an annual fixed fee payable calculated by multiplying the total contract area in km^2 by an annual rate per square km from first commercial production and offset against the royalty payable for that year.

47 *Supra* note 23, at Part XI, DR 59 (1) and (5).

48 *Supra* note 34, at paragraph 3.

The London workshop took place in December 2016 and principally focused on identifying the components (including capex, opex and financing costs as well as revenues) of a financial model that is needed by the ISA to assist in finalising the exploitation payment regime, particularly the proposed royalty regime.[49] Environmental 'fees' in the form of potential funds and bonds were also discussed including an environmental liability trust fund, seabed sustainability fund and an environmental bond.

A further workshop took place in Singapore in April 2017 during which a working financial model for nodules exploitation in the CCZ developed by GSR and UKSRL was introduced.[50] This model has since been provided to the ISA Secretariat that we understand will soon arrange for the model to be verified by expert economists. Points to note in respect of the model include the following: it assumes, amongst other things, an *ad valorem* royalty regime transitioning from a light to full royalty in eight years, an exploitation contract application fee of USD one million, an annual contract administration fee of USD 100,000 and a Sponsoring State corporate tax rate of 25%. Of note too is the likely 18% IRR hurdle rate for privately owned contractors which some consider they will need in order to move forward and green light development of a nodules exploitation project in the CCZ.

As mentioned earlier, the financial terms applicable to an exploitation contract have been moved to Part VII of the Revised Draft Exploitation Regulations. Significant changes from the Initial Draft Exploitation Regulations include:

- the removal of the proposed annual contract administration fee;[51]
- the inclusion of a proposed formulae for calculating the *ad valorem* royalty payment which takes into account the average metal grade per tonne of nodules and the average listed price of the relevant metal on a recognised international metal exchange that is "published" by the ISA;[52]
- the introduction of a "first commercial production period" and "second commercial production period" (the length of which is currently not specified)[53]

49 For a summary of the workshop see http://www.resolv.org/site-dsm/files/2016/10/DSM-PRW2-Workshop-Summary-FINAL1.pdf.

50 Deep Seabed Mining—Payment Regime Workshop (PRW) #3, pg. 2–4 available at https://www.isa.org.jm/files/documents/EN/Regs/DraftExpl/DSM-PRW-3.pdf.

51 But see the increase in the number of other administrative fees for particular matters in Appendix II of the Revised Draft Exploration Regulations.

52 *Supra* note 42, at DRs 51, 52, 53 and 54; note also that the use of a listed price reduces the risk of gaming the system through transfer pricing, non-arm's length sale contracts etc.

53 *Supra* note 42, DRs 51 and 52.

during which the royalty rate will be fixed and not subject to change without the agreement of the relevant sponsored Contractor;[54] and
- that the "system of payments" (which the author considers should be defined for clarity) applied to a particular exploitation contract will not change during the initial term of the contract without the agreement of the relevant sponsored Contractor.[55]

The author anticipates that these changes will be broadly welcomed by Contractors, particularly privately owned Contractors looking for a period of stability during which the royalty will not change (the impact of which will therefore be capable of being modelled accurately prior to the project's final investment decision).

Observers should note (but not be concerned) that the revised draft does not contain any of the environmental related fund(s) and other incentives that have been mooted during the consultation process, discussed at a number of payment workshops and included in the Draft Environmental Regulations. The LTC has noted that the environmental provisions in the Revised Draft Exploitation Regulations provide the environmental "framework" and therefore one can presume that these will be added into the document as it further develops. Unsurprisingly, there is broad consensus from those Contractors participating in the payment workshops that these should be funded by revenue received by the ISA through the royalty rather than as separate payments made by Contractors.

Lastly, it is important to note that although the revised payment terms broadly reflect discussions at the payment workshops, it is only "offered as a basis for further discussion, without any specific endorsement at the current stage"[56] from the Secretariat. The Secretariat will therefore prepare and issue a further consultation paper on the payment regime in March 2018 for stakeholder comment.

3.5.3 Regulatory Stability

As one would expect, regulatory stability and predictability is a key concern of Contractors, particularly those that are privately owned. One of the key challenges facing the ISA is that they are developing a regulatory regime for an industry that currently does not exist as there are currently no nodules, sulphides, crusts DSM mining projects in existence, so some flexibility to change

54 *Supra* note 42, DR 73.

55 *Supra* note 42, DR 72.

56 ISA, Note by the Secretariat "*Draft Regulations on exploitation of mineral resources in the Area*", ISBA/23/LTC/6 at Section II F.

the regulations within the short to medium term is clearly needed. The question is what should trigger changes and whether changes to regulations that occur after the award of an exploitation contract should apply to that contract. The proposal for the exploration regime (at least the view taken by a number of contractors with whom the author has spoken) is that changes in the exploitation regulations that take place after the entry into of an exploration contract do not apply to that contract unless agreed by the Contractor.

The Initial Draft Exploitation Regulations dealt with this issue in DR 59 (Review of the Authority's Regulations) by providing an obligation on the ISA Council to review the regulations five years after their approval by the ISA Assembly or at any time thereafter with a right to amend the regulations in light of that review. In addition, this review by the Council could also be carried out at any time at the request of any State Party to the Convention, the LTC or any contractor (through its sponsoring State) if "in the light of improved knowledge or technology, it becomes apparent that the regulations are not adequate".[57] It was however not entirely clear whether or not these changes would apply to contracts already entered into without the agreement of the relevant sponsored Contractor, particularly as the standard contract referred to complying with the regulations of the Authority but did not make clear whether this included complying with regulations as amended from time to time.[58]

A number of contractors including UKSRL, OMS, NORI and GSR were concerned about the uncertainty this created with recommendations including providing for a legally enforceable stabilisation mechanism to compensate contractors for post contract regulatory changes, which have a material negative financial impact on contractors,[59] making it clear that such changes do not apply to contracts already entered into,[60] or "having clear periods of stability in the regulations absent particular justifications or changes in circumstances".[61]

Whilst the regulatory review related provisions were amended in the Revised Draft Exploitation Regulations, the amendments now provide the ISA with more, rather than less, flexibility to make regulatory changes. The revised draft now makes clear that the Contractor must comply with the regulations as amended from time to time.[62] In addition, a significant number of provisions

57 *Supra* note 23, DR 59(2).

58 *Supra* note 23, DR 59(3) (second sentence); DR 59(4) and Annex VII Section 2.1.

59 GSR and OMS submissions; Note also that the standard contract terms do provide for stabilization related contract amendment but this requires the agreement of both parties to the exploitation contract.

60 NORI submission.

61 *Supra* note 42 (UKSRL submission).

62 *Supra* note 42, Annex X, Section 3.1.

that were previously contained in the standard contract terms have been moved into the regulations in order to provide the ISA with the flexibility to make changes to them post contract award without requiring the agreement of the Contractor. This is likely to be a concern for Contractors, particularly the provisions (now proposed regulations) listed below:

- Insurance;[63]
- Maintenance of commercial production;[64]
- Temporary suspension of production;[65] and
- Change of control.[66]

It will be interesting to see how Contractors react to these changes although the author does expect that they will take some comfort from the stability included in the revised payment regime with the proposal to fix the royalty rate for the first and second commercial production periods.

4 Conclusions

If the ISA fails to provide a workable exploitation framework that encourages early movers to invest the billions of dollars necessary to proceed to exploitation, a massive opportunity for this nascent industry will be lost.

While there is still much work to be done, the latest draft is a welcome step in the right direction for the ISA. Of course, there are still a significant number of issues that remain to be addressed in the exploitation regulations, but it is clear that the ISA's drafting team has tried to address legitimate concerns of key stakeholders, particularly those of States and sponsored Contractors. Of those key areas identified, the author considers that much will depend on how the final regime addresses environmental risks and sponsored Contractor operational flexibility while at the same time providing contractors (and their investors) with sufficient regulatory certainty to make the investments necessary to get these exploitation projects up and running. Crucial to this will be agreeing on a payment regime which provides an equitable sharing of the royalties payable to the ISA for the mining of resources also known as the common heritage of mankind.

The recent increase in the ISA's budget and the size and breadth of expertise of the LTC, the appointment at the end of 2016 of Michael Lodge as the

63 *Supra* note 42, DR 26.
64 *Supra* note 42, DR 30.
65 *Supra* note 42, DR 32.
66 *Supra* note 42, DR 43.

Secretary General of the ISA and the re-scheduling of ISA organ meetings/increase in ISA Council meetings for 2018 and 2019 all show that the ISA is very much heading in the right direction. As this paper notes, a significant amount of work has occurred in the development of the draft regulatory regime since the Initial Draft Exploitation Regulations was published and the author hopes that this momentum is sustained over the next few years.

The revised schedule recently proposed by the Secretariat provides for the Council to approve the draft exploitation regime in 2020, which, if the momentum created over the last 18 months is maintained, is a very achievable target date. Contractors need to continue participating in workshops and ISA consultations but States, including Sponsoring States, need to up their game and increase their level of contribution to the process. There is a wealth of experience in developing countries of mineral resources related regulation and now is the time to step up and participate in the process.

PART 4

Marine Pollution and Coastal Ecosystems

CHAPTER 14

Marine Debris in Indonesia: State of Understanding and Ongoing Efforts in Reducing Its Impacts on Marine Habitat

Brahmantya Satyamurti Poerwadi and Abdul Muhari***

Abstract

This article describes the present understanding of marine debris' impact on marine habitats and efforts that have been done in Indonesian marine and fisheries sectors. We first explain briefly the complexity of the marine litter problem: sources, factors, stakeholders, processes and its impact on the ecosystem. From that state of understanding, we show development of practical-oriented policy based on national and technical regulations to establish a national action plan for marine debris prevention. A series of activities were conducted at the national and local level to indicate the commitment of the Ministry of Marine Affairs and Fisheries to combating marine litter. Obstacles during the activities are described and, lastly, we draw conclusions and ways to move ahead.

1 Introduction

The coastal and marine areas are places where diverse activities are undertaken such as fishing, mining, tourism, and transport. In contrast to the terrestrials, coastal and marine areas receive impact either directly or indirectly from the activities conducted on land and at sea, particularly things related to debris and litter. Of the total urban population living in the coastal area of around 130 million people, it is estimated that at least 38 metric tons (MT) of garbage are generated annually and 17 MT per year of which is uncollected waste. In this amount of total uncollected waste, 45 percent of which was

* Director General of Marine Spatial Management, Ministry of Marine Affairs and Fisheries, Republic of Indonesia. The authors' PowerPoint is available at http://www.virginia.edu/colp/pdf/yogya-poerwadi-muhari.pdf.

** Head of Coastal Disaster Mitigation Section, Ministry of Marine Affairs and Fisheries, Republic of Indonesia.

 | DOI:10.1163/9789004366619_019

dumped into the rivers, and 30 percent of which is plastic, is burned and dumped into places that do not have the means of processing waste (Jambeck *et al.* 2015).

In Indonesian waterways alone, composition of the waste can be drawn as is given in Figure 14.1, where at most cases 44 percent of the total waste is organic or household garbage, 21 percent is diapers, 16 percent is plastic bags in addition to the other 5 percent of plastic packaging and the other 14 percent is other plastics, bottles, glass and metal.

From a global and regional point of view, Indonesia contributes about 0.48 – 1.29 million metric tons (MMT) of plastic debris, which is the second largest amount on earth after China (Jambeck *et al.* 2015).

Trash stranded in coastal areas or washed into the ocean and drowned into the sea floor has brought serious impacts on ecosystems such as entanglement

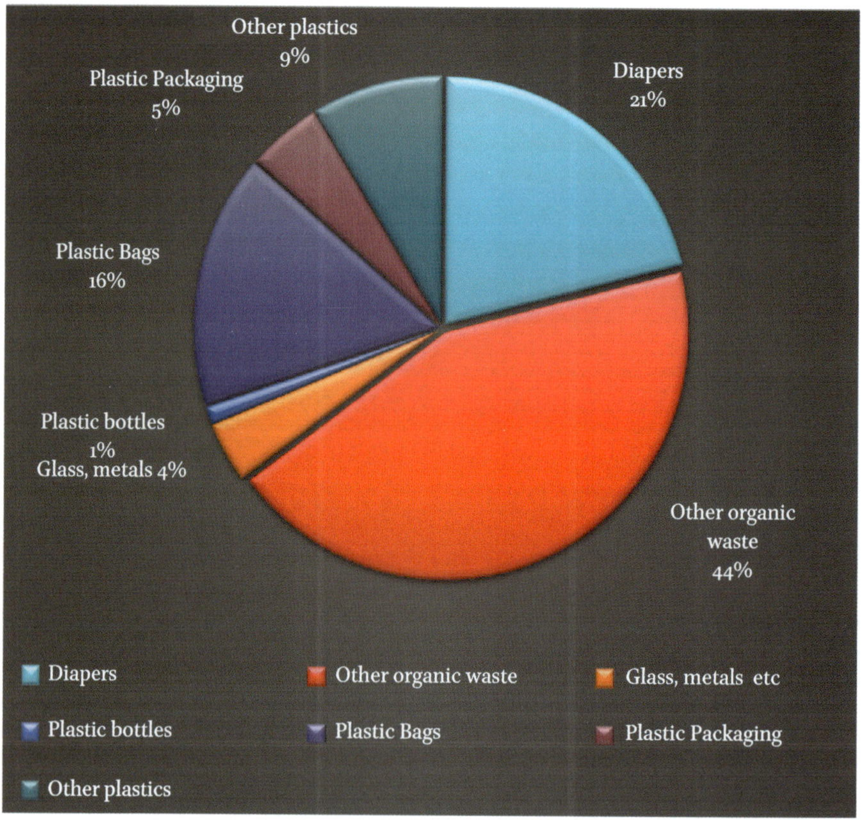

FIGURE 14.1 Composition of waste found in waterways.
SOURCE: WORLD BANK, 2018

in mangrove roots. This affects aeration and disturbs mangrove growth that leads to mortality. In the ocean, plastic covers and interferes with the respiratory processes that eventually leads to mortality. Moreover, micro plastic can contaminate water, where it is consumed by marine biota, which in turn causes of suffocation and intestinal blockage and increases the exposure of marine biota to toxic materials.

2 Policies and Regulations

Indonesian government efforts to address marine litter started in 2008 when the Law No. 18/2008 about Trash Management was established and the Law No. 1/2014 –which is the updated Law No. 27/2007 about the Coastal and Small Area Management. These main regulations have been derived into a more technical regulation such as Governmental Decree No. 81/2012, about Household Garbage Management, and some other local government regulations. The key in implementing those regulations is to initiate commitment and activities inter-ministries, within ministries and local governments as well as non-governmental organizations.

Implementation of the above-mentioned regulations is mainly dedicated to establish grass-root level awareness through education, campaigns and socialization, providing facilities and encouraging innovation on trash processing that have economic value, which can be managed and utilized by communities to increase their daily economic income.

The national government therefore introduced the National Action Plan for Marine Debris Prevention, which in general consists of four main strategies:

> *Strategy 1*: Increase public awareness, law enforcement and stakeholder participation;
> *Strategy 2*: Control plastic waste in terrestrial and coastal areas;
> *Strategy 3*: Control plastic waste in the marine area; and
> *Strategy 4*: Establish mechanisms of funding and institutional building to control marine plastic waste.

In implementing the national strategy, the Ministry of Marine Affairs and Fisheries committed to applying the strategies in coastal and fishing communities through:

a. Conducting continuous public awareness campaigns for fishermen to reduce ghost net, garbage and other plastic waste from fishing activities;

b. Improving waste processing facilities and infrastructure in fishery ports under the control of the Ministry of Marine Affairs and Fisheries;
c. Encouraging community groups under the coordination of local representative of MMAF in mangrove area management, to make regulations to gradually reduce plastics litter into the sea;
d. Playing an important role in controlling plastic litter from fishing and cultivation activities on the coast and sea; and
e. Improving hygiene facilities and infrastructure in small islands.

Some practical experiences and lessons from the activities are described below.

3 Efforts and Activities

Under the Directorate General of Marine and Spatial Management, the Ministry of Marine Affairs and Fisheries has been conducting various efforts to reduce marine litter in a comprehensive way. The activities are framed in a national campaign called the "Love the Sea" movement. The movement is strengthening by the Director General Decree No. 11/2015 about Technical Guideline of Culinary Tourism Waste in Coastal Area (DG of Marine Spatial Planning, 2015).

Under the umbrella of the national movement, locally based activities are conducted such as beach cleanups, local and national campaigns, education and socialization on the threat of marine debris, technical assistance in plastic waste processing as well as the provision of waste processing equipment and composters (Figures 14.2, 14.3 and 14.4).

FIGURE 14.2 Beach condition in a fishery port before and after the cleanup.
PHOTO SOURCE: MINISTRY OF MARINE AFFAIRS AND FISHERIES, 2017

FIGURE 14.3 Community education and socialization on marine debris.
PHOTO SOURCE: MINISTRY OF MARINE AFFAIRS AND FISHERIES, 2017

FIGURE 14.4 Materials for socializations and education.
PHOTO SOURCE: MINISTRY OF MARINE AFFAIRS AND FISHERIES, 2017

4 Conclusions

We have described problems, policies and practical efforts in dealing with marine debris in Indonesia. While the socialization, education, beach cleanups, and training on plastic waste processing are now on going, obstacles in ensuring its sustainability on a community basis are arising, such as the economic value of the products they produce from plastic recycling and the market cost of plastic shredders that do not cover production cost. The Ministry of Marine

Affairs, therefore will continue to encourage the use of biodegradable plastic in fisheries activities and training to enhance the quality of reusable plastic products. We believe that reducing the use of plastic materials in fishery activities will in the long-term significantly reduce the amount of plastic litter in the ocean.

References

Director General of Marine Spatial Management (2015), Decree No. 11/2015 about Technical Guideline of Tourism Waste in Coastal Area.

Jambeck, J.R., R. Geyer, C. Wilcox, T.R. Siegler, M. Perryman, A. Andrady, R. Narayan, K.L. Law (2015). "Plastic waste inputs from land into the ocean," *Science* 347 (6223), pp. 768–771.

CHAPTER 15

A Perspective on Marine Pollution

*Stephen A. Macko**

Abstract

With a large and growing portion of the world's population living only a short distance from coasts, the potential for human impact is easily seen. Rivers and terrestrial run off are thought to transport perhaps 80 percent of the materials that are typically introduced into the coastal environment and which can be characterized as pollutants. Commonly included in this category are materials like plastics, toxins such as DDT, PCBs, PAHs, heavy metals, and petroleum. These materials, some of which have limited water solubility, are able to be incorporated in lower trophic levels and will be magnified up the food chain. More recently, with an increasing awareness of more subtle, nonpoint source introduction, inorganic nitrogen from wastewater or fertilizers, or endocrine disrupting materials has been documented. Release of excess nitrogen into coastal environments is associated with terrestrial practices utilizing the inorganic fertilizers used in agriculture and clearly causes nearshore eutrophication. These excesses are capable of establishing water columns that are categorized as "dead zones" owing to the near absence of oxygen in the water. The impacts of future coastal seabed mining are only beginning to be understood. While introduction of significant amounts of petroleum hydrocarbons through accidents still occurs, it has been declining; the greatest amount of petroleum contaminants released into marine environments is associated with nonpoint sources from rivers and the land, especially urban locations. Macroscale plastics are increasingly disposed of into the coastal environment and affect all trophic levels of the ecosystem. It is anticipated that within the next 25 years the amount of discarded plastic may exceed the amount of fish in the ocean. Only recently has the sheer volume of microplastic debris been given the increased awareness of its potential impact. The possibility also exists of even greater influence on coastal areas in locations like the High Arctic, a result of oil exploration and transport in locations with diminished ice cover. Together, heightened levels of pollutants will reduce coastal

* Professor, Department of Environmental Sciences, University of Virginia. The author's PowerPoint presentation is available at http://www.virginia.edu/colp/pdf/yogya-macko.pdf.

 | DOI:10.1163/9789004366619_020

biodiversity and productivity and result in a reduction of food resources. Information and public awareness of the sources, and detrimental effects of these materials, has to be understood if this problem can be coherently addressed.

The vast size of the ocean, covering 71% of the Earth's surface, along with the lack of adequate understanding of the scope of pollutants entering the ocean, makes it difficult to evaluate long-term implications for the introduction of the thousands of compounds that are released into the marine environment (Matthews, 1973). Marine pollution occurs when potentially harmful materials gain entry into the ocean (Farmer, 1997). These materials include, among others, petroleum, chemicals, and wastes derived from agricultural and industrial sources. Rivers and terrestrial runoff are thought to transport perhaps 80 percent of the materials that are typically introduced into the coastal environment and which can be characterized as pollutants (Daoji, 2004). Increasingly, sources from atmospheric transport are seen to be a contributing factor for the introduction of pollutants into the ocean. Land and air pollution have been shown time and again to have harmful impact on the ocean environment. The scope of this present manuscript covers some of the most prevalent types of marine pollution including the release of petroleum, excess nutrients, and the establishment of area of hypoxia, chemicals and marine plastic debris. (Issues related to atmospheric carbon dioxide and climate (IPCC, 2007; NAS, 2008) are addressed elsewhere in this volume.)

Common pollutants, including toxins such as DDT, PCBs, PAHs, heavy metals, or endocrine disrupting materials have been documented as being introduced into the marine environment (Gerlach, 1975). These materials are able to be incorporated in lower trophic level organisms and will be magnified up the food chain. For example, when pesticides are released in agricultural practices and are washed into rivers, they eventually enter into marine food webs. Once in the food web, the pesticides can accumulate and be biomagnified to concentrations that can be harmful to humans along with other organisms of the food chain. Releases of heavy metals, like lead and mercury, have also been documented to having been released into coastal marine food webs, causing changes to behavior, biochemistry, and reproduction. Chemicals accumulate in organisms, becoming concentrated in their bodies and in the surrounding water or soil. These animals are in turn eaten by larger animals, which can travel large distances with an increased chemical load. People become contaminated either directly from household products or by eating contaminated seafood and animal fats. Evidence is mounting that a number of man-made chemicals can cause serious health problems – including cancer, damage to the immune system, behavioral problems, and reduced fertility. Classic examples of such events are the release of mercury in Minimata, Japan, or lead derived from leaded-fuels.

More recently, awareness has increased of subtle, nonpoint source introduction of inorganic nitrogen (Carpenter *et al.* 1998). Nutrient pollution emanates from diverse sources including agricultural runoff, sewage, and wastewater discharges. Release of excess nitrogen into coastal environments is associated with terrestrial practices utilizing the inorganic fertilizers used in agriculture or on lawns, or through urban runoff, as well as from sewage and wastewater discharges. The heightened nitrogen has caused nearshore eutrophication, overloading marine environments with high concentrations of nutrients, which can produce large algal blooms. The death and decomposition of this large abundance of productivity consumes the oxygen, which in turn creates hypoxic, or oxygen depleted areas, where fish and other marine life cannot thrive. The loss of oxygen is capable of establishing water columns that are categorized as "dead zones" owing to the near absence of oxygen in the water. Eutrophication has created enormous dead zones in several parts of the world, including the Gulf of Mexico, the Chesapeake Bay, and the Baltic Sea. An estimated 500 dead zones now exist in the world and many more areas suffer adverse effects of high-nutrient pollution (Figure 15.1). Some algal blooms contain toxic substances, and may harm local economies as they can severely disrupt the fisheries and tourism upon which many communities depend. Eutrophication can be seen to be related to the increase in nutrients, which are typically compounds containing nitrogen or phosphorus. It can result in an increase in the ecosystem's primary productivity (excessive plant growth and decay), and further effects including lack of oxygen and severe reductions in water quality, fish, and other animal populations. The most common locations are where rivers empty into the ocean, and with it the associated chemicals used as fertilizers in agriculture, but it may also be associated with the introduction of waste from livestock and humans. Any excess of oxygen-depleting chemicals in the water can lead to hypoxia and the creation of the dead zone.

Estuaries tend to be naturally eutrophic because land-derived nutrients are concentrated where runoff enters the marine environment in a confined channel. The World Resources Institute has identified nearly 400 hypoxic coastal zones around the world, concentrated in areas in Western Europe, the Eastern and Southern coasts of the United States, and East Asia, particularly in Japan. In the ocean, there are frequent red tide algae blooms that kill fish and marine mammals, and cause respiratory problems in humans and some domestic animals when the blooms reach close to shore (Carlisle, 2017).

In addition to land runoff, atmospheric anthropogenic fixed nitrogen can enter the open ocean. This addition could account for around one third of the

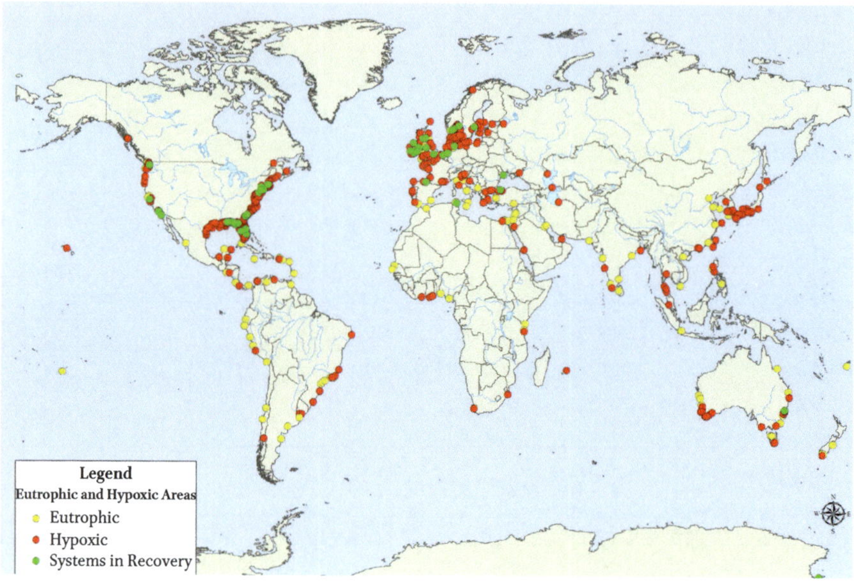

FIGURE 15.1 Coastal locations of areas of diminished oxygen or "Dead Zones."
SOURCE: WORLD RESOURCES INSTITUTE (HTTP://WWW.WRI.ORG/SITES/DEFAULT/FILES/WORLD_21DEC2010.JPG), WASHINGTON, DC.

ocean's external (non-recycled) nitrogen supply and up to three per cent of the annual new marine biological production. It has been suggested that accumulating reactive nitrogen in the environment may have consequences as serious as putting carbon dioxide in the atmosphere (Selman, 2007).

To address this type of pollution, mitigation may include more efficient use of fertilizers, improved wastewater management, and other techniques to minimize nutrient run-off. Proposed solutions to eutrophication in estuaries also include restoration of shellfish populations, such as oysters (Newell *et al.* 2002). Oyster reefs remove nitrogen from the water column and filter out suspended solids, subsequently reducing the extent of harmful algal blooms and anoxic conditions (Burkholder *et al.* 2011). Filter feeding by shellfish is beneficial to water quality, controlling phytoplankton density and sequestering nutrients, which can be removed from the system through shellfish harvest, burial in the sediments, or lost through denitrification (Kaspar *et al.* 1985).

A potential for a new source of ocean pollution is deep sea mining (Nath and Sharma, 2000). This activity is a relatively new mineral retrieval process that takes place on the ocean floor. Ocean mining sites are usually around large areas of polymetallic nodules or active and extinct hydrothermal vents at about

1,400–3,700 meters below the ocean's surface. The vents create polymetallic sulfide deposits, which contain strategic and precious metals including silver, gold, copper, manganese, cobalt, and zinc. The deposits could be mined using either hydraulic pumps or bucket systems that take ore to the surface to be processed. As with all mining operations, deep sea mining raises questions about environmental damages to the surrounding areas.

Because deep sea mining is a relatively new field, the complete consequences of full-scale mining operations are unknown (Halfar and Fujita, 2007). However, it is likely certain that removal of parts of the sea floor will result in disturbances to the benthic layer, increased toxicity of the water column, and sediment plumes from tailings. Removing parts of the sea floor disturbs the habitat of benthic organisms, possibly, depending on the type of mining and location, causing permanent disturbances (Sharma, 2005). Aside from the direct impact of mining the area, leakage, spills, and corrosion would alter the mining area's chemical makeup (Glasby, 2000).

Among the impacts of deep sea mining, sediment plumes could have the greatest impact. Plumes are caused when the tailings from mining (usually fine particles) are dumped back into the ocean, creating a cloud of particles floating in the water. Plumes occur when the tailings are pumped back down to the mining site. The floating particles increase the turbidity, or cloudiness, of the water, clogging filter-feeding apparatuses used by benthic organisms.

Surface plumes cause a more serious problem. Depending on the size of the particles and water currents, the plumes could spread over vast areas. The plumes could impact zooplankton and light penetration, in turn affecting the food web of the area.

1 Petroleum and Oil Spills

Oil is both a natural and unnatural pollutant. Oil seeps are naturally occurring places along the ocean floor where oil migrates upward through fractures in the sea floor and throughout the ocean (NOAA, 2016). Oil seeps are the primary source of oil entering the ocean; however, because the oil seepage is slow the environment is able to adapt and the negative impacts are not as great (Figure 15.2; NOAA, 2016).

In comparison, unnatural oil spills are extremely devastating to oceans with many more negative impacts. Small oil spills from ships and deep-water drilling are common, with some large oil spills occurring occasionally (Figure 15.3). The most notable recent spills are the *Deepwater Horizon* Oil Spill in 2010,

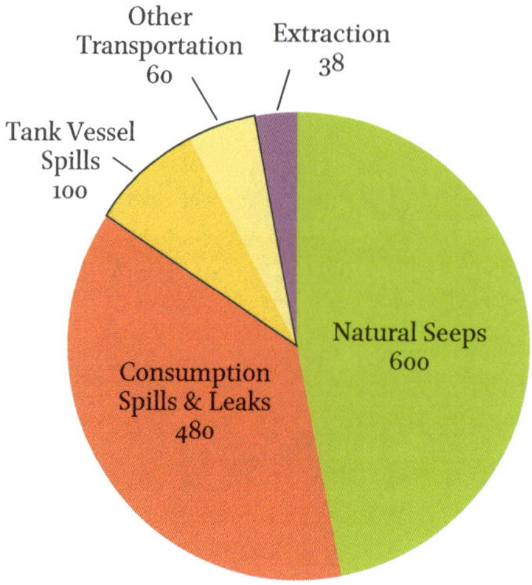

FIGURE 15.2
Global sources of oil to the ocean. In thousands of metric tonnes. Natural seepage exceeds the secondmost source, that being leaks and consumption spillage. Sources related to tanker accidents are a distant third, followed by other introductions from transport and those related to extraction.
SOURCE: BASED ON DATA FROM *OIL IN THE SEA III* FROM THE NATIONAL RESEARCH COUNCIL (2002).

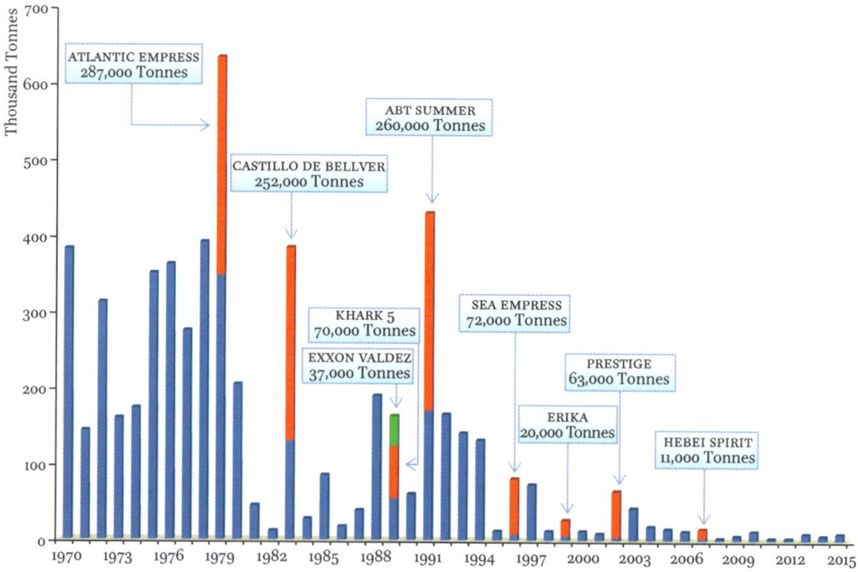

FIGURE 15.3 Declining amount of oil spilled in the ocean from tanker accidents Accidental oil spills from tankers presently constitute a diminishing portion of oil finding its way into the marine environment.
SOURCE: INTERNATIONAL TANKER OWNERS POLLUTION FEDERATION OR ITOPF.

which released almost five million barrels of crude oil into the Gulf of Mexico, and the *Exxon Valdez* oil spill in 1989 in Alaska. While oil spills such as those just listed are most commonly heard about in relation to the ocean, they are responsible for only about 12 percent of the oil entering the oceans (NAS, 2017).

Oil spillage on land is a much greater problem, with 36 percent of the oil entering the ocean coming from cities and industry (NAS, 2017). This is not difficult to imagine with the number of automobiles leaking oil on the streets every day; oil that is then susceptible to runoff with the next rain.

Oil spills can have huge and immediate economic, social, and environmental impacts. Near a spill, people can lose their livelihoods as fisheries and tourism areas may be closed; the clean-up costs are enormous; and tens of thousands of marine animals may be killed or harmed. This damage can continue long after the spill has ended. The chemicals used to break up the oil can be toxic, and it is impossible to remove all the spilled oil. Even after an area has been cleaned up, it can take a decade or more to fully recover. Some of the oil can sink with the ship, and can continue to contaminate the seabed and marine organisms. This oil can also resurface. In 2001, a cyclone off the island of Yap in Micronesia disturbed the oil tanker *USS Mississinewa*, which was sunk during World War II. For two months, thousands of liters of oil and gasoline leaked out of the rusted shipwreck onto the beaches of the atoll, stopping the 700 islanders from fishing. There are numerous other shipwrecked tankers around the world.

Animals and plants can be negatively affected by not only the oil spill itself, but also the cleanup process (Office of Response and Restoration, 2017). The location and type of oil spilled is very important for determining the effects on the environment and the life in the ocean (Office of Response and Restoration, 2017). Light oils, such as gasoline, evaporate very quickly when spilled, thus decreasing the time factor. However, they can easily catch fire, and are very toxic to touch or inhale (Office of Response and Restoration, 2017). In comparison, heavy oils, such as those used for ships, are significantly less toxic, but can remain in the ocean for months, lead to long-term health defects, and inhibit an animal's ability to regulate its temperature (Office of Response and Restoration, 2017). Oil spills spread out over the surface of the ocean creating a sheen that can prevent sunlight from passing through and thus also prevent processes, such as photosynthesis, that depend on light (Conserve Energy Future, 2017). For marine animals, oil spills are also extremely toxic and often affect their ability to float, regulate their temperature, find food, and in the worst cases, their ability to respire (Conserve Energy Future, 2017). Some areas with a high degree of biodiversity, like the straits near or around Indonesia, and a large amount of shipping traffic, could be seen as constantly threatened with a major impact should a tanker spill happen.

Oil spills cause huge damage to the marine environment, but in fact are responsible for only around 12% of the oil entering the seas each year. According to a study by the US National Research Council, the majority of oil entering the ocean, 36%, comes down drains and rivers as waste and runoff from cities and industry.

2 The "Other Carbon Dioxide Problem": Ocean Acidification

Carbon dioxide levels in the atmosphere have now reached 400ppm. These levels are a direct consequence of the increased levels of carbon dioxide emissions by human activities, including the burning of fossil fuels (Doney, 2006). With the increased atmospheric concentration of carbon dioxide have come increased levels of dissolved carbon dioxide in the ocean as marine waters scavenge the gas out of the atmosphere, thus increasing the amount dissolved in the ocean (Caldeira and Wickett, 2003). Over the past 100 years, the oceans have absorbed about one third of the carbon dioxide emitted by anthropogenic sources (Raven *et al.* 2005). This scavenging of the gaseous carbon dioxide can have great and predictable responses in the ocean water chemistry, affecting carbonate ion concentrations, calcite, and aragonite mineral saturation levels, and eventually influencing the pH of the ocean water. The ocean carbon cycle involves two forms of carbon: organic carbon and inorganic carbon (Feely *et al.* 2004, 2008). The inorganic carbon cycle is particularly relevant when discussing ocean acidification for it includes the many forms of dissolved CO_2. When CO_2 dissolves, it reacts with water to form ions from the dissolved carbon dioxide: carbonic acid (H_2CO_3), bicarbonate (HCO_3^-) and carbonate (CO_3^{2-}). The relative abundance of these species depends on factors such as seawater temperature and alkalinity (Turley, 2008; Tyrrell, 2008).

Although the natural absorption of CO_2 by the world's oceans has helped mitigate the atmospheric climatic effects of anthropogenic emissions of CO_2, it is believed that increased levels in the ocean have caused a decrease in pH of approximately 0.15 units on the pH scale (Doney, 2006; Table 15.1), or a 30% increase in acidity since this scale is logarithmic (Ridgwell *et al.* 2007). This increase will likely have negative consequences, primarily for oceanic calcifying organisms. These span the food chain from autotrophs to heterotrophs and include organisms such as coccolithophores, corals, foraminifera, echinoderms, crustaceans, and molluscs. The "skeletons" of these organisms are composed of calcite and aragonite (mineral forms of calcium carbonate) and are stable in surface waters since the carbonate ion is at supersaturating concentrations.

TABLE 15.1 Declining pH of the Ocean: increasing acidity, declining carbonate

Year	1750	2000	2100?
CO_2 ppm	275	375 (1.36x)	750 (2.73x)
pH of the Ocean	8.24	8.13 (1.29x H^+)	7.87 (2.35x H^+)

However, as ocean pH falls further, so does the concentration of this ion, and when carbonate becomes undersaturated, structures made of calcium carbonate are vulnerable to dissolution (Feely *et al.* 2004). The ocean is approaching pH levels not seen in millions of years. Increasing ocean acidity will affect a vast majority of marine life (either directly or indirectly), but some of the first to feel the effects are shellfish, such as oysters. The slightest decrease of 0.1 units of pH has adverse repercussions on calcifiers. Acidification results in the water becoming unstable for calcium carbonate minerals that shellfish produce to make their shells. Without their protective shells, oysters are vulnerable and simply cannot live. Although the corrosion of their defensive shell seems enough to rapidly increase their death rate, the devastation of oysters from acidification goes even further (Waldbusser *et al.* 2011).

An important aspect of this "other carbon dioxide problem" is that, unlike models of climatic warming which are based on complex models of many forcings and feedbacks, heightened acidity, or lower pH of the ocean, is fairly predictable. The mechanisms for increasing acidity are well-established, physical chemical processes: increasing of carbon dioxide in the atmosphere will increase the amount dissolved in the ocean. The pH of the ocean is dependent on the amount of the dissolved CO_2. The "unknowns" are simply the levels that atmospheric carbon dioxide will reach, and the rate at which the surface ocean attains equilibrium with that level. As fossil fuels continue to contribute carbon dioxide to the atmosphere, the pH of the ocean will continue to decline.

Most studies have found that coccolithophores, a type of planktonic algae, coralline algae, corals, shellfish, foraminifera, and pteropods all experience reduced calcification or increased dissolution under lower pH or elevated CO_2 (Raven *et al.* 2005). However, a few studies have suggested that with ocean acidification, the direction of the response, enhanced or declining, varies between species. While the full ecological consequences of these changes in calcification are still uncertain, it appears likely that many calcifying species will be adversely affected. Lower pH also appears to negatively impact non-calcifying larvae during planktonic stages, affecting hardening of chitin and resulting in increased mortality.

Aside from calcification stress, organisms may suffer other adverse effects, either directly as reproductive or physiological effects, including CO_2-induced acidification of body fluids, or indirectly through negative impacts on food resources. With diminished calcifying planktonic organisms, the entire food resource may be disrupted, with a cascading effect up the food chain, should no other primary food source be readily available (Kleypas *et al.* 2006). A change in any part of the food web may have consequences on the rest of the food web, ocean biogeochemistry and the whole ecosystem. Such a modification has already been observed in the Antarctic: in the Southern Ocean GLOBEC study with diminishing krill, predators of krill have turned to alternate foods, with associated potential loss of energy from longer food chains or foods not supplying appropriate levels of essential biochemical nutrients. More acidic conditions would hinder growth of calcium carbonate shells and skeletons by many other marine plants and animals.

Ocean acidification may also force some organisms to reallocate metabolic energy away from feeding and reproduction in order to maintain internal cell pH. It has even been suggested that ocean acidification will alter the acoustic properties of seawater, allowing sound to propagate further, increasing ocean noise and impacting animals that use sound for echolocation or communication. However, as with calcification, as yet there is not a full understanding of these processes in marine organisms or ecosystems. Leaving aside direct biological effects, it is expected that ocean acidification in the future will lead to a significant decrease in the burial of carbonate sediments for several centuries, and even the dissolution of existing carbonate sediments (Ridgewell *et al.* 2007; Turley, 2008). Inclusion of biological effects suggests that the ecosystem we know as the world's ocean, an environment that provides one sixth of the protein consumed by human, is dramatically changing. Millions of species of marine organisms will be affected directly by acidity, others by modification of the food chains on which they depend. At the extreme, large numbers of those species could be lost.

Ocean acidification has been seen to destroy ecosystems for marine life and the detrimental impacts are evident when looking at oysters, but, it is still not clear how capable other carbonate organisms will be in responding to the heightened acidity. Some organisms may have a higher resilience against a rise in pH, and therefore may still thrive at least for a while. A coral reef in the western Pacific suggests that some calcifiers may be able to adapt. On Palau in the western Pacific there exists high acidification, low aragonite saturation and yet from all appearances, a stable coral reef (Orr *et al.* 2005).

Despite low pH, carbonate, and aragonite, coral reefs at Palau show high coral calcification, diversity, and cover. Calcification in different areas of this reef are shown to be comparable to reefs with both high and low calcification rates, demonstrating that even under the stress induced by these conditions, calcification can occur for at least one of the reef building species (Shamberger *et al.* 2014).

Similar studies on the resilience of other reefs around the globe are also being conducted, including studies in the Eastern Tropical Pacific, Hawaii and near volcanic vents. Studies in Eastern Tropical Pacific reefs confirmed a general consensus that there would be reduced resilience of coral reefs in response to increasing carbon dioxide concentration (Kleypas, 2006). The eastern Tropical Pacific reefs are in zones of upwelling with high CO_2 and nutrient concentrations. However, research found that there was an abnormally low saturation of carbonate in this ecosystem, and as a result, these reefs would be more susceptible to bioerosion (Manzello *et al.* 2008).

A further consequence of ocean acidification will be an impact on humans through declining fish harvests resulting in diminishing captures for nutrition, and also lower revenues from those captures of shellfish or finfish, as well as associated habitat loss including that which results in ecotourism benefits in areas like coral reefs. A study of US commercial fisheries (Cooley and Doney, 2009) attempted to constrain the economic effects of ocean acidification using anticipated increases in atmospheric carbon dioxide. The annual domestic commercial harvest of molluscs alone could ultimately be impacted sufficiently to lower revenues by billions of dollars.

3 Plastic Debris

Marine debris is mainly discarded human rubbish which floats on, or is suspended in the ocean. Perhaps 90 percent of marine debris is plastic, a component that has been rapidly accumulating since the end of World War II. With the continued release of plastic debris, presently thought to be around 10 million tons per year, the total mass of plastic in the oceans will continue to accumulate, and has been suggested that it will soon become as massive as the weight of the fish in the ocean.

Discarded plastic bags, six pack rings, and other forms of plastic waste which end up in the ocean present dangers to wildlife and fisheries. Aquatic life can be threatened through entanglement, suffocation, and ingestion. Fishing nets, usually made of plastic, can be left or lost in the ocean by fishermen. Known as ghost nets, these entangle fish, dolphins, sea turtles, sharks, dugongs, crocodiles, seabirds, crabs, and other creatures, restricting movement, causing starvation, laceration, infection, and, in those that need to return to the surface to breathe, suffocation.

Macroscale plastics are increasingly disposed of into the coastal environment and affect all trophic levels of the ecosystem. While it is anticipated that within the next 25 years the mass of this discarded plastic may exceed that of fish in the ocean, only recently has the additional sheer volume of microplastic and nanoplastic debris been given the increased awareness of its potential impact.

The marine debris – all of the trash and other solid material that enters the ocean – threatens wildlife and marine habitats, and presents health and safety concerns for humans. Trash and debris in coastal and ocean ecosystems also pose safety hazards, and impose significant direct and indirect costs to society. Packaging and plastics consistently make up a significant percent of all marine trash. Preventing trash from entering the ocean presents a huge challenge due to the many sources, including municipal storm sewers, poor trash management by industrial and waste management facilities, disposal from vessels and offshore platforms, and littering by individuals in coastal and inland areas. The challenge is heightened by the fact that effective trash and litter prevention strategies require significant changes in the behavior of businesses, government entities, and individual citizens.

Plastic refuse constantly threatens our global ecosystems. Entanglement, ingestion, and exposure to persistent organic pollutants are primary threats to biota. Although numerous reports of plastic threatening macroorganisms exist, concerns are mounting that species further down on the food chain face a similar fate with microplastics. The source of oceanic microplastics has previously been speculated to result from inland sources, often dismissing floating detritus material breakdown due to UV attenuation and lower temperatures in ocean waters. There is now evidence for a rapid, low-temperature abiotic process producing micron-sized plastic particulates due to interactions with UV radiation and seawater. These results are contrary to the current theory that the breakdown of plastic debris into smaller microplastic occurs over the course of perhaps 100 to 1000 years. Since plastic may have been floating in the oceans for decades, it is alarming to consider that the amount of microplastics and nanoplastics present in the oceans are grossly underestimated. Global mass production of plastic has increased exponentially over the past several decades and shows no signs of slowing down, with approximately 300 million tonnes currently produced. Modern societies have not been able to handle the enormous wastes resulting from a dependence on these synthetic materials and as a result, an increased flux of plastic debris enters aquatic environments. Numerous investigations have recently highlighted the ingestion and entanglement of these polymers with marine and terrestrial macro and microorganisms. Speculation exists that some oceanic plastic debris is degraded by microbial communities due to observations of pitting along the surface of retrieved samples and an increase in bacteria present on samples compared to the surrounding waters. Plastic polymers have been previously considered resistant to biological attack, owing to their high molecular weight and degree of crystallinity. Physical properties of these molecules include a high degree of water repellency and hydrophobicity. Of all the different plastic polymers, polyethylene has been considered the least reactive because of the lack of functional groups available for interaction with biological enzymes that reduces their likelihood of microbial degradation. Solid garbage also makes its way to the ocean. If not disposed of correctly, almost everything thrown away can reach the sea, including plastic bags, balloons, glass bottles, shoes, and packaging material.

Plastic garbage, which decomposes very slowly, is often mistaken for food by marine animals. High concentrations of plastic material, particularly plastic bags, have been found blocking the breathing passages and stomachs of many marine species, including whales, dolphins, seals, puffins, and turtles. Plastic six-pack rings for drink bottles can also choke marine animals. This garbage can also come back to shore, where it pollutes beaches and other coastal habitats. Many animals that live on or in the sea consume flotsam by mistake, as it often looks similar to their natural prey. Plastic debris, when bulky or tangled, is difficult to pass, and may become permanently lodged in the digestive tracts of these animals. Especially when evolutionary adaptions make it impossible for the likes of turtles to reject plastic bags, which resemble jellyfish when immersed in water, as they have a system in their throat to stop slippery foods from otherwise escaping, thereby blocking the passage of food and causing death through starvation or infection.

Plastics accumulate because they do not biodegrade in the way many other substances do. They will photodegrade on exposure to the sun, but they do so properly only under dry conditions, and water inhibits this process. In marine environments, photodegraded plastic disintegrates into ever-smaller pieces while remaining polymers, even down to the molecular level. When floating plastic particles photodegrade down to zooplankton sizes, jellyfish attempt to consume them, and in this way the plastic enters the ocean food chain. Many of these long-lasting pieces end up in the stomachs of marine birds and animals, including sea turtles, and black-footed albatrosses.

Plastic debris of all sizes tends to accumulate at the center of ocean gyres. In particular, the Great Pacific Garbage Patches have high levels of plastic particulates, suspended in the upper water column. In samples taken in 1999, the mass of plastic exceeded that of zooplankton (the dominant animal life in the area) by a factor of six. Midway Atoll, in common with all the Hawaiian Islands, receives substantial amounts of debris from the garbage patch. Ninety percent plastic, this debris accumulates on the beaches of Midway where it becomes a hazard to the bird population of the island. Midway Atoll is home to two-thirds (1.5 million) of the global population of Laysan albatrosses. Nearly all of these albatrosses have plastic in their digestive system and one-third of their chicks die.

Toxic additives used in the manufacture of plastic materials can leach out into their surroundings when exposed to water. Waterborne hydrophobic pollutants collect and magnify on the surface of plastic debris, thus making plastic far more deadly in the ocean than it would be on land. Hydrophobic contaminants are also known to bioaccumulate in fatty tissues, biomagnifying up the food chain and putting pressure on apex predators (Oehlmann, *et al.* 2009). Some plastic additives are known to disrupt the endocrine system when consumed, others can suppress the immune system or decrease reproductive rates. Floating debris can also absorb persistent organic pollutants from seawater, including PCBs, DDT, and PAHs. Aside from toxic effects, when ingested, some of these affect animal brain cells similarly to estradiol, causing hormone disruption in the affected wildlife.

A growing concern regarding plastic pollution in the marine ecosystem is the use of microplastics. Microplastics are little beads of plastic less than five mm wide, and they are commonly found in hand soaps, face cleansers, and other exfoliators (Moore, 2001; Moore, 2003). When these products are used, the microplastics go through the water filtration system and into the ocean, but because of their small size they are likely to escape capture by the preliminary treatment screens on wastewater plants. These beads are harmful to the organisms in the ocean, especially filter feeders, because they can easily ingest the plastic and become sick. The microplastics are such a concern because it is difficult to clean them up due to their size, so humans must avoid using these harmful plastics by purchasing products that use environmentally-friendly materials.

Plastic microfibers are a specific subset of a larger category of plastic pollutants called microplastics. Microplastics include plastic beads, formerly popular in personal care products, and small pieces of plastic that come from larger plastic items that have begun to degrade (Fendall and Sewell, 2009; Bomgardner, 2017). Owing to often undetectable size and long lifetimes, a large amount of these plastics are ending up in environments around the world. Natural phenomena carry these pollutants, like many others, to waterways and eventually the oceans. It is well-known that much of the plastic products that are used end up in huge floating trash islands, such as the Texas-sized "Great Pacific Garbage Patch," centered in ocean gyres. However, the undetectability of microplastics has allowed this problem to be unreported until recently (O'Connor, 2017). Their small size also causes them to be mistaken for food by many marine animals. An irregular shape, compared to plastic beads or other forms of plastic pollution, poses a threat to smaller, lower trophic level organisms. This causes microplastics to be spread through food chains, and perhaps even end up being consumed by humans (O'Connor, 2017).

Additionally, plastic microfibers may be less than five mm long, and on the scale of tens of microns wide (Bomgardner, 2017). They are the synthetic, fibrous materials that are shed from synthetic clothing materials, like nylon, acrylic, and polyester (O'Connor, 2017). Microfibers are entering the environment through regular washing. The fibers escaping from washing machines are essentially the lint found trapped by dryers, however, washers have no method of capturing these fibers. Microfibers are washed down the drain, where they enter waste treatment plants. The filters in most waste water treatment plants, although they do collect 65 to 92 percent of microfiber pollution, are not fine enough to prevent substantial amounts of the smallest microfibers from passing through (O'Connor, 2017). This wastewater is then released to the surrounding environment where many plastics make their way into rivers and oceans (Derraik, 2002). There in fact exists evidence that even microplastics that are filtered out may make their way into the environment. The rationale behind this claim is that this wastewater is often used as agricultural fertilizer. Washing of synthetic apparel is just one route plastics take to the ocean. Microfibers could also be released into the atmosphere, where they could be brought to the ocean by rain, and there could be other industries besides apparel that contribute to microplastic pollution (Bomgardner, 2017).

4 Summary

The variety, scope and effects of marine pollution requires that we remain vigilant for protecting the ever-ongoing threats to the diverse and fragile ecosystems of the ocean. Not only do we need to be aware of the present activities, but also insightful of new techniques and technologies which could introduce materials which could have devastating impacts at a minimum, locally, but potentially globally (Jenssen, 2003). We depend on the ocean as a primary source of nutrition with 16% of our protein derived from its fisheries. The ocean is crucial with maintaining the heat budget of the planet. Changes to the pH of the ocean suggest that the very base of the food chain, the calcium carbonate containing phytoplankton, will be seriously impacted and thus affect organisms higher on the food chain. As we search for future energy on, or transport materials across the continental shelves, especially in the high Arctic, we need to minimize the chance of accidents and must be prepared for the near certainty of an accidental release of petroleum. Such an event would be devastating to the ecosystems, potentially forever, or be the cause for destruction of the last refuges of stressed organisms on the brink of extinction. And for some current perturbations, we must address and find solutions while we still can: removing the ocean as a receptacle for our one-use plastics and reducing the excess nutrients from coastal areas which cause the eutrophication and nearshore anoxia. This is not the challenge to accomplish these tasks. It is a requirement. We have no other option.

References

Bomgardner, M. (2017). "The Great Lint Migration." *Chemical and Engineering News* 95:16–17.

Burkholder, J.M. and S.E. Shumway (2011). "Bivalve shellfish aquaculture and eutrophication." In: *Shellfish Aquaculture and the Environment*. Eds. Sandra E. Shumway. John Wiley & Sons.

Caldeira, K. and M.E. Wickett (2003). "Anthropogenic carbon and ocean pH." *Nature* 425 (6956): 365–365.

Carlisle, E. (2017). "The Gulf of Mexico Dead Zone and Red Tides." http://www.tulane.edu/~bfleury/envirobio/enviroweb/DeadZone.htm.

Carpenter, S.R.; N.F. Caraco; D.L. Correll; R.W. Howarth; A.N. Sharpley; V.H. Smith (1998). "Nonpoint Pollution of Surface Waters with Phosphorus and Nitrogen." *Ecological Applications* 8 (3): 559–568. doi:10.1890/1051-0761(1998)008[0559:NPOSWW]2.0.CO;2.

Cooley, S.R. and S. Doney (2009). "Anticipating ocean acidification's economic consequences for commercial fisheries." *Environmental Research Letters* 4 DOI 10.1088/1748-9326/4/2/024007/.

Conserve Energy Future. (2017). Oil Spill. http://www.conserve-energy-future.com/effects-of-oil-spills.php.

Daoji, L. and D. Daler (2004). "Ocean pollution from land-based sources: East China Sea, China." *Ambio* 33: 1–2. Royal Swedish Academy of Sciences.

Derraik, J.G.B. (2002). "The pollution of the marine environment by plastic debris: a review." *Marine Pollution Bulletin* 44: 842–852.

Doney, S.C. (2006) "The Dangers of Ocean Acidification." *Scientific American* 294: 58–65.

Farmer, A. (1997). *Managing Environmental Pollution*. Psychology Press.

Feely, R.A.; C.L. Sabine; J.M. Hernandez-Ayon; D. Ianson; B. Hales (2008). "Evidence for Upwelling of Corrosive 'Acidified' Water onto the Continental Shelf." *Science* 320 (5882): 1490–2.

Feely, R.A., C.L. Sabine, K. Lee, W.L. Berelson, J. Kleypas, V.J. Fabry, and F.J. Millero (2004). "Impact of anthropogenic CO_2 on the $CaCO_3$ system in the oceans," *Science* 305 (5682): 362–366.

Fendall, L.S.; M.A Sewell. (2009). "Contributing to marine pollution by washing your face: Microplastics in facial cleansers." *Marine Pollution Bulletin* 58 (8): 1225–1228. PMID19481226. doi:10.1016/j.marpolbul.2009.04.025.

Gerlach, S.A. (1975) *Marine Pollution*. Springer, Berlin.

Glasby, G.P. (2000). "Economic Geology: Lessons Learned from Deep-Sea Mining." *Science* 289 (5479): 551–553. PMID17832066. doi:10.1126/science.289.5479.551.

Halfar, J.; R.M. Fujita. (2007). "Ecology: Danger of Deep-Sea Mining." *Science* 316 (5827): 987. PMID17510349. doi:10.1126/science.1138289.

Intergovernmental Panel on Climate Change (2007). Fourth Assessment Report: Climate Change, 2007. Cambridge University Press, Cambridge, United Kingdom and New York, NY, USA.

Jenssen, B.M. (2003). "Marine pollution: the future challenge is to link human and wildlife studies" (PDF). *Environ. Health Perspect*ive 111 (4): A198–A199.

Kaspar, H.F.; P.A. Gillespie; I.C. Boyer; A.L. MacKenzie (1985). "Effects of mussel aquaculture on the nitrogen cycle and benthic communities in Kenepuru Sound, Marlborough Sounds, New Zealand." *Marine Biology* 85 (2): 127–136.

Kleypas, J.A., R.A. Feely, V.J. Fabry, C. Langdon, C.L. Sabine and L.L Robbins (2006). Impacts of ocean acidification on coral reefs and other marine calcifiers: A guide for future research. Report of a workshop held 18–20 April 2005, St. Petersburg, FL, sponsored by NSF, NOAA, and the U.S. Geological Survey, 88 pp.

Manzello, D.P., J.A. Kleypas, D.A. Budd, C.M. Eakin, P.W. Glynn, and C. Langdon (2008). "Poorly cemented coral reefs of the eastern tropical Pacific: Possible insights into reef development in a high-CO2 world," *Proc. Natl. Acad. Sci.* U.S.A., 105: 450–510.

Matthews, G. (1973). "Pollution of the oceans: An international problem?" *Ocean Management.* 1: 161–170. doi:10.1016/0302-184X(73)90010-3.

Moore, C.J.; S.L. Moore; M.K. Leecaster; S.B. Weisberg (2001). "A Comparison of Plastic and Plankton in the North Pacific Central Gyre." *Marine Pollution Bulletin* 42 (12): 1297–1300. PMID11827116. doi:10.1016/S0025-326X(01)00114-X.

Moore, C. (2003). "Across the Pacific Ocean, plastics, plastics, everywhere." http://www.naturalhistorymag.com/htmlsite/master.html?http://www.naturalhistorymag.com/htmlsite/1103/1103_feature.html.

Nath, B.N.; R. Sharma (2000). "Environment and Deep-Sea Mining: A Perspective." *Marine Georesources & Geotechnology* 18 (3): 285–294. doi:10.1080/10641190009353796.

National Academies of Sciences (2008). *The Ecological Impacts of Climate Change.* NAS, Washington, DC.

National Academies of Sciences (2003). *Oil in the Sea.* National Academies Press. http://dels.nas.edu/Report/Inputs/10388.

National Ocean and Atmospheric Administration (NOAA). 2016. "What Is an Oil Seep?" http://oceanservice.noaa.gov/facts/oilseep.html.

Newell, R.I.E.; J.C. Cornwell; M.S. Owens (2002). "Influence of simulated bivalve biodepositon and microphytobenthos on sediment nitrogen dynamics, a laboratory study." *Limnology & Oceanography* 47 (5): 1367–1379.

O'Connor, M.C. (2017). "Microfibers are polluting our food chain." https://www.theguardian.com/sustainable-business/2017/feb/12/seafood-microfiber-pollution-patagonia-guppy-friend.

Oehlmann, J.; U. Schulte-Oehlmann; W. Kloas; O. Jagnytsch; I. Lutz; K.O. Kusk; L. Wollenberger; E.M. Santos; G.C. Paull; K.J.W. Van Look; C.R. Tyler (2009). "A Critical Analysis of the Biological Impacts of Plasticizers on Wildlife." *Philosophical Transactions of the Royal Society B: Biological Sciences*. DOI: 10.1098/rstb.2008.0242.

Office of Response and Restoration (NOAA). (2017). How Oil Harms Animals and Plants in Marine Environments. http://response.restoration.noaa.gov/oil-and-chemical-spills/oil-spills/how-oil-harms-animals-and-plants-marine-environments.html.

Orr, J.C.; V.J. Fabry; O. Aumont; L. Bopp; S.C. Doney; R.A. Feely; A. Gnanadesikan; N. Gruber; A. Ishida; F. Joos; R.M. Key; K. Lindsay; E. Maier-Reimer; R. Matear; P. Monfray; A. Mouchet; R.G. Najjar; G.K. Plattner; K.B. Rodgers; C.L. Sabine; J.L. Sarmiento; R. Schlitzer; R.D. Slater; I.J. Totterdell; M.F. Weirig; Y. Yamanaka; A. Yool (2005). "Anthropogenic ocean acidification over the twenty-first century and its impact on calcifying organisms." *Nature* 437 (7059): 681–686.

Raven, J.A. *et al.* (2005). *Ocean acidification due to increasing atmospheric carbon dioxide*. Royal Society, London, UK. The Clyvedon Press Ltd.

Ridgwell, A.; I. Zondervan; J.C. Hargreaves; J. Bijma; T.M. Lenton (2007). "Assessing the potential long-term increase of oceanic fossil fuel CO_2 uptake due to CO_2-calcification feedback." *Biogeosciences* 4: 481–492.

Selman, M. (2007) *Eutrophication: An Overview of Status, Trends, Policies, and Strategies*. World Resources Institute.

Sharma, R. (2005). "Deep-Sea Impact Experiments and their Future Requirements." Marine *Georesources & Geotechnology*. 23 (4): 331–338. doi:10.1080/10641190500446698.

Turley, C. (2008). "Impacts of changing ocean chemistry in a high-CO_2 world." *Mineralogical Magazine* 72(1): 359–362.

Tyrrell, T. (2008). "Calcium carbonate cycling in future oceans and its influence on future climates." *J. Plankton Res.* 30: 141–156.

US Department of Commerce, National Oceanic and Atmospheric Administration (USDA NOAA). "What Is the Biggest Source of Pollution in the Ocean?" http://oceanservice.noaa.gov.

Waldbusser, G.G.; E.P. Voigt, H. Bergschneider; M.A. Green & R.I.E. Newell 2011. "Biocalcification in the Eastern oyster (*Crassostrea virginica*) in relation to long-term trends in Chesapeake Bay pH." *Estuaries and Coasts* 34: 221–231.

CHAPTER 16

A Human Rights Perspective on the Protection of Coastal Ecosystems from Land-Based Marine Pollution under the United Nations Convention on the Law of the Sea

*Chie Kojima**

Abstract

This paper examines the potential of the United Nations Convention on the Law of the Sea (UNCLOS) as a living instrument, which is capable of safeguarding human rights in relation to the protection of coastal ecosystems from land-based marine pollution. It discusses how general obligations of States to prevent, reduce and control land-based marine pollution can be linked with rights to life, health and food. It proposes that a human rights perspective may reinforce UNCLOS obligations on the prevention, reduction and control of land-based marine pollution, which are often criticized as vague and aspirational.

1 Introduction

Adverse effects of land-based marine pollution on coastal ecosystems have significant impacts on food security and human rights of coastal populations. The loss of marine living resources in the coastal area threatens food security, in particular of developing countries and indigenous people. The contamination of coastal seafood affects lives and health of people who rely on them for living. The Minamata disease, a mercury poisoning caused by eating coastal fish and shellfish in the 1950s and 1960s, was an early example in which contaminated coastal seafood severely affected human health. The control of land-based marine pollution remains unsolved and Goal 14 of the Sustainable Development Goals (SDGs) includes the goal of sustainably managing and protecting coastal

* Professor of International Law, Faculty of Law, Musashino University, Tokyo, Japan; Ph.D. (Chuo), LL.M. (Yale) and J.S.D. (Yale). The author would like to thank Mr. Julian Hinz for reviewing this chapter.

 | DOI:10.1163/9789004366619_021

ecosystems to avoid significant adverse impacts for maintaining healthy and productive oceans.[1]

The 1982 United Nations Convention on the Law of the Sea (hereinafter referred to as UNCLOS)[2] is the only treaty at the global level, which stipulates the general obligation of States to prevent, reduce and control land-based marine pollution. Although UNCLOS is regarded as the "Constitution of the Oceans",[3] the provisions of UNCLOS concerning the prevention, reduction and control of land-based marine pollution do not contain detailed environmental standards and are considered to be weak[4] and "largely aspirational".[5] The reasons why the global regulation of land-based pollution remains weak are complex: they include the States' reluctance to restrict their economic development, the complexity of sources and substances to be regulated, the need to address different geographical features such as enclosed or semi-enclosed coastal sea areas, and the gaps between developed and developing countries in technology and economy.[6]

This paper takes a fresh look at UNCLOS as the only global binding instrument for the prevention, reduction and control of land-based marine pollution, and as a living instrument[7] that is capable of safeguarding human rights.

1 14.2 of Goal 14 states: "By 2020, sustainably manage and protect marine and coastal ecosystems to avoid significant adverse impacts, including by strengthening their resilience, and take action for their restoration in order to achieve healthy and productive oceans". Transforming our world: the 2030 Agenda for Sustainable Development, UN Doc. A/70/L.1(18 Sep. 2015). Such adverse impacts may include not only land-based marine pollution but also global climate change, invasions of exotic species, overfishing, and illegal, unreported and unregulated fishing.

2 United Nations Convention on the Law of the Sea, *adopted* 10 December 1982, 1833 UNTS 3 (*entered into force* 16 November 1994) [UNCLOS].

3 United Nations, *A Constitution for the Oceans, Remarks by Tommy T.B. Koh, of Singapore, President of the Third United Nations Conference on the Law of the Sea*, Final Session, Montego Bay, 6th and 11th of December 1982. Available at: www.un.org/depts/los/convention_agreements/texts/koh_english.pdf (accessed 2 August 2018).

4 Y. Tanaka, "Regulation of Land-Based Marine Pollution in International Law: A Comparative Analysis between Global and Regional Legal Frameworks", 66 *Zeitschrift für ausländisches öffentliches Recht und Völkerrecht (Heidelberg Journal of International law)* 535 (2006), at 547. See generally, D.L. VanderZwaag & A. Powers, "The Protection of the Marine Environment from Land-Based Pollution and Activities: Gauging the Tides of Global and Regional Governance", 23 *International Journal of Marine and Coastal Law* 423 (2008); E.A. Kirk, "Noncompliance and the Development of Regimes Addressing Marine Pollution from Land-Based Activities", 39 *Ocean Development and International Law* 235 (2008); D. Hassan, "International Conventions Relating to Land-Based Sources of Marine Pollution Control: Applications and Shortcomings", 16 *Georgetown International Environmental Law Review* 657 (2003–2004).

5 VanderZwaag & Powers, *id.*, at 425.

6 Tanaka, *supra* note 4, at 548–549.

7 The interpretation and application of UNCLOS are considered flexible and responsive to emerging problems as a living instrument. See, A.E. Boyle, "Further Development of the Law of the Sea Convention: Mechanisms for Change" 54 *International and Comparative*

First, it gives a brief look at the Minamata disease as an unsolved issue today. Second, it provides an overview of relevant UNCLOS provisions and discusses how these provisions can evolve through different techniques integrating other international instruments. Third and finally, it explores possibilities to integrate human rights considerations into the obligation to protect coastal ecosystems from land-based marine pollution.

2 Lessons from Minamata

Minamata disease is a neurological syndrome caused by methylmercury-containing effluents discharged from chemical plants[8] in Kumamoto and Niigata Prefectures in the 1950s and 1960s. The direct cause of this disease was methylmercury concentrated in fish and shellfish, which were taken and consumed by the local population. Those who were diagnosed with Minamata disease suffered from sensory disturbance, ataxia, concentric constriction of the visual field, and auditory disorders.[9]

The Japanese government's failure to take control or implement prevention measures after the recognition of the first instance of Minamata disease in Kumamoto Prefecture in 1956 led to the outbreak of the same disease in Niigata Prefecture in 1965. It was only in 1968 that Chisso, the corporation operating the chemical plant in Minamata, ended the production of acetaldehyde, which involved the use and discharge of large amounts of mercury. The failure of the Japanese government to regulate Chisso's economic activities effectively demonstrates the difficulties in implementing the goals of sustainable development in legal, political and societal contexts. According to the report of the Ministry of the Environment, "the failure to prevent the spread of Minamata disease was attributable to the major impact the Chisso Minamata Plant had on the local economy in respect to employment and tax revenues, and to the concerns that any actions on the plant could adversely affect Japan's strong economic growth at the time".[10]

Both the responsible companies and the Japanese government took a series of relief measures for the victims, but a number of lawsuits have proven these

Law Quarterly 563–584 (2005); See also: J. Barrett and R. Barnes, *Law of the Sea: UNCLOS as a Living Treaty* (London: British Institute of International and Comparative Law, 2016).

8 The plants were operated by Shin Nippon Chisso Hiryo K.K. ("Chisso") in Kumamoto Prefecture and Showa Denko K.K. ("Showa Denko") in Niigata Prefecture.

9 Ministry of the Environment, *Lessons from Minamata Disease and Mercury Management in Japan* (2013) at 3 ["MOE Report"]. Available at: https://www.env.go.jp/chemi/tmms/pr-m/mat01/en_full.pdf (accessed 2 August 2018).

10 *Id.*, 4, at 6. Chisso produced the largest amount of acetaldehyde in Japan at that time.

measures to be inappropriate.[11] Since the 2004 Supreme Court decision[12] and until the end of July 2010, the number of applicants for the disease certification amounted to 8,282 people from which 2,806 people sought compensation before courts.[13] In order to settle increased claims, the government enacted the Act on Special Measures Concerning Relief for Victims of Minamata Disease and Solution to the Problem of Minamata Disease[14] in 2009. Under the Act, 48,327 applications for lump-sum payment and 16,824 applications for Minamata Disease Victims' Passbook have been filed by July 2012. The number of these applications surpassed the number of patients of Minamata disease who were already certified and entitled to receive payments.[15]

The Minamata disease is only one example of land-based marine pollution affecting coastal ecosystems, which eventually harmed the enjoyment of human rights of coastal populations. Mercury is still being discharged from various sources worldwide. According to the UNEP Global Mercury Assessment, aquatic releases of mercury originate from industrial sites such as power plants or factories, contaminated sites including old mines, landfills, and waste disposal locations, artisanal and small-scale gold mining sites, and deforestation.[16] It is assessed that anthropogenic releases to waters are estimated at least 1000

---

11 Under the 1969 Act on Special Measures Concerning Relief for Health Damage by Pollution (Act No. 90 of 1969) succeeded by the 1973 Act Concerning Compensation and prevention of Pollution-Related Health Damage (Act. No. 111 of 1973), the victims of Minamata disease could receive payments to cover their medical expenses once certified by the prefectural governor or city mayor. From 1967 to 1969, first lawsuits were brought against Showa Denko and Chisso before courts in Niigata and Kumamoto respectively for compensation to the victims, which upheld the rights of the patients to receive compensation for damage suffered. Subsequently, compensation agreements were concluded between the responsible companies and the group of patients, which stipulated that each certified patient would receive a consolation payment, medical expenses and an annuity and that patients certified after the date of the signing of the agreements could claim for compensation under the agreements. The increasing disputes concerning the certification and compensation for damage led to a political settlement in 1995, which led to the suspension of a series of lawsuits and the sharp decrease of applications under the 1973 Act for the Minamata disease patient certification.

12 The Supreme Court decision in 2004, which upheld the entitlement of those who had not been certified under the 1973 Act, made the number of applicants for the disease certification increased again. MOE Report, *supra* note 9, at 12.

13 *Id.* at 13.

14 Act No. 81 of 2009.

15 As of the end of May 2013, the number of certified patients amounted to 2,977 including those who passed away. Among the certified are 1,784 in Kumamoto Prefectures, 491 in Kagoshima Prefecture, and 702 in Niigata Prefecture. MOE Report, *supra* note 9, at 9–10.

16 United Nations Environment Programme (UNEP), Global Mercury Assessment 2013, at 7.

tonnes per year.[17] East Asia is the largest contributor to mercury deposition in almost all aquatic regions, ranging from 25 to 53 percent.[18] These statistics indicate that threats of mercury accumulation in living marine resources continue today. The following sections examine the obligation of States to protect coastal ecosystems from land-based marine pollution and its relation to human rights.

3 Protection of Coastal Ecosystems from Land-Based Marine Pollution under UNCLOS

UNCLOS codifies a number of customary rules on one hand, but also provides "umbrella or framework provisions [...] capable of being implemented at a later date".[19] For instance, the obligation of States to ensure that activities under their jurisdiction do not cause damage by pollution to other States and their environment under Article 194(2) of UNCLOS is a customary rule recognized in the *Trail Smelter Arbitration*.[20] The obligation of States to take all measures to prevent, reduce and control all sources of marine pollution under Article 194(1) and (3) of UNCLOS can be characterized as a framework provision which can adopt changes in society such as technological or scientific developments.

While UNCLOS itself is not a so-called framework treaty followed by protocols and annexes,[21] it has means to integrate the development of international rules and standards through various techniques, including references to "international rules and standards established through competent international organizations",[22] the adoption of implementing agreements,[23] and the

17 *Id.*

18 Arctic Monitoring and Assessment Programme, Global Mercury Modelling: Update of Modelling Results in the Global Mercury Assessment 2013 (2015), at 6.

19 C.S.G. Jefferies, *Marine Mammal Conservation and the Law of the Sea* (Oxford: Oxford University Press, 2016), 37.

20 *Trail Smelter Case* (*United States* v. *Canada*), III RIAA 1905 (11 Mar. 1941). The Arbitral Tribunal stated that "under the principles of international law, as well as the law of the United States, no State has the right to use or permit the use of its territory in such a manner as to cause injury by fumes in or to the territory of another or the properties or persons therein when the case is of serious consequence and the injury is established by clear and convincing evidence" (at 1965).

21 Boyle, "Further Development of the Law of the Sea Convention", *supra* note 7, at 564.

22 E.g., UNCLOS, Arts. 211, 213–214, 216–218, 222.

23 Agreement Relating to the Implementation of Part XI of the United Nations Convention on the Law of the Sea of 10 December 1982, *adopted* 28 July 1994, 1836 UNTS 3 (entered

interpretation and application of existing provisions in view of other treaty regimes.[24] These techniques can be utilized to reinforce the general obligations of States to prevent, reduce and control land-based marine pollution under Articles 194 and 207 of UNCLOS.

3.1 *UNCLOS and Land-Based Marine Pollution*

Article 194(1) of UNCLOS provides that States have an obligation to take, individually or jointly, all measures that are necessary to prevent, reduce and control pollution of the marine environment from all sources including "the release of toxic, harmful or noxious substances, especially those which are persistent, from land-based sources, from or through the atmosphere or by dumping".[25] States can fulfill this obligation by deciding best practical means by themselves and "in accordance with their capabilities".[26] Necessary measures prescribed under Article 194(1) include the adoption of laws and regulations to prevent, reduce and control pollution of the marine environment from land-based sources, including rivers, estuaries, pipelines and outfall structures.[27] Such laws and regulations involve measures to minimize the release of toxic, harmful or noxious substances, especially those which are persistent.[28]

Article 207(4) of UNCLOS obliges States to cooperate to establish global and regional rules, standards and recommended practices and procedures to prevent, reduce and control pollution of the marine environment from land-based sources. In doing so, States must take into account "characteristic regional features, the economic capacity of developing States and their need for economic development". Article 207(3) of UNCLOS also provides that States are under obligation to harmonize their policies at the appropriate regional level, but what exactly this provision aims to achieve is unclear. Although a number of regional treaties regulating land-based marine pollution, which address the particular circumstances of individual States and regions,[29] have been concluded, regulatory measures under these regional treaties are different from

in force provisionally 16 November 1992, definitively 28 July 1996); United Nations Agreement for the Implementation of the Provisions of the United Nations Convention on the Law of the Sea of 10 December 1982 relating to the Conservation and Management of Straddling Fish Stocks and Highly Migratory Fish Stocks, *adopted* 4 December 1995, 2167 UNTS 88 (*entered in force* 11 December 2001).

24 Boyle, "Further Development of the Law of the Sea Convention", *supra* note 7, at 566. Boyle writes that "UNCLOS is a treaty which functions within a larger legal system".

25 UNCLOS, Art. 194(3)(a).

26 UNCLOS, Art. 194(1).

27 UNCLOS, Art. 207(1).

28 UNCLOS, Art. 207(5).

29 Tanaka, *supra* note 4, at 549–550.

each other[30] and do not control every substance in a horizontal way.[31] Other challenges—although this is not limited to the control of land-based marine pollution—include the lack of financial and human resources as well as political will to ratify and implement those agreements domestically.

One of the techniques for the evolution of UNCLOS used in those provisions is the following: When adopting laws and regulations, States must take into account "internationally agreed rules, standards and recommended practices and procedures";[32] when establishing global and regional rules, standards and recommended practices, States are encouraged to act "through competent international organizations and diplomatic conferences".[33] In the context of land-based marine pollution, UNEP serves as one of such an international organization. For example, the UNEP Global Programme of Action for the Protection of the Marine Environment from Land-Based Activities (GPA),[34] created in 1995 as a guidance for national and regional authorities in implementing sustained action to prevent, reduce, control and eliminate marine degradation from land-based activities, serves as a reference for States in implementing UNCLOS provisions domestically.[35]

3.2 *UNCLOS and Coastal Ecosystems*

Part XII of UNCLOS not only provides the obligations of States to prevent, reduce and control marine pollution, but also to protect and conserve coastal ecosystems. Recent international decisions such as the Arbitral Tribunal decision in *Chagos Marine Protected Area*,[36] held that Part XII, Article 194 of UNCLOS is "not limited to measures aimed strictly at controlling marine pollution and extends to measures focused primarily on conservation and the preservation of ecosystems".[37] The Arbitral Tribunal in the *South China Sea*

30 For instance, mercury and cadmium are included in the grey list in the 1974 Helsinki Convention, while they are categorized in the black list in the 1974 Paris Convention, the 1980 Athens Protocol and the 1983 Quito Protocol.

31 Tanaka, *supra* note 4, pp. 555–556.

32 UNCLOS, Art. 207(1).

33 UNCLOS, Art. 207(4).

34 UNEP, Global Programme of Action for the Protection of the Marine Environment from Land-Based Activities, UN Doc. UNEP (OCA)/LBA/IG.2/7 (5 December 1995). Available at: http://www.cep.unep.org/pubs/meetingreports/LBS%20ISTAC%20III/english/whole gpa.pdf (accessed 2 August 2018).

35 *Id.*, para. 14.

36 *Chagos Marine Protected Area Arbitration* (*Mauritius* v. *United Kingdom*), Award of 18 March 2015, 211, para. 538.

37 *Id.*, para. 320.

Arbitration[38] also reiterated that measures under Part XII of the Convention include those that are necessary to protect and preserve rare or fragile ecosystems as well as habitats of endangered species.[39]

UNCLOS, however, does not define the term "ecosystem". The term "ecosystem" appears only in Article 194(5) of UNCLOS as follows: "The measures taken in accordance with this Part shall include those necessary to protect and preserve rare or fragile ecosystems as well as the habitat of depleted, threatened or endangered species and other forms of marine life". The Arbitral Tribunal in the *South China Sea Arbitration* therefore interpreted the meaning of "ecosystems" under Article 194(5) in the light of Article 2 of the 1992 Convention on Biological Diversity (CBD).[40] In the Award, the Tribunal emphasized the general character of CBD by expressing the definition of ecosystem under CBD as the "internationally accepted definition".[41] This serves as another technique for the evolution of UNCLOS, an interpretation of its generic terms through the principle of "systematic integration"[42] expressed under Article 31(3)(c) of the Vienna Convention on the Law of Treaties.[43] Article 31(3)(c) permits references to "any relevant rules of international law applicable in the relations between the parties" in interpreting a treaty.

These developments demonstrate that the protection of coastal ecosystems is inherent in the general obligation to protect and preserve the marine environment under Article 192 and that necessary measures under Article 194(1) should include those to protect and preserve coastal ecosystems including rare and fragile ecosystems under Article 194(5).

---

38 *South China Sea Arbitration* (*Republic of the Philippines* v. *People's Republic of China*), PCA Case No 2013-19, Judgment 12 July 2016.

39 *Id.*, para. 945.

40 *Convention on Biological Diversity*, *adopted* 5 June 1992, 1760 UNTS 79 (*entered into force* 29 December 1993) (CBD). Article 2 defines an ecosystem as "a dynamic complex of plant, animal and micro-organism communities and their non-living environment interacting as a functional unit".

41 *South China Sea Arbitration*, *supra* note 38, para. 945.

42 International Law Commission, Fragmentation on International Law: Difficulties Arising from the Diversification and Expansion of International Law, Report of the Study Group of the International Law Commission Finalized by Martti Koskenniemi, UN Doc. A/CN.4/L.682 (13 April 2006), available at legal.un.org/ilc/documentation/english/a_cn4_l682.pdf (accessed 2 August 2018), paras 410–423.

43 Vienna Convention on the Law of Treaties, *adopted* 23 May 1969, 1155 UNTS 331 (*entered into force* 27 January 1980). See C. Kojima, "South China Sea Arbitration and the Protection of the Marine Environment: Evolution of UNCLOS Part XII through Interpretation and the Duty to Cooperate", 21 *Asian Yearbook of International Law* 166 (2015).

4 UNCLOS and Human Rights

UNCLOS is not a human rights instrument, but "[r]ules of the Law of the Sea are sometimes inspired by human rights considerations and may or must be interpreted in light of such considerations".[44] Human rights considerations in UNCLOS include, for instance, Article 73(3) on the prohibition of imprisonment for fisheries violations in the exclusive economic zone and Article 230 on the requirement that monetary penalties should only be imposed for violations related to marine pollution. Even if such considerations cannot be implied in UNCLOS provisions, international courts and tribunals are increasingly interpreting UNCLOS in the light of human rights principles.[45] The International Tribunal for the Law of the Sea stated, in the *Saiga No. 2 Case*, that "[c]onsiderations of humanity must apply in the law of the sea, as they do in other areas of international law".[46] Such interpretation is in accordance with Article 31(3)(c) of the Vienna Convention on the Law of Treaties as well as Article 293 of UNCLOS which permits international courts and tribunals to apply other rules of international law not incompatible with UNCLOS. Considering these trends, there are no reasons why the protection and preservation of the marine environment under Part XII of UNCLOS should not be interpreted in the light of human rights law.

The discussion on the relationship between human rights and the environment is not new. It was as early as 1972 when the participating States at the Stockholm Conference on the Human Environment recognized that environmental degradation hampers the enjoyment of human rights.[47] While the importance of protecting the environment for human well-being is recognized, international conventions do not always recognize a human right to the environment.[48] Some regional human rights conventions guarantee a right to

44 T. Treves, "Human Rights and the Law of the Sea", 28 *Berkeley Journal of International Law* 1 (2010), at 12.

45 *Id.* at 5.

46 *M/V Saiga (No. 2) (St. Vincent v. Guinea)*, ITLOS Reports 1999, at para. 155.

47 D. Shelton, "Human Rights and the Environment: Substantive Rights", in: *Research Handbook on International Environmental Law* (Fitzmaurice, Ong and Merkouris eds., 2011), at 265.

48 A. Boyle, "Human Rights and the Environment: Where Next?" 23(3) *The European Journal of International Law* 613 (2012), at 616. Boyle writes: "the relationship between human rights and environmental protection in international law is far from simple or straightforward".

the environment,[49] but many other conventions do not include such a right.[50] However, cases brought before the European Court of Human Rights, which invoked the right to life, the right to privacy and family life, and the right to protection of property from environmental harm, indicate that those human rights should be protected whether or not a right to the environment is guaranteed in an international convention.[51] The situation in domestic law is not much different: more than 100 constitutions in the world guarantee a right to clean and healthy environment,[52] but the laws of the rest of the world do not recognize such a right.

With regard to States' obligation, the Human Rights Council adapted a resolution in 2011, which recognizes that environmental damage can have negative implications, both direct and indirect, for the effective enjoyment of human rights and urges States to consider human rights when developing their environmental policies.[53] This recommendation, although non-binding, can be a useful reference when interpreting Article 193 of UNCLOS, which provides that "States have the sovereign right to exploit their natural resources pursuant to their environmental policies and in accordance with their duty to protect and preserve the marine environment". In the light of the 2011 resolution, the phrase "pursuant to their environmental policies" under Article 193 should include considerations to human rights.

The 2017 Report of the Special Rapporteur on the issue of human rights obligations relating to the enjoyment of a safe, clean, healthy and sustainable environment took a step further and examined the question on the dependence of human rights on biodiversity as follows:

> The full enjoyment of human rights, including the rights to life, health, food and water, depends on the services provided by ecosystems. The provision of ecosystem services depends on the health and sustainability

49 Article 24 of the African Convention on Human and Peoples' Rights provides that "[a]ll peoples shall have the right to a general satisfactory environment favorable to their development". Article 11 of the Additional Protocol to the American Convention on Human Rights in the Area of Economic, Social and Cultural Rights provides as follows: "1. Everyone shall have the right to live in a healthy environment and to have access to basic public services. 2. The States Parties shall promote the protection, preservation and improvement of the environment". Article 24(2)(c) of the Convention on the Rights of the Child obliges its States Parties to take appropriate measures "to combat disease and malnutrition [...] taking into consideration the dangers and risks of environmental pollution".

50 E.g., European Convention for the Protection of Human Rights and Fundamental Freedoms.

51 Shelton, "Human Rights and the Environment", *supra* note 47, at 267.

52 *Id.*

53 Human Rights Council, Human Rights and the Environment, UN Doc. A/HRC/RES/16/11 (12 April 2011).

> of ecosystems, which in turn depend on biodiversity. The full enjoyment of human rights thus depends on biodiversity, and the degradation and loss of biodiversity undermine the ability of human beings to enjoy their human rights.[54]

This statement clearly indicates that not only marine pollution itself but also the loss of biodiversity in coastal ecosystems caused by pollution can also affect the enjoyment of human rights.

These discussions suggest that States' obligations to protect and preserve the marine environment and their obligations to protect human rights are intertwined with each other. Certain provisions of UNCLOS also allude to rights to health[55] and food.[56] Article 1(1)(4) of UNCLOS for example defines marine pollution as "hazards to human health". The notion of sustainability enshrined under UNCLOS, in particular in provisions related to the sustainable management of fisheries,[57] is linked to the notions of the right to adequate food and food security,[58] implying that living marine resources should be accessible for both present and future generations. Even if the link to human rights is not clear from the text of UNCLOS provisions, international courts and tribunals may interpret them in the light of human rights principles.[59]

In this way, human rights, especially human rights to life, health, and food[60] could be of particular relevance when interpreting the obligations of States

---

54 Human Rights Council, Report of the Special Rapporteur on the issue of human rights obligations relating to the enjoyment of a safe, clean, healthy and sustainable environment, UN Doc. A/HRC/34/49 (19 January 2017), at 3.

55 The UN Committee on Economic, Social and Cultural Rights states that the right to health "extends to the underlying determinants of health, such as food and nutrition, housing, access to safe and potable water and adequate sanitation, safe and healthy working conditions, and a healthy environment". CESCR General Comment No. 14 (2000): The Right to the Highest Attainable Standard of Health (Art. 12), para. 4.

56 The UN Committee on Economic, Social and Cultural Rights states that the core content of the right to adequate food implies "the availability of food in quantity and quality sufficient to satisfy the dietary needs of individuals, free from adverse substances, and acceptable within a given culture". CESCR, *Substantive Issues Arising in the Implementation of the International Covenant on Economic, Social, and Cultural Rights: General Comment 12, The Right to Adequate Food (Art. 11)*, U.N. ESCOR, 20th Sess., U.N. Doc. E/C.12/1999/5 (1999), at 3.

57 E.g., UNCLOS Arts. 63, 64 and 118.

58 CESCR, "Substantive Issues", *supra* note 56, at 3.

59 Treves, *supra* note 44, at 5.

60 Article 3 of the Universal Declaration of Human Rights and Article 6 of the International Covenant on Civil and Political Rights provide the right to life. The Constitution of the World Health Organization and Article 12 of the International Covenant on Economic, Social and Cultural Rights recognize the right to the highest attainable standard of physical and mental health. Article 25 of the Universal Declaration of Human Rights and

to prevent, reduce and control marine pollution and to protect and conserve coastal ecosystems under Part XII of UNCLOS. Human rights conventions and non-binding instruments that human rights bodies adopted deserve to be a part of "internationally agreed rules, standards and recommended practices and procedures"[61] that States must take into account when adopting laws and regulations on the prevention of land-based marine pollution.

5 Conclusion

UNCLOS Part XII is drafted to interact with other treaty regimes. The fundamental values enshrined under UNCLOS such as sustainability of living marine resources and the protection of the marine environment from pollution are intertwined with the notion of human rights. It follows that human rights obligations and the obligations to protect and preserve the marine environment under UNCLOS must be fulfilled and enforced simultaneously. This proposition must be distinguished from traditional human rights enforcement mechanisms where individuals claim infringement of their human rights directly before domestic or regional courts; rather, it engages States' responsibility to protect the marine environment for the benefit of individuals. Eventually, this human rights perspective on land-based marine pollution could "be used to compel governments to regulate environmental risks, enforce environmental laws, or disclose environmental information"[62] and induce more effectiveness to the vague and aspirational provisions of UNCLOS on the prevention, reduction and control of land-based marine pollution.

Article 11 of the International Covenant on Economic, Social and Cultural Rights recognize the right to an adequate standard of living, including the right to food.

61 UNCLOS, Art. 207(1).

62 Boyle, "Human Rights and the Environment", *supra* note 48, at 615.

PART 5

Climate Change and the Oceans

CHAPTER 17

Ocean Acidification and Sustainable Development Goal 14: A Goal but No Target?

*Karen N. Scott**

Abstract

This paper explores the theme of climate change and the oceans with a focus on the "other" side of climate change: ocean acidification. It is estimated that the acidity of the oceans has increased by about 30 percent since the beginning of the Industrial era and the rate of change is particularly significant in colder Polar waters. Until recently, ocean acidification has received little attention at the international level and the particular threat posed by carbon dioxide (as opposed to other greenhouse gases) to the oceans is not specifically acknowledged within the climate change regime. By contrast, the United Nations General Assembly and other fora, such as the 1992 Convention on Biological Diversity, are increasingly acknowledging the threat posed by ocean acidification to the health of our oceans. Sustainable Development Goal 14, among other objectives, calls upon states to "minimize and address the impacts of ocean acidification, including through enhanced scientific cooperation at all levels" (14.3). Nevertheless, specific action, including targets to limit ocean acidification, have yet to be established by any institution. This chapter examines the interaction between law of the sea institutions and the climate regime in order to assess progress to date and potential for future action on ocean acidification and conclude with a number of proposals for reform.

1 Introduction

Sustainable Development Goal (SDG) 14.3 urges states to "minimise and address the impacts of ocean acidification, including through enhanced scientific cooperation at all levels."[1] Ocean acidification refers to the lowering of ocean pH as a consequence of changes in ocean chemistry arising from increased

* University of Canterbury, New Zealand. The author's PowerPoint presentation is available at http://www.virginia.edu/colp/pdf/yogya-scott.pdf.

1 UNGA Res. 70/1, Transforming our world: the 2030 Agenda for Sustainable Development (15 September 2015).

 | DOI:10.1163/9789004366619_022

levels of carbon dioxide (CO_2) being drawn down into the oceans from the atmosphere.[2] Although the processes of ocean acidification are increasingly understood, the impact of pH changes on species and ecosystems is uncertain. Nevertheless, research indicates that ocean acidification may have significant detrimental impacts on calcifying organisms and reef ecosystems for example,[3] and attempts have been made to quantify its implications for the blue economy.[4] Despite improvements in the understanding of ocean acidification as a major threat to ocean ecosystems, it "has remained a footnote in the development of climate change and related environmental policy at both international and national levels."[5] Largely ignored by the climate change regime established under the auspices of the 1992 United Nations Framework Convention on Climate Change (UNFCCC)[6] despite its principal cause being CO_2 emissions, ocean acidification is also not directly addressed by instruments regulating marine pollution. Bridging and, to a very limited extent, filling the gaps between these regimes are an increasing number of soft commitments and targets such as SDG 14.3. This contrast between climate change targets (however unsatisfactory) and ocean acidification goals implies the latter is of lesser importance than the former.

This chapter will explore the regulatory landscape for ocean acidification, which can also be described as a regime complex: overlapping but non-integrated regimes. It will examine in particular, regimes for the protection of the atmosphere, the oceans, biodiversity and soft obligations as they apply to ocean acidification. In Section 5 the absence of meaningful targets associated

2 R. Zeebe, J. Cachos, K. Caldeira *et al.*, "Carbon Emissions and Acidification" 321 *Science*, 2011, 51.

3 See the literature cited in Section 2, below.

4 See for example L. Rodrigues, J. van den Bergh and A. Ghermandi, "Socio-economic impacts of ocean acidification in the Mediterranean Sea," 38 *Marine Policy*, 2013, 349; A. Speers, E. Besedin, J. Palardy *et al.*, "Impacts of climate change and ocean acidification on coral reef fisheries: An integrated ecological-economic model," 128 *Ecological Economics*, 2016, 33; J. Ekstrom, L. Suatoni, S. Cooley *et al.*, "Vulnerability and adaptation of US shellfisheries to ocean acidification," 5 *Nature Climatic Change*, March 2015, 207; A. Punt, D. Poljak, M. Dalton *et al.*, "Evaluating the impact of ocean acidification on fishery yields and profits: The example of red king crab in Bristol Bay," 285 *Ecological Modelling*, 2014, 39; D. Narita and K. Rehdanz, "Economic impact of ocean acidification on shellfish production in Europe," 60 *Journal of Environmental Planning and Management*, 2017, 500; L. Mathis, S. Cooley, N. Lucey *et al.*, "Ocean acidification risk assessment for Alaska's fishery sector," 136 *Progress in Oceanography*, 2015, 71.

5 J. Ekstrom and B. Crona, "Institutional misfit and environmental change: A systems approach to address ocean acidification," 576 *Science of the Total Environment*, 2017, 599.

6 1992 United Nations Framework Convention on Climate Change 1771 UNTS 107, entered into force 21 March 1994 (UNFCCC).

with ocean acidification will be examined and contrasted with the various targets established in respect of climate change under the UNFCCC. The Chapter will conclude with some brief remarks as to other measures that should be implemented to support CO_2 emissions reductions as well as a suggestion designed to support and facilitate the coordination and integration of the ocean acidification regime complex.

2 Ocean Acidification: Causes and Consequences

Stable for over 800,000 years,[7] ocean pH has decreased by 0.1 pH units since the Industrial Revolution,[8] representing an increase of 30 percent in ocean acidity to date.[9] Ocean acidification is largely a consequence of increased levels of carbon dioxide (CO_2) in the atmosphere[10]—400 parts per million (ppm) in 2015 in contrast to 280 ppm in 1750[11]—of which about 30 percent is absorbed annually by the oceans, the most important sink for CO_2.[12] Recent research however, has highlighted that other greenhouse gases such as SO_x, NO_x and NH_3 may also contribute to ocean acidification.[13] A secondary source, comprising land-based runoff, primarily from agriculture, may have significant localised impacts.[14] In addition to decreasing ocean pH, ocean acidification also decreases

7 C. Turley and J. Gattuso, "Future biological and ecosystem impacts of ocean acidification and their socioeconomic-policy implications," 4 *Current Opinion in Environmental* Sustainability, 2012, 278 at p. 278.

8 V. Rérolle, C. Floquet, and M. Mowlem, "Seawater-pH measurements for ocean-acidification observations," 40 *Trends in Analytical Chemistry*, 2012, 146 at p. 146.

9 S. Dupont and H. Pörner, "A snapshot of ocean acidification research," 160 *Mar. Biol.* 2013, 1765 at p. 1765.

10 S. Doney, V. Fabry, R. Feely *et al.*, "Ocean Acidification: The Other CO_2 Problem," *Annual Review of Marine Science*, 2009, 169 at p. 170.

11 WMO, *Greenhouse Gas Bulletin* No. 12, 24 October 2016 available online at: https://ane4bf-datap1.s3-eu-west-1.amazonaws.com/wmocms/s3fs-public/GHG_Bulletin_12_EN_web_JN161640.pdf?aZaKZhdpDfJdmHvtbSvLwbj6zb_PWwdz.

12 M. Rhein, S.R. Rintoul, S. Aoki *et al.*, "Observations: Ocean" in T.F. Stocker, D. Quin, G.K. Platter *et al.*, *Climate Change 2013: The Physical Science Basis. Contribution of Working Group I to the Fifth Assessment Report of the Intergovernmental Panel on Climate Change* (Cambridge, Cambridge University Press, 2013) 255–315, p. 260 [hereinafter, IPCC Fifth Assessment Report].

13 K. Hunter, P. Liss, V. Surapipith *et al.*, "Impacts of anthropogenic SOx, NOx and NH3 on acidification of coastal waters and shipping lanes," 38 *Geophysical Research Letters*, 2011, L13602.

14 W. Cai, X. Hu, W. Huang *et al.*, "Acidification of subsurface coastal waters enhanced by eutrophication," 4 *Nature Geoscience*, November 2011, 766.

the saturation of calcium carbonate ($CaCO_3$), which is the compound integral to the shells and skeletons of many marine species.[15] Although ocean acidification is undoubtedly a global phenomenon, its impact in terms of regions and species is highly variable. Thus far its impact is thought to be greatest at the surface of the oceans[16] and, as CO_2 is absorbed more easily at lower temperatures, higher latitude oceans are particularly vulnerable.[17] Moreover, polar species are thought to have less capacity to adapt to changes in ocean acidity owing to their low energetic cost adaptations, which enable them to survive at low temperatures.[18] Species most likely to be impacted by ocean acidification are calcifying organisms such as pteropods[19] and shelled molluscs.[20] Coral reef ecosystems may also suffer from decalcification as a consequence of ocean acidification,[21] and in some experiments lower pH levels has also been linked to the phenomenon of coral bleaching,[22] which is primarily linked to ocean temperature rise. The impact of ocean acidification on fish species more generally is under-researched but some experiments indicate that low pH levels may damage key organs in the larvae of yellowfin tuna,[23] slow the development

---

15 B. Hönisch, A. Ridgwell, D. Schmidt *et al.*, "The Geological Record of Ocean Acidification," 335 *Science*, 2012, 1058–1059.

16 G.G. Tarling, V. Peck, P. Ward *et al.*, "Effects of acute ocean acidification on spatially-diverse polar pelagic foodwebs: Insights from on-deck microcosms," 127 *Deep-Sea Research II*, 2016, 75 at p. 75.

17 *Ibid.* at p. 76.

18 P. Thor, A. Bailey, C. Halsband *et al.*, "Seawater pH Predicted for the Year 2100 Affects the Metabolic Response to Feeding in Copepodites of the Arctic Copepod *calanus glacialis*," *PLOS ONE* (December 19, 2016) DOI:10.1371/journal.pone.0168735; P. Matson, T. Martz and G. Hofmann, "High-frequency observations of pH under the Antarctic sea ice in the southern Ross Sea," 23 *Antarctic Science*, 2011, 607 at p. 612.

19 J. Orr, V. Fabry, O. Aumont *et al.*, "Anthropogenic ocean acidification over the twenty-first century and its impact on calcifying organisms," 437 *Nature*, 2005, 681; N. Bednaršek, C. Harvey, I. Kaplan *et al.*, "Pteropods on the edge: Cumulative effects of ocean acidification, warming, and deoxygenation," 145 *Progress in Oceanography*, 2016, 1.

20 J. Ekstrom, L. Suatoni, S. Cooley *et al.*, "Vulnerability and adaptation of US shellfisheries to ocean acidification," 5 *Nature Climate Change*, March 2015, 207.

21 O. Hoegh-Guldbert, P.J. Mumby, A.J. Hooten *et al.*, "Coral Reefs Under Rapid Climate Change and Ocean Acidification," 318 *Science*, 2007, 1737. For example, calcification of the Great Barrier Reef has decreased by around 21 percent between 1988 and 2003. See S. Doney, V. Fabry, R. Feely *et al., supra* n. 10 at p. 175.

22 K.R.N. Anthony, D.I. Kline, G. Diaz-Pulido *et al.*, "Ocean acidification causes bleaching and productivity loss in coral reef builders," 105 (45) *PNAS*, 2008, 17442; P.L. Jokiel, K.S. Rodgers, I.B. Kuffner *et al.*, "Ocean acidification and calcifying reef organisms: a mesocosm investigation," 27 *Coral Reefs*, 2008, 473.

23 A. Frommel, D. Margulies, J. Wexler *et al.*, "Ocean acidification has lethal and sub-lethal effects on larval development of yellowfin tuna, *Thunnus albacares,*" 482 *Journal of Experimental Marine Biology and Ecology*, 2016, 19.

of embryos and larvae,[24] negatively impact the ability of fish to sense predators[25] or distinguish between predator and non-predator odours,[26] and change behaviour, including becoming less predator-adverse.[27] More generally, it is thought that a lower pH ocean environment has reduced capacity to absorb carbon dioxide, weakening the ocean as a natural (or unnatural) sink for CO_2.[28]

In contrast to climate change predictions, actual and modelled changes in ocean pH are relatively certain, and without substantial decreases in CO_2 emissions, the current pH of 8.1 could decrease to between 7.9 and 7.7 by 2100.[29] The impacts of such a change on ecosystems and individual species is, by contrast, much less certain. Research to date is primarily based on laboratory experiments, limited mesocosms and observations drawn from naturally acidic environments as well as historic high pH environments.[30] Experiments where dramatic impacts have been observed typically assume very high levels of CO_2 emissions, beyond that which is predicted for 2100 on even worst-case scenarios.[31] Most research to date indicates that impacts are likely to be mixed with significant variations in the way species,[32] including corals, respond.[33] Some experiments suggest that phytoplankton, for example, are

24 K. Verkaik, J. Hamel and A. Mercier, "Impact of ocean acidification on reproductive output in the deep-sea annelid *Ophryotrocha sp.* (Polychaeta: Dorvilleidae)," 137 *Deep-Sea Research II*, 2017, 368.

25 P. Munday, D. Dixson, M. McCormick *et al.*, "Replenishment of fish populations is threatened by ocean acidification," 107 *PNAS*, July 2010, 12930.

26 T. Branch, B. DeJoseph, L. Ray *et al.*, "Impacts of ocean acidification on marine seafood," 28 *Trend in Ecology and Evolution*, 2013, 178 at p. 180.

27 P. Munday, M. McCormick, M. Meekan *et al.*, "Selected mortality associated with variation in CO_2 tolerance in a marine fish," *Ocean Acidification*, 2013 1.

28 V. Rérolle, C. Floquet, M. Mowlem, *supra* n. 8 at p. 146.

29 C. Turley and J. Gattuso, *supra*, n. 7 at p. 278.

30 *Id.*

31 For example, in the experiment carried out by Munday, Dixson, McCormick *et al.* in which the authors demonstrated a dramatic decrease in the survival of larval fish an assumption of CO_2 levels of between 700 and 850ppm was made. See P. Munday, D. Dixson, M. McCormick *et al.*, *supra* n. 25.

32 I.E. Hendriks, C.M. Durart, M. Álvarez, "Vulnerability of marine biodiversity to ocean acidification: A meta-analysis," 86 *Estuarine, Coastal and Shelf Science*, 2010, 157 at p. 161.

33 J. Kavousi, J. Reimer, Y. Tanaka *et al.*, "Colony-specific investigations reveal highly variable responses among individual corals to ocean acidification and warming," 109 *Marine Environmental Research*, 2015, 9; J. Kavousi, Y. Tanaka, K. Nishida *et al.*, "Colony-specific calcification and mortality under ocean acidification in the branching coral *Montipora digitata*," 119 *Marine Environmental Research*, 2016, 161; John M. Pandolfi, Sean R. Connolly, Dustin J. Marshall, "Projecting Coral Reef Futures Under Global Warming and Ocean Acidification," 333 *Science*, 2011, 418.

not particularly affected by pH changes of the order predicted[34] or are adaptable to a higher CO_2 environment,[35] and even some corals[36] and Antarctic sea urchins[37] have been found resilient to pH changes. In one experiment whilst one group of copepod species decreased in biomass under elevated CO_2 conditions, another group increased biomass by around 30 percent under the same conditions.[38] Moreover, research has suggested that there may be positive impacts arising from ocean acidification, including an increase in the capacity of seagrass to sequester CO_2[39] and a reduction in the release of nitrous oxide (N_2O) into the atmosphere, a greenhouse gas 300 percent more damaging than CO_2.[40]

Nevertheless, as Widdicombe and Spicer point out, the fact that individuals survive higher levels of CO_2 in experiments does not mean that there are no other negative impacts relating to range, behaviour or ecosystem function.[41] Moreover, species not impacted in short-term experiments may suffer longer-term effects that have yet to be documented.[42] It is acknowledged that there is significant spatial variability in the drawdown of CO_2,[43] and pH dynamics, particularly in the polar regions, are poorly understood.[44] The limits of current research are universally acknowledged. For example, there are more species of

34 T. Eberlein, S. Wohlrab, B. Rost *et al.*, "Effects of ocean acidification on primary production in a coastal North Sea phytoplankton community," *PLoS ONE 12(3) e0172594*, 2017.

35 K. Lohbeck, U. Riebesell and T. Reusch, "Adaptive evolution of a key phytoplankton species to ocean acidification," 5 *Nature Geoscience*, 2012, 346.

36 P. Edmunds and A. Yarid, "The effects of ocean acidification on wound repair in the coral *Porites* spp.," 486 *Journal of Experimental Marine Biology and Ecology*, 2017, 98.

37 E. Jones, M. Fenton, M. Meredith *et al.*, "Ocean acidification and calcium carbonate saturation states in the coastal zone of the West Antarctic Peninsula," 139 *Deep-Sea Research II*, 2017, 181, at p. 188.

38 J. Taucher, M. Haunost, T. Boxhammer *et al.*, "Influence of ocean acidification plankton community structure during a winter-to-summer succession: An imaging approach indicates that copepods can benefit from elevated CO_2 via indirect food web effects," *PLoS ONE 12(2): e0169737*, 2017.

39 S. Garrard and N. Beaumont "The effect of ocean acidification on carbon storage and sequestration in seagrass beds; a global and UK context," 86 *Marine Pollution Bulletin*, 2014, 138.

40 A. Rees, I. Brown, A. Jayakumar *et al.*, "The inhibition of N_2O production by ocean acidification in cold temperature and polar waters," 127 *Deep-Sea Research II*, 2016, 93.

41 S. Widdicombe and J. Spicer, "Predicting the impact of ocean acidification on benthic biodiversity: What can animal physiology tell us?" 366 *Journal of Experimental Marine Biology*, 2008, 187 at p. 188.

42 J. Godbold and M. Solan, "Long-term effects of warming and ocean acidification are modified by seasonal variation in species responses and environmental conditions," 368 *Phil Trans R Soc B* 20130186, 2013.

43 E. Jones, M. Fenton, M. Meredith *et al.*, *supra*, n. 37 at p. 182.

44 P. Matson, T. Martz and G. Hofmann, *supra*, n. 18 at p. 607.

coral found below, in contrast to above, a depth of 50 metres, and these are largely unmonitored so the impacts of ocean acidification on these species are simply not known.[45] There is no doubt however, that "the current rate of (mainly fossil fuel) CO_2 release stands out as capable of driving a combination and magnitude of ocean geochemical changes potentially unparalleled in at least the last ~300 My of Earth history, raising the possibility that we are entering an unknown territory of ecosystem change."[46]

3 Ocean Acidification and the Regime Complex

Ocean acidification as a source of marine pollution and a threat to biodiversity is thus far not subject to global binding regulatory control under any international regime. It is an issue, which simultaneously cuts across and falls between international regimes with mandates to protect the atmosphere, the oceans and biodiversity. Ocean acidification is an archetypal issue subject to what scholars have described as a regime complex:[47] "a loosely coupled set of specific regimes" which "pertain to the same issue domain or spatially defined area … and interact with one another in the sense that the operation of each affects the performance of the others."[48] Others have termed the regulatory framework for ocean acidification as a polycentric order "in the sense that they include the self-organizing relationship between many centers of decision-making that are formally independent of each other."[49] The challenge for international regulators and policy-makers is creating meaningful linkages and connections between these regimes, developing a network or web, or regulatory control that effectively addresses ocean acidification.

3.1 *The Atmosphere*

The objective of the 1992 United Nations Framework Convention on Climate Change is the "stabilization of greenhouse gas concentrations in the

45 J. Roberts and S. Cairns, "Cold-water corals in a changing ocean" 7 *Current Opinion in Environmental Sustainability*, 2014, 118, at p. 124.

46 B. Hönisch, A. Ridgwell, D. Schmidt *et al.*, *supra*, n. 15 at p. 1062.

47 M. Simons and T. Stephens, "Ocean Acidification: Addressing the other CO_2 Problem," 12 *Asia Pacific Journal of Environmental Law*, 2009, 1, 6; R. Baird, M. Simons and T. Stephens "Ocean Acidification: A Litmus Test for International Law," 4 *Carbon and Climate Law Review*, 2009 459, 463.

48 O.R. Young, "Building an International Regime Complex for the Arctic: Current Status and Next Steps," 2 *The Polar Journal*, 2014, 391, 394.

49 V. Galaz, B. Crona, H. Österblom *et al.*, "Polycentric systems and interacting planetary boundaries—Emerging governance of climate change-ocean acidification-marine biodiversity," 81 *Ecological Economics*, 2012, 21, 22.

atmosphere at a level that would prevent dangerous anthropogenic interference with the climate system."[50] Climate system, which is defined as "the totality of the atmosphere, hydrosphere, biosphere and geosphere and their interactions,"[51] undoubtedly includes the oceans. However, the focus of the UNFCCC is on the atmosphere, and this is demonstrated by the definition of climate change in the Convention itself as "a change of climate which is attributed directly or indirectly to human activity that alters the composition of the *global atmosphere* and which is in addition to natural climate variability observed over comparable time periods."[52] Until recently the impact of CO_2 on the oceans has been largely ignored by the UNFCCC regime. Under the 1997 Kyoto Protocol[53] Annex I parties can choose from a "basket" of six (now seven)[54] greenhouse gases to meet their emission commitments and no specific targets are set in relation to CO_2, which is the principal greenhouse gas that is the source of ocean acidification. Even greater flexibility is afforded to the parties to the 2015 Paris Agreement[55] in which states merely commit to "nationally determined contributions"[56] in support of the Agreement's objective to hold temperatures below 2° C or, preferably, 1.5° C.[57]

At recent meetings of the UNFCCC the oceans-climate connection has, however, been more widely recognised. Over 40 ocean-related events were held during COP 21 in December 2015 for example and, in 2016, the ocean was designated as one of nine Global Climate Action Events. Also at COP 22 in 2016 the Global Ocean Acidification Observing Network (GOA-ON), which is a collaborative network of institutions undertaking research on ocean acidification processes in order to inform policy development, was created.[58] Simultaneously, a Strategic Action Roadmap on Oceans and Climate: 2016–2021 was released,

---

50 1992 UNFCCC, Article 2.

51 1992 UNFCCC, Article 1(3).

52 1992 UNFCCC, Article 1(2) (emphasis added).

53 1997 Protocol to the United Nations Framework Convention on Climate Change, 2303 UNTS 214, entered into force 16 February 2005 (Kyoto Protocol).

54 The list of greenhouse gases under the 1997 Kyoto Protocol was amended in Doha in 2012. See Decision 1/CMP.8 *Amendment to the Kyoto Protocol pursuant to its Article 3, paragraph 9 (the Doha Amendment)* (2012).

55 2015 Paris Agreement (2016), 55 ILM 743 (in force 4 November 2016).

56 2015 Paris Agreement, Article 4.

57 2015 Paris Agreement, Article 2.

58 See *Global Climate Action Event: Oceans* (12 November 2016) available at: http://climateaction.unfccc.int/media/1055/gca-oceans-programme.pdf.

which sets out six policy recommendations relating to the role of the oceans in the climate, mitigation, adaptation, displacement, financing and capacity development.[59] Nevertheless, the 2016 Marrakech Action Proclamation for our Climate and Sustainable Development makes no specific reference to the oceans[60] and, more generally, no target comparable to the Paris commitment to limiting temperature rise to less than 2° C has been created in relation to CO_2 emissions or ocean pH change.

The only instrument adopted thus far that directly addresses ocean acidification is the 1999 Protocol to Abate Acidification, Eutrophication and Ground-level Ozone[61] to the 1979 Convention on Long-range Transboundary Air Pollution,[62] which applies to Europe and North America. The objective of the Protocol is to control and reduce emissions of sulphur, nitrogen oxides, ammonia, volatile organic compounds and particulate matter which cause adverse effects on health and the environment, including acidification.[63] The Protocol requires parties to establish critical loads for acidity,[64] which are defined as "the maximum amount of acidifying deposition an ecosystem can tolerate in the long term without being damaged" and which in the long term "will not cause adverse effects to the structure and functions of ecosystems"[65] and obligations to reduce emissions are set out in Annex II. This is important in the context of local ocean acidification but notably provides no obligations to reduce CO_2 more generally, which, as noted above, is the principal cause of ocean acidification.

59 B. Cicin-Sain, J. Barbiere, H. Terashima *et al.*, *Toward a Strategic Action Roadmap on Oceans and Climate: 2016–2021* (Washington, DC: Global Ocean Forum 2016) available at: https://globaloceanforumdotcom.files.wordpress.com/2013/03/strategic-action-roadmap-on-oceans-and-climate-november-2016.pdf.

60 Available at: https://unfccc.int/files/meetings/marrakech_nov_2016/application/pdf/marrakech_action_proclamation.pdf.

61 1999 Protocol to Abate Acidification, Eutrophication and Ground-level Ozone to the 1979 Convention on Long-range Transboundary Air Pollution (as amended in 2012) (Gothenburg) available at: http://www.unece.org/env/lrtap/multi_h1.html (entered into force 17 May 2005).

62 1979 Convention on Long-range Transboundary Air Pollution, 18 ILM 1442 (entered into force 16 March 1983).

63 1999 Protocol to Abate Acidification, Eutrophication and Ground-level Ozone, Article 2.

64 *Ibid.*

65 1999 Protocol to Abate Acidification, Eutrophication and Ground-level Ozone, Annex I. Methodologies for establishing critical loads are set out in Annex I and vary according to location (EMEP parties and parties in North America).

3.2 *The Oceans*

The 1982 United Nations Convention on the Law of the Sea (LOSC)[66] creates a general obligation to protect and preserve the marine environment[67] and to take individually or jointly all necessary measures to "prevent, reduce and control pollution of the marine environment from any source."[68] Pollution is defined as "the introduction by man, directly or indirectly, of substances or energy into the marine environment, including estuaries, which results or is likely to result in such deleterious effects as harm to living resources and marine life."[69] Anthropogenic sources of CO_2 indirectly (or even directly, as a consequence of geoengineering) introduced into the marine environment thus will constitute pollution provided it reaches the harm threshold as stipulated in Article 1(4) of the LOSC. The LOSC establishes obligations with respect to pollution arising from land-based sources and the atmosphere,[70] offshore activities,[71] dumping[72] and vessels.[73] These obligations largely rely on standards established under external instruments such as the 1979/83 MARPOL[74] (in respect of vessels), the 1972 London Convention[75] and 1996 Protocol[76] (in respect of dumping), and none have as a principal objective the mitigation of ocean acidification.

The least developed rules relating to pollution under the LOSC are set out in Article 212 in respect of atmospheric pollution. Although parties are required to adopt laws and regulations to prevent, reduce and control pollution of the marine environment from or through the atmosphere, and must endeavour to establish global or regional rules to this effect, no minimum standards are specified. It is arguable that the global standards (such that they are) established

66 1982 United Nations Convention on the Law of the Sea 1833 UNTS 397, entered into force 16 November 1994 (LOSC).
67 1982 LOSC, Articles 192 and 193.
68 1982 LOSC, Article 194(1).
69 1982 LOSC, Article 1(4).
70 1982 LOSC, Articles 207 and 212.
71 1982 LOSC, Articles 208 and 209.
72 1982 LOSC, Article 201.
73 1982 LOSC, Article 211.
74 International Convention for the Prevention of Pollution from Ships as Modified by the Protocol of 1978 relating thereto 1340 UNTS 62, entered into force 2 October 1983 (MARPOL 73/78).
75 1972 Convention for the Prevention of Marine Pollution by Dumping from Ships and Aircraft, 11 ILM (1972), 262 (London Convention) (in force 30 August 1975).
76 1996 Protocol to the London Convention on the Prevention of Marine Pollution by Dumping of Wastes and Other Matter (1997), 35 ILM 1 (in force 8 November 2006).

under the 1992 UNFCCC regime may be regarded as the relevant rules and regulations but, in contrast to other provisions under the LOSC, there is no obligation on the parties to comply with these external standards under Article 212 of the Convention. Moreover, as noted above, the 1992 UNFCCC regime pays little heed to the particular problem of ocean acidification and neither the standards relating to emissions reductions nor to temperature rise have been developed with the oceans in mind. The one source of atmospheric emissions that is subject to binding regulation is air pollution from ships. The 1973/8 MARPOL was amended in 1997 to add Annex VI, which sets out air pollution standards for vessels with particular focus on sulphur oxide (SOx) and nitrogen oxide (NOx). Subsequent amendments have strengthened emissions standards,[77] addressed ship energy-efficiency more generally[78] and designated four Emission Control Areas (ECAS) where emission standards are higher.[79] Although these standards are generally applicable to all LOSC parties by virtue of Article 211(2) of the Convention, they make only a small contribution to addressing ocean acidification, which is largely caused by land-based sources of CO_2.

Unfortunately, Article 207 of the LOSC, which addresses land-based sources of pollution, is similarly under-developed. Parties are subject to a general obligation to prevent, reduce and control land-based sources of pollution, and are encouraged to establish global and regional standards but, as for atmospheric pollution, there is no obligation to adhere to those standards when adopting domestic legislation. Moreover, no globally legally binding standards have been adopted to date. Instead, the 1995 Global Programme for Action (GPA)[80] establishes soft, non-binding targets to limit the release of land-based pollutants, including nutrients, which can exacerbate local sources of acidification.[81] At the regional level, binding targets in respect of land-based sources of pollution have

77 Resolution MEPC.132(53) (adopted on 22 July 2005).

78 For example, all ships must develop a Ship Energy Efficiency Management Plan (SEEMP). See Resolution MEPC.203(62) (adopted on 15 July 2011).

79 These comprise: Baltic Sea area (SO_x only); North Sea area (SO_x only); North American area (SO_x, NO_x and PM); and United States Caribbean Sea area (SO_x, NO_x and PM).

80 See https://www.unep.org/gpa/.

81 Ocean acidification resulting from land-based activities was identified as a threat to vulnerable developing states in the 2006 Beijing Declaration on furthering the implementation of the Global Programme of Action for the Protection of the Marine Environment from Land-based Activities adopted at the Intergovernmental Review Meeting on the Implementation of the Global Programme of Action for the Protection of the Marine Environment from Land-based Activities, Second session, Beijing, 16–20 October 2006 (UNEP/GPA/IGR.2/7) (23 October 2006), Annex V.

been adopted in areas such as the Mediterranean[82] and North East Atlantic[83] for example, but these are similarly more focused on localised pH abatement as opposed to addressing ocean acidification more generally.

By contrast, the rules relating to dumping at sea are both detailed and binding. Article 210 requires both flag and port states to apply global rules and standards to activities that constitute dumping at sea. Those international rules and standards are developed in detail under the 1972 London Convention and the 1996 Protocol. Currently the 1996 Protocol is of indirect application to ocean acidification pollution. The Protocol was amended in 2006 to permit subseabed sequestration of CO_2 streams from captured CO_2 processes[84] but this should not contribute to ocean acidification unless the CO_2 is accidentally released into the marine environment. Of greater significance is the 2013 amendment to the Protocol,[85] not yet in force, which prohibits ocean fertilization activities (designed to stimulate ocean biological productivity and increase draw-down of CO_2 from the atmosphere into the ocean) unless those activities are for scientific purposes and comply with the conditions of the Protocol including the Risk Assessment Framework.[86] If deployed as a climate change mitigation measure,[87] ocean fertilization is deliberately designed to increase the level of CO_2 in the oceans (and correspondingly decrease atmospheric levels of CO_2) and this will inevitably lead to a reduction in ocean pH.[88] The 2013 amendment prohibits ocean fertilization activities for purposes other than scientific research and this prohibition has been agreed to by the parties to both the 1972 Convention and Protocol in a non-binding resolution[89] as well

82 1996 Protocol for the Protection of the Mediterranean Sea against Pollution from Land-Based Sources and Activities, available at: http://wedocs.unep.org/bitstream/handle/20.500.11822/7096/Consolidated_LBS96_ENG.pdf?sequence=5&isAllowed=y (in force 18 May 2006).

83 1992 Convention for the Protection of the Marine Environment of the North East Atlantic (OSPAR Convention), 2354 UNTS 67 (in force 25 March 1998), Annex I.

84 1996 Protocol, Annex 1, [1(8)] and [4].

85 Resolution LP.4(8) *On the amendment of the London Protocol to Regulate the placement of matter for ocean fertilization and other marine geoengineering activities* (18 October 2013).

86 1996 Protocol, Annexes 4 and 5.

87 References for ocean fertilization as a climate change mitigation measure.

88 Unless action were taken to counter a reduction in ocean pH through artificial means such as weathering. See L.D.D. Harvey, "Mitigating the atmospheric CO_2 increase and ocean acidification by adding limestone powder to upwelling regions," 113 *Journal of Geophysical Research*, 2008, C04028; H. Kheshgi, "Sequestering Atmospheric Carbon Dioxide by Increasing Ocean Alkalinity," 20 *Energy*, 1995, 915.

89 Resolution LC.LP.1 (2008) *on the Regulation of Ocean Fertilization.*

as by the parties to the 1992 Convention on Biological Diversity[90] in several decisions.[91] Nevertheless, for the purposes of the 1982 LOSC, it is worth noting that at present the global rules and standards referred to under Article 210 of the Convention relate to the 1972 London Convention rather than the 1996 Protocol,[92] so the Protocol rules associated with CO_2 sequestration and ocean fertilization are arguably yet to be of global application by virtue of the LOSC.

3.3 *The Protection of Species and Ecosystems*

Increasingly, states should be considering the implications of ocean acidification on vulnerable species and ecosystems as part of decision-making in relation to the environment more generally. The 1992 Biodiversity Convention for example, requires parties to identify processes and activities likely to have an adverse impact on biodiversity[93] and to develop plans and programmes for the conservation of biodiversity as well as integrating conservation values into relevant sectoral or cross-sectoral plans and programmes, such as energy production and transport.[94] However, whilst states are increasingly considering climate change or the impacts of climate change as part of oceans and (particularly) coastal planning,[95] fewer regional seas regimes expressly require ocean acidification to be incorporated into planning and conservation measures. In the North East Atlantic for example, outside of the regulation of sub-seabed CO_2 sequestration,[96] the OSPAR Commission merely encourages parties to gather baseline data on pH levels for monitoring purposes.[97] A similar focus on monitoring and assessment underpins the approach of Arctic states as set out in the 2013 Kiruna Declaration of the Eighth Ministerial Meeting of the

---

90 1992 Convention on Biological Diversity, 31 *ILM* (1992) 818, entered into force 29 December 1993 (CBD).

91 See CBD Decision IX/16 (2008) *Biodiversity and Climate Change* at C.4; CBD Decision X/33 (2010) *Biodiversity and Climate Change* at [8(w)]; CBD Decision XI/20 (2012) *Climate Related Geoengineering*; CBD Decision XIII/14 *Climate-related geoengineering* (2016).

92 R.R. Churchill and A.V. Lowe, *The Law of the Sea*, 3rd ed. (Manchester University Press 1999), 369.

93 1992 Biodiversity Convention, Article 7(c).

94 1992 Biodiversity Convention, Article 6.

95 For example, parties to the 2008 Protocol on Integrated Coastal Zone Management in the Mediterranean (ref) must integrated climate change and its impacts into coastal planning in the region. See also UNEP/MAP, Mediterranean Strategy for Sustainable Development 2016–2025, objective 4.

96 *OSPAR Guidelines for Risk Assessment and Management of Storage of CO_2 Streams in Geological Formations* (2007–12).

97 OSPAR Commission, *JAMP Guidelines for Monitoring Chemical Aspects of Ocean Acidification* HASEC 14/14/1, Annex 6 (2014). See also OSPAR Commission and ICES, *Final Report to OSPAR of the Joint OSPAR/ICES Ocean Acidification Study Group* (SGOA) (2014).

Arctic Council.[98] The Baltic Sea states go a little further in that ocean acidification is expressly referred to in Recommendation 35/1 (2014), which sets out targets and the regulatory framework for establishing a system of MPAs in the Baltic[99] but it falls short of fully integrating ocean acidification into ocean planning processes. Regional Fisheries Management Organisations (RFMOs) have been slower to respond to concerns over ocean acidification than the impacts of climate change more generally. The most robust response to date has been taken by the Commission established under the Convention for the Conservation of Antarctic Marine Living Resources[100] (CCAMLR), which in 2009 adopted Resolution XXVIII *Climate Change*, expressly recognising the threat posed by ocean acidification to the Southern Ocean and urging the Commission to consider both climate change impacts and ocean acidification in its decision-making.[101] An Intersessional Correspondence Group (ICG) designed to develop approaches for integrating consideration of the impacts of climate change into the work of the Commission was established in 2015.[102] The ICG is tasked with making recommendations as to how the Commission can use information relating to climate change in its decision-making under Article II of the Convention, and how the Commission can consider climate change impacts across its agenda.[103]

4 Ocean Acidification and the Creation of Soft Goals

To date, the approach adopted to connect and bridge the gaps between regimes has been to adopt soft commitments relating to oceans acidification. The most high profile commitment to date has been made in Sustainable Development Goal 14.3, which calls upon states to "minimise and address the impacts of ocean acidification, including through enhanced scientific cooperation at all levels." This builds on the call for collective action to prevent further ocean

98 The Declaration is available at: https://oaarchive.arctic-council.org/bitstream/handle/11374/93/MM08_Final_Kiruna_declaration_w_signature.pdf?sequence=1&isAllowed=y.

99 HELCOM Recommendation 35/1 (2014) *System of Coastal and Marine Baltic Sea Protected Areas (HELCOM MPAs)*. The Helsinki Commission was established under the 1992 Convention on the Protection of the Marine Environment of the Baltic Sea Area (Helsinki Convention). Reference.

100 1980 Convention on the Conservation of Antarctic Marine Living Resources, 1329 UNTS 47 (entered into force 7 April 1982).

101 CCAMLR Resolution XXVIII Climate Change preamble and [1].

102 The terms of reference of the ICG are set out in CCAMLR XXXIV, *Report of the Thirty-fourth Meeting of the Commission* (Hobart, Australia 19–30 October 2015), Annex 8.

103 *Ibid.*, Annex I, [7].

acidification and to take steps to promote ecosystem resilience as set out in UNGA Resolution 66/288 The Future We Want (2012).[104] The UN General Assembly has a relatively long-standing interest in ocean acidification, which has been regularly noted as an issue of serious concern in resolutions on the oceans and the law of the sea since 2007.[105] Part one of the 2017 UN Secretary General's Report on Oceans and the Law of the Sea is devoted to the topic of climate change and ocean acidification, and it emphasises the synergies and need to connect instruments with mandates relating to climate change, the oceans, and sustainable development.[106] The report was designed to facilitate discussions at the eighteenth meeting of the United Nations Open-ended Informal Consultative Process on Oceans and the Law of the Sea held in May 2017, on the theme "the effects of climate change on oceans."[107] At this meeting states emphasised the importance of implementing commitments agreed at Paris in 2015,[108] noting that ocean acidification comprised one of many cumulative threats to the ocean environment[109] and advocated strong inter-agency coordination relying on UN-Oceans to promote coordination and coherence of UN agencies with oceans-related mandates.[110]

The 1992 Biodiversity Convention has also provided a forum for soft targets with respect to ocean acidification. The threat is regularly highlighted in the decisions of the conference of the parties on ocean and coastal biodiversity[111] and two synthesis reports on ocean acidification have been commissioned to date.[112] More importantly, Target 10 of the 2010 Aichi Biodiversity Targets aims

104 UNGA Resolution 66/288 The Future We Want (2012) [166].

105 For example in UNGA Res. 71/257 (2016) Oceans and the Law of the Sea, ocean acidification was noted as an issue of serious concern and states were urged to take global and regional action to combat its impacts (see the preamble and [186], [187] and [189]).

106 Report of the Secretary General, *Oceans and the Law of the Sea* (A/72/70) (6 March 2017).

107 *Report on the work of the United Nations Open-ended Informal Consultative Process on Oceans and the Law of the Sea at its eighteenth meeting* A72/95 (16 June 2017). In 2013 ocean acidification comprised the topic for the fourteenth United Nations Open-ended Informal Consultative Process on Oceans and the Law of the Sea. See *Report of the work of the United Nations Open-ended Informal Consultative Process on Oceans and the Law of the Sea at its fourteenth meeting* A68/159 (17 July 2013).

108 *Report of the eighteenth meeting*, [18].

109 *Ibid.*, [21].

110 *Ibid.*, [112–114].

111 See for example Decision XII/23 (2014) *Ocean and Coastal Biodiversity* [6–10].

112 Secretariat of the Convention on Biological Diversity (2014), *An Updated Synthesis of the Impacts of Ocean Acidification on Marine Biodiversity* (Eds. S. Hennige, J.M. Roberts & P. Williamson). Montreal, Technical Series No. 75; Secretariat of the Convention on Biological Diversity (2009). Scientific Synthesis of the Impacts of Ocean Acidification on Marine Biodiversity. Montreal, Technical Series No. 46.

to minimise the impacts of climate change or ocean acidification on coral reefs and other vulnerable ecosystems so as to maintain their integrity and ecosystem function by 2015.[113]

Regionally the focus of regional seas conventions and regional fisheries management organisations (RFMOs) has been on the impacts of climate change rather than on ocean acidification *per se*. A notable exception is the Arctic Council, which in 2013 formally welcomed the Arctic Ocean Acidification Assessment,[114] approved its recommendations, noted with concern the potential impacts of ocean acidification on marine life and people, recognised that carbon dioxide emission reductions are the only effective way of mitigating ocean acidification and requested that Arctic states take action to mitigate and adapt and continue to monitor and assess the state of Arctic ocean acidification.[115]

Soft targets and processes for collaboration and capacity building have also been developed by scientific and other non-governmental organisations. In 2008 for example, 122 scientists approved a declaration in Monaco urging states to develop plans to cut emissions drastically and to improve communication between policymakers and scientists in order to improve understanding of the impacts of ocean acidification.[116] The Global Network of Scientific Academies (IAP) adopted a statement on ocean acidification in 2009, and recommended that CO_2 should be reduced by 50 percent of 1990 levels by 2050 in addition to action designed to reduce other stressors on ocean ecosystems.[117] Following the 2012 Rio+20 conference the Ocean Acidification International Coordination Centre based in Monaco and supported by the International Atomic Energy Agency (IAEA) was established with the principal mandate to support capacity building and communication of science associated with ocean acidification.[118] Finally, a Global Ocean Acidification Observing Network was established following workshops held in 2012 and 2013 and aims to improve understanding of global ocean acidification conditions, improve understanding of ecosystem responses to ocean acidification and acquire and exchange

113 Decision X/2 (2010) *Strategic Plan for Biodiversity 2011–2020*, Annex.

114 AMAP, 2013. AMAP Assessment 2013: Arctic Ocean Acidification. Arctic Monitoring and Assessment Programme (AMAP), Oslo, Norway.

115 2013 Kiruna Declaration of the Eighth Ministerial Meeting of the Arctic Council.

116 Monaco Declaration, adopted at the Second International Symposium on the Ocean in a High CO_2 World, Monaco, October 6–9, 2008. Available at: https://www.iaea.org/nael/docrel/MonacoDeclaration.pdf.

117 2009 IAP Statement on Ocean Acidification available at: http://www.interacademies.net/File.aspx?id=9075.

118 See: https://www.iaea.org/ocean-acidification/page.php?page=2178.

data and knowledge necessary to optimise modelling for ocean acidification and its impacts.[119]

5 A Goal but No Target: Does This Matter?

Ocean acidification is a problem concurrent with rather than a consequence of climate change. Nevertheless, a reduction in CO_2 emissions is integral to the mitigation of both threats, so the climate change regime is arguably best placed to take the institutional lead in responding to ocean acidification. Article 2 of the 2015 Paris Agreement establishes a target to keep the global average temperature increase to below 2^0 C above pre-industrial levels and encourages states to purse efforts to limit the temperature increase to no more than 1.5^0 C above pre-industrial levels. As noted above, parties to the Agreement have committed to nationally determined contributions designed to meet this target and these contributions build on the previous emissions targets applied to developed states under the 1997 Kyoto Protocol. By contrast, the UNFCCC regime establishes no targets designed to mitigate ocean acidification. There is no ocean pH change target comparable to the 2^0 C limit on global temperature rise and neither the Kyoto Protocol nor the Paris Agreement establish targets in relation to CO_2 emissions. Outside of the UNFCCC regime, as noted above, a number of soft goals have been set with the aim of addressing ocean acidification but these goals lack specific targets and (with the exception of 2010 Aichi Biodiversity Target 10) time frames.

This divergence in approach—to the mitigation of climate change and the mitigation of ocean acidification—has significant symbolic as well as practical implications. It implies that ocean acidification is less important and a lower priority than climate change, and this is both wrong and dangerous. A target, even an imperfect target, is symbolic. It demonstrates a need for urgent action, it provides an incentive for change and operates as a focal point around which collective action at all levels may coalesce. Thus it is advocated that the UNFCCC adopt a target to stabilise global pH change in the oceans. This target, would naturally support the Paris 2^0 C target and efforts to more generally combat climate change, but would require states to reduce CO_2 emissions directly in order to address ocean acidification. States may for example, develop nationally

119 See: http://www.goa-on.org/ApproachAndGoals.php and Newton, J.A., Feely, R.A., Jewett, E.B., Williamson, P. & Mathis, J., 2015. *Global Ocean Acidification Observing Network: Requirements and Governance Plan.* 2nd ed., GOA-ON, http://www.goa-on.org/docs/GOA-ON_plan_print.pdf.

determined contributions in order to meet such a target or to address climate change and the oceans more generally. There are no doubt scientific challenges in establishing a pH target, particularly in light of the significant regional and seasonal variation in ocean pH and the incomplete understanding of ocean processes and indeed the likely impacts of ocean acidification on species and ecosystems. However, many environmental targets are to an extent arbitrary: articulating a need to take action but not necessarily sufficient in actually mitigating the threat. The emissions target of a 5 percent reduction below 1990 levels by 2012 established under the Kyoto Protocol[120] and indeed the maximum 2^0 C temperature rise codified by the Paris Agreement can be characterised as arbitrary targets. In light of the evident risks posed by ocean acidification and the uncertainty surrounding its impacts, taking action and establishing a binding target implements and is arguably mandated by the precautionary approach, which underpins the climate change regime,[121] the law of the sea[122] and, arguably, customary international law.[123]

6 Integrating Ocean Acidification into Planning and Decision-making

Establishing a global target to stabilise ocean pH levels represents an important but a partial response to mitigating and responding to ocean acidification. States also need to take national action to reduce as far as possible local contributions to and amplifiers of pH change through agriculture and other land-based pollutants.[124] Ocean acidification should be expressly considered as part

120 1997 Kyoto Protocol, Article 3(1).

121 1992 UNFCCC, Article 3(3).

122 See Case No. 17, *Responsibilities and Obligations of States Sponsoring Persons and Entities with Respect to Activities in the Area,* Advisory Opinion, Seabed Disputes Chamber of the International Tribunal for the Law of the Sea (2011), [135].

123 See for example the decision of the International Court of Justice in the *Pulp Mills on the River Uruguay* Case (Argentina v. Uruguay), ICJ Reports (2010) p. 14, where the Court concluded that the precautionary approach may be relevant in the interpretation of the statute in dispute between Argentina and Uruguay relying implicitly on Article 31(3)(c) of the 1969 Vienna Convention on the Law of Treaties (at [164]).

124 For discussion of local action and associated recommendations in the United States see J. Ekstrom and B. Crona, "Institutional misfit and environmental change: A systems approach to address ocean acidification," 576 *Science of the Total Environment*, 2017, 599; S. Weisbert, N. Bednaršek, R. Feely *et al.*, "Water quality criteria for an acidifying ocean: Challenges and opportunities for improvement," 126 *Ocean and Coastal Management*, 2016, 31; C. Logan, "A Review of Ocean Acidification and America's Response," 60 *BioScience*, 2010, 819; R. Craig, "Dealing with Ocean Acidification: The Problem, The Clean Water Act, and

of environmental impact assessments, strategic assessment processes and marine spatial planning at national and regional levels. Similarly, as a cumulative pressure, the risk of ocean acidification needs to be incorporated into fisheries management alongside the impacts of climate change more generally. More specifically, at risk ecosystems such as coral reefs must be managed so as to minimise other pressures such as destructive fishing practices and mining. The resilience of vulnerable ecosystems must be supported through active conservation measures such as the designation of marine protected areas or other forms of area-based management tools.

7 Concluding Remarks

Ocean acidification in many ways demonstrates the disconnect between environmental regimes. Within a regime, complex institutions and treaties intersect and overlap but do not necessarily leverage their connectivity to create coherent and strong obligations.[125] Filling the gaps between regimes and, to an extent, bridging those gaps, are an increasing plethora of soft goals and other initiatives designed to promote and disseminate scientific research, and to incentivise state action. Nevertheless, with no binding targets comparable to those developed in respect of climate change under the UNFCCC regime, ocean acidification is thus far relegated to a secondary environmental concern. The adoption of a binding target under the auspices of the UNFCCC regime relating to ocean pH would address this concern. However, more needs to be done to connect the mandates of other regimes (such as those relating to land-based pollution, dumping and regional air pollution) in order to develop a more integrated and coherent response. One option is for the UN General Assembly to adopt a resolution on oceans and climate change, setting out priorities and principles for coordinated action at global and regional levels, in effect, creating an overarching framework to manage the ocean acidification regime complex.

State and Regional Approaches," 90 *Washington Law Review*, 2015, 1583; R. Kelly and M. Caldwell, "Ten Ways States can Combat Ocean Acidification (And Why they Should)," 37 *Harvard Environmental Law Review*, 2013, 57.

125 See more generally K. Scott, "International Environmental Governance: Managing Fragmentation through Institutional Connection," 12 *Melbourne Journal of International Law*, 2011, 177.

CHAPTER 18

Sea Level Rise and the Law of the Sea: Maritime Zones and Maritime Boundaries

*J. Ashley Roach**

Abstract

This paper discusses the ongoing work of two International Law Association (ILA) Committees, one on Baselines under International Law of the Sea and a second on International Law and Sea Level Rise, through September 2017. It explores the legal issues raised by the effects of sea level rise on maritime zones and maritime boundaries and proposed solutions. The legal issues were explored at the 18th meeting of the UN's Informal Open-ended Informal Consultative Process on Oceans and the Law of the Sea in May 2017. The ILA Sea Level Rise Committee gave further consideration to them at an intersessional meeting in Lupod, Croatia, 14–16 September, and is expected to give further consideration to them during a third intersessional meeting in Singapore during the week of 12 March 2018 for preparation of its report to the August 2018 Sydney biennial ILA conference. The International Law Commission may undertake a study of the issues.

1 Introduction

The International Law Association (ILA), founded in 1873 and based in London, operates through committees composed of members of various branches who volunteer to consider international law issues of concern. The law of the sea has regularly been considered by these committees, two of which are currently

* Captain J. Ashley Roach, JAGC, USN (retired) is Visiting Senior Principal Research Fellow at the Centre for International Law, National University of Singapore. He was an attorney advisor in the Office of the Legal Advisor, US Department of State, from 1988 until he retired at the end of January 2009, responsible for law of the sea matters. He serves as chair of the ILA Committee on Baselines under the International Law of the Sea and is a member of the ILA Committee on International Law and Sea Level Rise. This paper is presented in his personal capacity but draws heavily on reports of these two ILA committees. He has taught, advised and published extensively on a wide range of law of the sea issues. The author's PowerPoint is available at < http://www.virginia.edu/colp/pdf/yogya-roach.pdf>.

 | DOI:10.1163/9789004366619_023

working on baselines and the effect of rising sea levels on maritime zones and maritime boundaries, and migration. The paper begins with work that was completed in 2012 on the ambulatory normal baseline.[1] It then recounts the establishment of a new committee on sea level rise and international law in 2013 and the work done so far on the effect of significant sea level rise on maritime zones and maritime boundaries. Of particular importance is the committee's review of proposals for the progressive development of the law to deal with those issues. The paper closes with a description of some of the major questions being dealt with by the committee in preparation for its full report to the 2018 Sydney biennial ILA conference and the possibility that the International Law Commission might undertake a study of these issues.

2 Ambulatory Normal Baseline

The issue of the so-called ambulatory baseline was carefully examined by the ILA Committee on Baselines under the International Law of the Sea in its 2012 Report to the biennial Sofia Conference.[2] The Report initially noted the writings of Professors Prescott and Schofield:

> [i]n practice it seems likely that the dispute over whether the charted or actual low-water line should prevail will only arise in two situations. The first is when a country realises that the actual line lies significantly seawards of the charted line. The second is when a foreigner, accused of improperly entering a maritime zone, realises that the actual line lies significantly landwards of the low-water line shown on the chart.[3]

The Report added:

> As demonstrated in maritime delimitation cases there is a third situation in which a coastal state may contest its neighbor's asserted baseline

1 Reference here is to article 5 of the UN Convention on the Law of the Sea (hereinafter LOS Convention).

2 Hereinafter Sofia Report. *Available* on the Baseline Committee's web page *at* https://ila.vettoreweb.com/Storage/Download.aspx?DbStorageId=1291&StorageFileGuid=983044b2-6of0-402a-8533-39659e008b33. The Sofia Report also appears in ILA, *Report of the Seventy Fifth Conference held in Sofia August 2012*, 385–428 (London 2012). Page references are to the Committee's online version of the Report unless otherwise noted.

3 V. Prescott & C. Schofield, *The Maritime Political Boundaries of the World* (2d ed., Leiden/Boston, Martinus Nijhoff, 2005), p. 101. (Adapted from the footnote in the Sofia Report.)

> when it realizes that the actual line lies significantly landwards of the charted line.
>
> Finally, there is a special category of coastal States with an existential interest in this issue. The Committee refers here to the low-lying, small island developing states that may be particularly vulnerable to the effects of sea level rise. It is possible that some of these States could lose the entirety of their territory to the sea, and thereby the basic qualifications of statehood itself.[4]

After extensive analysis the Committee's general conclusions were that

> the legal normal baseline is the actual low-water line along the coast at the vertical datum, also known as the chart datum, indicated on charts officially recognized by the coastal State. The phrase "as marked on large-scale charts officially recognized by the coastal State" provides for coastal State discretion to choose the vertical datum at which that State measures and depicts its low-water line. The charted low-water line illustrates the legal normal baseline, and in most instances and for most purposes the charted low-water line provides a sufficiently accurate representation of the normal baseline. As a matter of evidence for proving the location of the normal baseline the charted line appears to enjoy a strong presumption of accuracy. However, where significant physical changes have occurred so that the chart does not provide an accurate representation of the actual low-water line at the chosen vertical datum, extrinsic evidence has been considered by international courts and tribunals in order to determine the location of the legal normal baseline.[5]

In reaching those general conclusions, the Committee considered the effect of articles 6 (reefs) and 13 (low-tide elevations) where the features may appear or disappear with changes in sea level, as follows:

> reefs and low-tide elevations are highly susceptible to coastal change ... Article 13 provides:

4 S. Park, *Climate Change and the Risk of Statelessness: The Situation of Low-lying Island States* (UNHCR, May 2011), *available at* http://www.unhcr.org/4df9cb0c9.pdf ("The Intergovernmental Panel on Climate Change (IPCC) thus concluded that '[s]ea-level rise impacts on the low-lying Pacific Island atoll States of Kiribati, Tuvalu, Tokelau and the Marshall Islands may, at some threshold, pose risks to their sovereignty or existence'" [quoting IPCC, Climate Change 2007, Fourth assessment report, Report of the international working group II, *Impacts, adaptation and vulnerability* 736]). (Adapted from the footnote in the Sofia Report.)

5 Sofia Report, p. 25.

> 1. A low-tide elevation is a naturally formed area of land which is surrounded by and above water at low tide but submerged at high tide. Where a low-tide elevation is situated wholly or partly at a distance not exceeding the breadth of the territorial sea from the mainland or an island, the low-water line on that elevation may be used as the baseline for measuring the breadth of the territorial sea
>
> 2. Where a low-tide elevation is wholly situated at a distance exceeding the breadth of the territorial sea from the mainland or an island, it has no territorial sea of its own.

Under this provision, whether a naturally-formed feature may be included in the baseline depends on its height relative to the vertical datum and its horizontal location relative to the nearest mainland or island. In order to be included, some part of the feature must be above water at low tide as defined by the low-water datum. Absent this characteristic, the feature would not have a low-water line. Some drying part of the feature as defined by its low-water line must also be within the territorial sea generated from the nearest mainland or island. Without both of these qualities a low-tide elevation does not contribute to the normal baseline. Therefore, an otherwise unqualified low-tide elevation situated within the territorial sea of a qualified low-tide elevation does not contribute to the baseline.

Changes in sea level, or other natural processes, that lead to the total submersion of an otherwise qualified low-tide elevation could have an impact on the location of the normal baseline and of outer limits measured from that feature. Similarly, changes in the coastal configuration of the nearest mainland or island could have the effect of placing a previously qualified low-tide elevation outside the territorial sea and therefore unable to contribute to the generation of the territorial sea outer limit. Of course, accretion along a mainland or island coast could bring a low-tide elevation within the territorial sea. Alternatively, if a feature were to rise above water even at high tide, a low-tide elevation could transform into an island with its own baseline, irrespective of its distance from other territory.[6]

Article 6 contains an exception to the distance from mainland or island requirement for reefs and provides:

> In the case of islands situated on atolls or of islands having fringing reefs, the baseline for measuring the breadth of the territorial sea is the seaward

6 *See Maritime Delimitation and Territorial Questions between Qatar and Bahrain* (*Qatar v. Bahrain*), Merits, Judgment, [2001] ICJ Rep. 40 (Mar. 16). (Adapted from the footnote in the Sofia Report.)

> low-water line of the reef, as shown by the appropriate symbol on charts officially recognized by the coastal State.

As with Article 5, the low-water line is the normal baseline and charts may be used to depict or show that line. The reef need only rise above low-water,[7] but unlike low-tide elevations that do not meet the requirements of Article 6, the Article 6 reef need not be within the territorial sea or any specified distance from the reference island.

The analysis of the normal baseline generated by these features does not differ from the analysis of Article 5. The normal baseline is the actual low-water line, not the charted line. The baselines on these features may not be preserved solely through the publication of a chart depicting them.

The Report then went on to consider the effect of sea level rise and coastal erosion:

> In the Committee's view, it follows that if the legal baseline changes with human-induced expansions of the actual low-water line to seaward, then it must also change with contractions of the actual low-water line to landward. These contractions could occur, for example, from the actual loss of land through erosion or from rising sea levels that, over time, would submerge coastlines and associated territory. In theory, these contractions could occur to such an extent that the entire territory, and actual low-water line, of low-lying small island nations would be below the vertical datum, thereby eliminating entirely the normal baseline and any entitlement to maritime zones generated from the baseline. Even if such an extreme scenario were not to arise, the Committee considers that the likelihood that some offshore low-lying small islands will be completely submerged will still remain, which will give rise to debate as to whether a coastal state loses the totality of its entitlement to claim a normal baseline from territory that has become submerged. Islands presently above the water surface at high tide may, as a result of sea level rise, disappear at high tide and become low tide elevations, resulting in the feature being reclassified from an island to an insular feature.[8]
>
> This may be the unavoidable consequence of using the actual low-water line as the baseline from which maritime zones are measured, but

7 *See Eritrea/Yemen*, Award of the Arbitral Tribunal in the second stage of the proceedings between Eritrea and Yemen (Maritime Delimitation) (Eritrea/Yemen), XXII RIAA 335 (Dec. 17, 1999) at 368, *available at* http://legal.un.org/riaa/ (citing to the language of Article 6, the Tribunal concludes that "[a] reef that is not also a low-tide elevation appears to be out of the question as a base point …"). (Adapted from the footnote in the Sofia Report.)

8 C. Schofield & I.M.A Arsana, *Imaginary Islands? Options to Preserve Maritime Jurisdictional Entitlements and Provide Stable Maritime Limits in the Face of Coastal Instability*, 6th IHO-IAG

> it is a consequence that authors seem to agree reflects the existing law of the normal baseline. ....
>
> Unlike most of the scenarios considered above in which possible differences between the charted and actual low-water lines are small and the effects local – sling mud banks, deltaic accretion, other forms of accretion and erosion, land reclamation projects, or the construction of harbour works – the prospect of significant sea level rise carries with it problems of global scale and effect and serious existential implications for some states. Among these problems are negative impacts on maritime boundaries negotiated in reliance on normal baselines in existence at the time of a delimitation,[9] and the outer limits of a State's maritime zones proclaimed in reliance upon a normal baseline. Under these circumstances, a question arises as to whether the existing law of normal baselines would or should apply.

After noting the proposals by scholars to change existing international law, the Baseline Committee noted that "the implementation of these proposals would undoubtedly create serious challenges to fundamental principles of the law of the sea (such as, for example, the principle that 'the land dominates the sea'),[10] notwithstanding that the proposals seek to promote stability, particularly regarding the preservation of the outer limits of maritime zones". After considering in detail the various scholarly arguments, the Committee concluded:

> that the existing law of the normal baseline applies in situations of significant coastal change caused by both territorial gain and territorial loss. Coastal states may protect and preserve territory through physical reinforcement, but not through the legal fiction of a charted line that is unrepresentative of the actual low-water line.

ABLOS Conference, Oct. 25–27, 2010, at 6, *available at* http://www.iho.int/mtg_docs/com_wg/ABLOS/ABLOS_Conf6/ABLOS_Conf6.htm. (Adapted from the footnote in the Sofia Report.)

9 J.L. Jesus, "Rocks, New-born Islands, Sea Level Rise and Maritime Space", in Frowein *et al.* (eds.), *Negotiating For Peace – Liber Amicorum Tono Eitel* (Berlin, Springer, 2003), pp. 599, 602; C. Schofield, "Shifting Limits? Sea Level Rise and Options to Secure Maritime Jurisdictional Claims", 3 *Carbon & Climate L. Rev.* 2009, 405, 406; K. Houghton *et al.*, "Maritime Boundaries in a Rising Sea", 3 *Nature Geoscience*, 2010, 813 (discussing case studies). (Adapted from the footnote in the Sofia Report.)

10 J.G. Stoutenburg, "Implementing a New Regime of Stable Maritime Zones to Ensure the (Economic) Survival of Small Island States Threatened by Sea-Level Rise", 26 *Int'l J. Marine & Coastal L.*, 2011, 263, 271–275. (Adapted from the footnote in the Sofia Report.)

> All coastal States face the threat of territorial loss as a result of predicted sea level rise. When coastal territory submerges below the selected low-water datum, the normal baseline would retreat, and in extreme cases would be lost. As indicated in the proposal establishing this Committee, low-lying, small-island developing states are likely to be the most severely affected by this phenomenon. If current predictions of sea level rise are realized, some States will become completely submerged. The resulting deterritorialization will likely mean, among other things, a total loss of baselines and of the maritime zones generated by coastal territory and measured from those baselines. Should the issue of deterritorialization fall to be considered by the international community at least in part as a baseline issue, the existing law of the normal baseline does not offer an adequate solution.
>
> Here the Committee raises the possibility of deterritorialization in the context of Article 5 baselines, but the loss of a State's territory to rising sea levels is not primarily a baseline or law of the sea issue. Substantial territorial loss is a much broader issue encompassing concerns of statehood, national identity, refugee status, state responsibility, access to resources, and international peace and security. This issue requires consideration by a committee established for the specific purpose of addressing this range of concerns. The work of that committee should take into account the spirit of the modern law of the sea in which the interests of differently situated states are balanced. That committee should also recall the aims of the Convention: to strengthen peace, security, cooperation, and friendly relations among nations in conformity with the principles of justice and equal rights; to take account of the interests and needs of humankind as a whole; and to promote the economic and social advancement of all peoples of the world considering the special interests and needs of developing countries.[11]

Consequently, the Committee concluded:

> the normal baseline is ambulatory, moving seaward to reflect changes to the coast caused by accretion, land rise, and the construction of human-made structures associated with harbour systems, coastal protection and land reclamation projects, and also landward to reflect changes caused by erosion and sea level rise. Under extreme circumstances the latter category of change could result in total territorial loss and the consequent total loss of baselines and of the maritime zones measured from those baselines. The existing law of the normal baseline does not offer an adequate solution to this potentially serious problem.

11 Sofia Report, pp. 30–31.

> The Committee recommends that the issue of the impacts of substantial territorial loss resulting from sea level rise be considered further by a Committee established for the specific purpose of addressing the wide range of concerns it raises.

3 Establishment of New Sea Level Rise Committee

In response to this recommendation, in November 2012 the ILA Executive Committee established a new committee on sea level rise and international law. The Committee's mandate through November 2018 is to "study the possible impacts of sea-level rise and the implications under international law of the partial and complete inundation of state territory, or depopulation thereof, in particular of small island and low-lying states", and to "develop proposals for the progressive development of international law in relation to the possible loss of all or of parts of state territory and maritime zones due to sea-level rise, including the impacts on statehood, nationality, and human rights".[12] The Committee is chaired by Professor Davor Vidas (Norway).[13] There are two co-rapporteurs: Professors David Freestone (UK – law of the sea issues) and Jane McAdam (Australia – human rights issues).[14]

4 Initial Meetings of the Sea Level Rise Committee

The new committee held its first meeting at the 76th Biennial ILA Conference in Washington, D.C. 7–11 April 2014 and its first intersessional committee meeting in Oslo, Norway 12–13 June 2015. Reports of the 9 and 10 April 2014 meetings in Washington are available on the committee's webpage of the ILA Headquarters.[15]

---

12 *Available at* http://www.ila-hq.org/index.php/committee-single.

13 The Chair had addressed these issues in *Sea-Level Rise and International Law: At the Convergence of Two Epochs, Climate Law* 4 (2014), 70–84, DOI: 10-1163/18786561-00402006.

14 The background and initial work of the committee are recounted in an article by the Chair and Co-rapporteurs, "International Law and Sea Level Rise: The New ILA Committee", 22 *ILSA J. Int'l & Comp. L.* (2015) 397–408.

15 Documents of the International Law and Sea Level Rise Committee *available at* https://ila.vettoreweb.com/Storage/Download.aspx?DbStorageId=939&StorageFileGuid=33f6b360-58f1-496e-95f2-99b7889c4061; https://ila.vettoreweb.com/Storage/Download.aspx?DbStorageId=938&StorageFileGuid=d7369f16-f024-487a-bd0f-27baa35b145b; and https://ila.vettoreweb.com/Storage/Download.aspx?DbStorageId=937&StorageFileGuid=96b5aaa9-9d50-490f-bdf5-d156a20b81fa.

5 Sea Level Rise Committee's 2016 Johannesburg Report

The Sea Level Rise Committee's first report was considered at the Johannesburg 2016 biennial meeting. This interim report reviewed the evidence of sea level rise and of the related literature and set the stage for preparation of its 2018 report in Sydney. In particular the 2016 Sea Level Rise Committee report considered issues related to the effect of sea level rise on the outer limits of maritime zones and effects of coastline changes on maritime boundaries.[16] The report considered both the proposals identified in the Sofia Report and subsequent proposals. These are discussed in the next section.

6 Proposals for Progressive Development[17]

6.1 *Proposals Considered in the Sofia Report of the Baseline Committee*

The 2012 Sofia Baseline Committee report had summarized five proposals for progressive development of the law to deal with the potential major consequences of sea level rise.

Judge Jesus argued that "a substantial rise in sea level, whatever the cause, should not entail the loss of States' ocean space and their rights over maritime resources, already recognized by the 1982 Convention and by the community of nations".[18]

After acknowledging that States may protect their coasts through the creation or reinforcement of sea defenses, Professor Soons suggested that "[a] less

16 2016 Johannesburg Sea Level Rise Committee interim report at 13–18 (hereinafter Johannesburg Report). The Interim Report of the ILA Sea Level Rise Committee (2016) is available on the Committee's webpage at https://ila.vettoreweb.com/Storage/Download.aspx?DbStorageId=936&StorageFileGuid=6bc111aa-af31-4e3d-84bc-c77223adfdf8. The Report is also available to Members in ILA, *Report of the Seventy-Seventh Conference held in Johannesburg August 2016*, 842–874 (London, 2017), at https://ila.vettoreweb.com/storage/download.Aspx?DbStorageId=3988&StorageFileGuid=c13287f2-28c2-4ef8-ad60-e121f0c287e4. Reports of the meetings of the committee are *available at* https://ila.vettoreweb.com/Storage/Download.aspx?DbStorageId=935&StorageFileGuid=15649a87-b002-4c30-a781-05e784eac3e7, and https://ila.vettoreweb.com/Storage/Download.aspx?DbStorageId=934&StorageFileGuid=e998bf71-9c08-41f1-bf05-d2115d8a19e8.

Unless otherwise noted, references to the Sea Level Rise Committee's Interim Report (2016) are to the online version on the Committee's webpage (Johannesburg Report).

17 The proposals in this section are taken from the Baseline Committee's 2012 Sofia report, pp. 29–30.

18 J.L. Jesus, "Rocks, New-born Islands, Sea Level Rise and Maritime Space", in: Frowein *et al.* (eds.), *Negotiating for Peace – Liber Amicorum Tono Eitel* (Berlin, Springer, 2003), p. 601. (Adapted from the Sofia Report.)

expensive, but probably also less dependable means for these States to prevent negative consequences as a result of sea level rise is to contribute towards the creation of a new rule of customary international law which allows coastal States in case of sea level rise to maintain the original outer limits of their maritime zones".[19] In order to succeed, these States would need to "gain approval for this practice in the relevant international fora".[20]

Professor Caron noted that "[i]n the case of a rising sea level, the law of baselines gives rise to a legal feedback that increases the potential for the waste of resources as well as private and interstate conflict".[21] Like Soons, he acknowledged that states may "preserve their rights and entitlements by committing resources to stabilize that aspect of the physical world which is threatened by climatic change [that is, the baseline]".[22] He continued noting the inefficiencies of preserving baselines, "not because the aspect itself is valuable, but rather because the entitlements are valuable, and those entitlements, for purely conventional reasons, require its preservation".[23] In order to address the problems of inefficiency and conflict, Caron suggested modifying the rule and concluded "that states should move toward permanently fixing ocean boundaries".[24]

Professor Rayfuse also acknowledged that "[c]urrent international law does not adequately address the continued maintenance of [maritime] entitlements in the context of sea level rise",[25] and suggested that "[a]doption by the international community of a positive rule freezing existing baselines would address the concerns of all States by sea level rise".[26] She proposed as an alternative for states whose very existence is threatened, "recognition in international law of a new category of State, the 'deterritorialized State', ... would continue to exist as international legal entities ... and would maintain

---

19 A.H.A. Soons, "The Effects of a Rising Sea Level on Maritime Limits and Boundaries", 37 *Neth. Int'l L. Rev.*, No. 2, 1990, 207, 231. (Adapted from the Sofia Report.)

20 *Id.* (Adapted from the Sofia Report.)

21 D.D. Caron, "Climate Change, Sea Level Rise and the Coming Uncertainty in Oceanic Boundaries: A Proposal to Avoid Conflict", in: Hong & van Dyke (eds.), *Maritime Boundary Disputes, Settlement Processes, and the Law of the Sea* (Leiden, Martinus Nijhoff, 2009), pp. 1, 2. (Adapted from the Sofia Report.)

22 *Id.* 13. (Adapted from the Sofia Report.)

23 *Id.* (Adapted from the Sofia Report.)

24 *Id.* 14 (using the term "boundaries" to refer to the outer limits of maritime zones). (Adapted from the Sofia Report.)

25 R. Rayfuse, "Sea Level Rise and Maritime Zones: Preserving the Entitlements of 'Disappearing' States" in: M.B. Gerrard and G.E. Wannier (eds.), *Threatened Island Nations: Legal Implications of Rising Seas and a Changing Climate* (Cambridge, Cambridge University Press, 2013), p. 191.

26 *Ibid.* (Adapted from the Sofia Report.)

sovereignty over their maritime zones".[27] According to this author, a freezing of maritime zone outer limits "would be consistent with the LOSC objectives of peace, stability, certainty, fairness, and efficiency in ocean governance".[28] Rayfuse concluded that "for small island States whose physical existence may be threatened by sea level rise – our disappearing States – a positive rule freezing baselines would also assuage some concerns as to their future, in a manner fully consistent with the presumption of the continuity of States".[29]

Professors Schofield and Arsana considered that one option to resolve this issue would be to "legally fix or declare the location of normal baselines and/or the maritime limits derived from them".[30] While considering the potential for unilateral state practice in this area to develop new customary law, they asserted that a "preferable approach would be to seek multilateral agreement on, effectively, a revised legal regime applicable to normal baselines".[31]

6.2 *Proposals Considered in the 2016 Johannesburg Interim Report*

The 2016 Sea Level Rise Committee Report first reviewed the issues identified in the 2012 Sofia Baseline Committee Report on effect of sea level rise on the outer limits of maritime zones:

> The issue simply stated is that as sea levels rise the low water line along the coast, which marks the "normal baseline" for the purposes of Article 5, will usually move inland and some key geographical features used as base points may be inundated and lost.
>
> If, as a matter of international law, the coastal baseline is ambulatory (as the Baseline Committee has concluded)[32] then where a baseline moves inland[33] and critical basepoints from which maritime zones are

27 *Id.* 179. *But see* J. Lisztwan, "Stability of Maritime Boundary Agreements", 37 *Yale J. Int'l L.*, No. 1 (Winter), 2012, 153 (offering a competing view on the equities of freezing existing baselines considering that zones preserved under such rules would otherwise have reverted to the commons). (Adapted from the Sofia Report.)

28 Rayfuse, *supra* note 25, at 191. (Adapted from the Sofia Report.)

29 *Ibid.* (Adapted from the Sofia Report.)

30 C. Schofield & I.M.A. Arsana, *Imaginary Islands? Options to Preserve Maritime Jurisdictional Entitlements and Provide Stable Maritime Limits in the Face of Coastal Instability*, 6th IHO-IAG ABLOS Conference, Oct. 25–27, 2010, at 11, *available at* http://www.iho.int/mtg_docs/com_wg/ABLOS/ABLOS_Conf6/ABLOS_Conf6.htm. (Adapted from the Sofia Report.)

31 *Id.* (Adapted from the Sofia Report.)

32 *Baselines Committee Sofia Report*, p. 426 (online, p. 31). (Footnote in the Johannesburg Report.)

33 Also, note that because of changes in sediment flows rising sea may in some circumstances also have an opposite effect in some places – accreting sediments to push the low water line seaward. (Footnote in the Johannesburg Report.)

> measured are inundated, then the outer limits of the maritime zones which are measured from this baseline may also move landward.[34] Where key geographical features which are used as basepoints for the construction of straight baselines are totally inundated then these movements inland may be even more substantial. This same principle of course applies to the archipelagic baselines of archipelagic States, where the effect of losses of key basepoint features may appear to be even more pronounced.

After recalling the analysis by Judge Jesus quoted above, the Report noted the more recent caution by Professor Schofield:

> that, extreme as this issue may seem in theory, the outer limits of broader maritime zones, notably the EEZ, are often less susceptible to change as they are reliant on a restricted number of contributing or critical basepoints. Thus, even significant changes in the location of the coastline may have only limited impacts in terms of 200 nautical mile outer limits so long as key basepoints are not subject to change.[35]

The Johannesburg Report continued:

> the planet is already irrevocably committed to continued sea-level rises in the course of this and the coming centuries, and the actual amount and pace of sea-level rise is as yet unknown. Moreover, as IPCC reports in AR5,[36] the actual amount of sea-level rise is uncertain even in the course of the 21st century, and significant changes, especially in some most affected regions, cannot be excluded at this point.

---

34 Where the outer edge of the natural prolongation of the continental shelf is less than 200 nautical miles from the coastal baseline, then the outer limit of the shelf may extend to 200 nm from the baseline (Article 76(1), LOSC). Due consideration should be here also given to Article 76(9) LOSC, which states that the "coastal State shall deposit with the Secretary-General of the United Nations charts and relevant information, including geodetic data, permanently describing the outer limits of its continental shelf". (Footnote in the Johannesburg Report.)

35 C.H. Schofield, "Defining the 'Boundary' between Land and Sea: Territorial Sea Baselines in the South China Sea", in: R. Beckman, M.R. Page and L. Bernard (eds.), *UNCLOS and the South China Sea* (Cheltenham, Edward Elgar, 2014), pp. 21–54. (Footnote in the Johannesburg Report.)

36 J.A. Church *et al.*, "Sea Level Change", in: *Climate Change 2013: The Physical Science Basis. Contribution of Working Group I to the Fifth Assessment Report of the Intergovernmental Panel on Climate Change* (Cambridge University Press, 2013). (Adapted from Johannesburg Report.)

> This Committee considered the views of a number of scholars who have written on this issue. There was a common understanding that the unavoidable consequence of the impact of sea-level rise, using the existing law of the normal baseline, is the loss of maritime space. However, legal authors that have addressed this issue so far seem equally dissatisfied by this result, and have proposed different solutions de lege ferenda. Soons had pointed out in 1990 that coastal States may legally defend existing coastlines by artificial means but accepted that this was an extremely expensive option.[37] As the 2012 Baselines Committee Sofia Report summarized the issue: "Short of actual physical protection of the coast the authors do not find that the existing law provides for any other way to protect the maritime interests of States threatened with a total loss of territory".[38]

The Report then described the Committee's view on the way ahead:

> The Sea-level Rise Committee considered that this was an appropriate issue on which to make proposals for progressive development of international law. In order to preserve – at least provisionally – existing maritime zone entitlements, the Committee discussed two possible options for progressive development. The first option would be to propose a new rule freezing the existing baselines in their current position, using the "large scale charts officially recognized by the coastal State";[39] the second option would be to propose a new rule freezing the existing defined outer limits of maritime zones measured from the baselines established in accordance with the LOSC.
>
> The practical complications of divorcing the actual low water line from the "legal" low water line had already been explored by the Baselines Committee, which had also looked at State practice in some detail. Although it had found evidence that at least one State took the view that the charted line was the definitive baseline for maritime entitlements

37 A.H.A. Soons, "The Effects of a Rising Sea Level on Maritime Limits and Boundaries", *Netherlands International Law Review*, 37 (1990), p. 231. (Adapted from footnote in Johannesburg Report.)

38 *Baselines Committee Sofia Report*, p. 422 (online, p. 29). (Footnote in the Johannesburg Report.)

39 Article 5, LOSC. Note there is no requirement in the LOSC for the coastal State to notify these "normal baselines". Article 16(2) appears only to require coastal state to give "due publicity" to such charts when lines are "determined in accordance with articles 7, 9 and 10". (Footnote in the Johannesburg Report.)

> even where it might depart from the actual low water line,[40] this was found by the Baselines Committee to be at the extreme end of the spectrum of State practice studied, and the majority of State practice was to the contrary or hybrid,[41] while also "the preponderance of the scholarship in this area appears to support the view that charts are not determinative of the naturally ambulatory normal baseline".[42]

The second option of proposing the freezing of the outer limits of maritime zones was the option also discussed by a number of authors cited, most of whom now take part in the work of the Committee as its members. Soons had suggested the creation of a new rule of customary international law which allows coastal States in case of sea level rise to maintain the original outer limits of their maritime zones.[43] Like Soons, Caron had recognized early on the waste of resources that would be involved in physically defending basepoints simply to protect maritime interests[44] and has suggested that, in order to address the problems of inefficiency and conflict, "States should move toward permanently fixing ocean boundaries".[45] He, like Jesus, is in favour of freezing baselines, whereas Rayfuse has proposed that a freezing of maritime zone outer limits "would be consistent with, and would significantly assist in, the promotion and achievement of the LOSC objectives of peace, stability, certainty, fairness, and efficiency in ocean governance".[46] Schofield and Arsana had considered that

40 See *Baselines Committee Sofia Report*, p. 412 (online, p. 21), citing Netherlands Ministry of Defense press release of Dec. 22, 2009, stating: "A change in the actual coastline thus has no effect, until it is included in the nautical chart" (translation by the Baselines Committee members Oude Elferink, Soons, and Kwiatkowska). (Footnote in the Johannesburg Report.)

41 *Baselines Committee Sofia Report*, p. 413 (online, p. 21–22). (Footnote in the Johannesburg Report.)

42 *Baselines Committee Sofia Report*, p. 413 (online, p. 22). (Footnote in the Johannesburg Report.)

43 Soons, "The Effects of a Rising Sea Level on Maritime Limits and Boundaries", p. 231. (Footnote in the Johannesburg Report.)

44 D.D. Caron, "When Law Makes Climate Change Worse", p. 621. (Footnote in the Johannesburg Report.)

45 D.D. Caron, "Climate Change, Sea Level Rise and the Coming Uncertainty in Oceanic Boundaries: A Proposal to Avoid Conflict", in Seoung-Yong Hong and Van Dyke (eds.), *Maritime Boundary Disputes, Settlement Processes, and the Law of the Sea* (Boston/Leiden, Brill/Martinus Nijhoff, 2009), pp. 1, 14. (Footnote in the 2016 Report.)

46 R. Rayfuse, "Sea Level Rise and Maritime Zones: Preserving the Entitlements of 'Disappearing' States", in M.B. Gerrard & G.E. Wannier (eds.), *Threatened Island Nations: Legal Implications of Rising Seas and a Changing Climate* (Cambridge University Press, 2013) p. 191. (Adapted from footnote in the Johannesburg Report.)

one option would be to "legally fix or declare the location of normal baselines and/or the maritime limits derived from them".[47]

Hayashi thinks that "the idea of freezing, and thus permanently fixing, the baselines and consequently the outer limits of various maritime zones, has considerable merit".[48] However he contrasts the two possible approaches pointing out that "the fixing of baselines would mean that the future submerged area becomes internal waters, whereas fixing only the outer limits of maritime zones would result in expanding the breadth of the territorial sea landward to the extent of the shifting baselines" and he considers the proposal to fix baselines, in contrast to fixing outer limits, as having "the merit of not changing the rules on the breadth of the territorial sea and the EEZ as stipulated in Articles 3 and 57, respectively".[49]

This issue was the subject of a preliminary discussion at the Oslo intersessional meeting of the [Sea Level Rise] Committee (June 2015) and there was preference expressed for considering modalities de lege ferenda for freezing the outer limits of maritime zones. It was felt that the issue would benefit from wider discussion within the Committee in its work on the 2018 final report.[50] The legal consequences of "divorcing" coastal baselines (which would still be ambulatory) from the outer limits of maritime zones, by freezing the outer limit without freezing the baseline, has implications for some of the key provisions of the LOS Convention, including the breadth of the territorial sea[51]

47 C. Schofield and I.M.A. Arsana, *Imaginary Islands? Options to Preserve Maritime Jurisdictional Entitlements and Provide Stable Maritime Limits in the Face of Coastal Instability*, 6th IHO-IAG ABLOS Conference, Oct. 25–27, 2010, p. 6; online: www.iho.int/mtg_docs/com_wg/ABLOS/ABLOS_Conf6/ABLOS_Conf6.htm. While considering the potential for unilateral state practice in this area to develop new customary law, they assert that a "preferable approach would be to seek multilateral agreement on, effectively, a revised legal regime applicable to normal baselines". (Adapted from footnote in the Johannesburg Report.)

48 M. Hayashi, "Sea Level Rise and the Law of the Sea – Future Options", in: Vidas and Schei (eds.), *The World Ocean in Globalisation: Challenges and Responses* (Boston/Leiden, Brill/Martinus Nijhoff, 2011), p. 197. (Footnote in the Johannesburg Report.)

49 *Ibid.*, p. 196. (Footnote in the Johannesburg Report.)

50 Note that on July 16, 2015, seven leaders of Polynesian States and Territories signed at Papeete the Taputapu-ātea Declaration on Climate Change. This Declaration states (in para. 4) that, ahead of COP 21 in Paris, "the Polynesian Leaders Group call upon all State Parties to the UNFCCC to" … "with regards to the loss of territorial integrity … acknowledge, under the [UNCLOS], the importance of the Exclusive Economic Zones for Polynesian Island States and Territories whose area is calculated according to emerged lands and permanently establish the baselines in accordance with the UNCLOS, without taking into account sea level rise". (Footnote in the Johannesburg Report.)

51 Article 3, LOS Convention: "Every State has the right to establish the breadth of its territorial sea up to a limit not exceeding 12 nautical miles, measured from baselines determined in accordance with this Convention". (Footnote in the Johannesburg Report.)

and of the EEZ.[52] Hayashi has also highlighted the risks of freezing what might already be "excessive" claims where States have established baselines in violation of the rules of the Convention.[53] Moreover, a decision would also need to be made as to which specific point in time that any "freezing" might be permissible. Concerns have also been raised about the fundamental principles of the law of the sea, such as the principle of the land dominating the sea, rather than vice versa.[54] Although the motive for making such a proposal would be to facilitate stability and orderly relations between States, there might well be unintended negative consequences.

The Report then noted additional work the Committee has to do:

> The Sea-level Rise Committee has also yet to discuss in detail how any proposal de lege ferenda might best be implemented. Hayashi has systematically adumbrated a number of options, some of which he admits may not be practical – they include the development of customary international law,[55] a protocol to the UN Framework Convention on Climate Change,[56] utilization of the amendment provisions of the LOSC, a decision of the Meeting of the State Parties to the LOSC (SPLOS), or by means of a diplomatic conference open also for States non-parties to the LOSC, or by an agreement adopted by the UN General Assembly after negotiation in its subsidiary bodies or informal consultations.[57]

52 Article 57, LOS Convention: "The exclusive economic zone shall not extend beyond 200 nautical miles from the baselines from which the breadth of the territorial sea is measured". (Footnote in the Johannesburg Report.)

53 M. Hayashi, "Sea Level Rise and the Law of the Sea – Future Options", p. 196, citing T. Palmer, "Sea-level Change and Baselines", in: *Proceedings of the Canadian Hydrographic Conference and National Surveyors Conference, 2008, available at* https://hydrography.ca/wp-content/uploads/files/2008conference/session_2A/2A-2_Palmer.pdf, p. 4. (Adapted from footnote in the Johannesburg Report.)

54 *Baselines Committee Sofia Report* p. 423 (online, p. 29). (Footnote in the Johannesburg Report.)

55 As suggested by Soons, "The Effects of a Rising Sea Level on Maritime Limits and Boundaries", p. 255. (Footnote in the Johannesburg Report.)

56 As proposed in 1990 by the Coastal Zone Management Subgroup of the IPCC, reported by Freestone and Pethick, "Sea Level Rise and Maritime Boundaries: International Implications of Impacts and Responses", at p. 76. D. Freestone and J. Pethick, "Sea Level Rise and Maritime Boundaries: International Implications of Impacts and Responses", in G. Blake (ed.) *International Boundaries: Fresh Perspectives*, vol. 5 (Routledge, 1994), pp. 73–90, reprinted as *Maritime Boundaries: World Boundaries volume 5* (Routledge, 2016). (Adapted from footnote in the Johannesburg Report.)

57 All discussed further by Hayashi, "Sea Level Rise and the Law of the Sea – Future Options", pp. 200–206. (Footnote in the Johannesburg Report.)

There is also the question of how long such a "freezing" might be effective? Would it be regarded as a permanent solution or as a temporary provisional measure until such time as the international community might agree upon a more permanent approach?

6.3 *Effects of Coastline Changes on Maritime Boundaries*

The second major problem considered by the Committee is the effects of coastline changes on maritime boundaries. The 2016 Sea Level Rise Committee Report introduced the problem as follows:

> The second issue flagged by the Baselines Committee was what it termed the "negative impacts on maritime boundaries negotiated in reliance on normal baselines in existence at the time of delimitation". In other words, if the representatives of two (or possibly more) States negotiating a maritime boundary delimitation agreement have relied on the coastal baseline in order, for example, to measure an equidistance line, and then the coastal baseline changes as a result of inundation and resultant shift in the location of baselines landward caused by sea-level rise. Is it possible for one party then to argue that this is a fundamental change of circumstances which would undermine the validity of the maritime boundary agreement?

The Report then recalled the law on fundamental change of circumstances:

> The concept of "fundamental change of circumstances" or clausula rebus sic stantibus is specifically referred to in Article 62 of the 1969 Vienna Convention on the Law of Treaties. Article 62(1) reads:
>
> > A fundamental change of circumstances which has occurred with regard to those existing at the time of the conclusion of a treaty, and which was not foreseen by the parties, may not be invoked as a ground for terminating or withdrawing from the treaty unless:
> >
> > a) the existence of those circumstances constituted an essential basis of the consent of the parties to be bound by the treaty; and
> > b) the effect of the change is radically to transform the extent of obligations still to be performed under the treaty.
>
> However, Article 62(2) does specifically exclude two situations from the application of this doctrine. It provides that:

> A fundamental change of circumstances may not be invoked as a ground for terminating or withdrawing from a treaty: (a) if the treaty establishes a boundary[.][58]

The Report proceeds to examine what experts have written about what these provisions mean:

> Commenting on this provision in the Vienna Convention, Aust notes that "the principle cannot be invoked if the treaty establishes a boundary, that term being used so as to include treaties which cede territory, not merely delimit a boundary".[59] Some scholars have suggested that the application of this provision of the Vienna Convention to maritime boundaries is still an open issue. For instance, Caron has argued that "it is entirely plausible that a State might argue that circumstances had changed in that the parties had not foreseen such a rise in sea level".[60]
>
> Lisztwan has looked in detail at the travaux préparatoires of the Vienna Convention text – itself based on an International Law Commission draft prepared by Rapporteur Sir Humphrey Waldock (building on the previous 1957 report of Sir Gerald Fitzmaurice). Her research reveals that many types of boundary treaties were referred to in the context of the negotiation and she concludes that "[a]lthough limited, those comments, and the absence of any contrary statements, suggest that the State Parties intended maritime boundaries to fall within the Article 62(2)

58 The second exception is not relevant here: "(b) if the fundamental change is the result of a breach by the party invoking it either of an obligation under the treaty or of any other international obligation owed to any other party to the treaty". (Footnote in the Johannesburg Report.)

59 A. Aust, *Modern Treaty Law and Practice* (3rd ed., Cambridge, Cambridge University Press, 2013), p. 264. (Footnote in the Johannesburg Report.) Accord, D.H. Hollis, "Defining Treaties", in: D.H. Hollis (ed.), *The Oxford Guide to Treaties* (Oxford, Oxford University Press, 2012), p. 36; J. Crawford, "*Brownlie's Principles of Public International Law*", (8th ed., Oxford, Oxford University Press, 2012), p. 393.

60 D. Caron, "Climate Change, Sea Level Rise and the Coming Uncertainty in Oceanic Boundaries: A Proposal to Avoid Conflict", pp. 13–14; cited by J. Lisztwan, "Stability of Maritime Boundary Agreements", *Yale Journal of International Law*, 37, No. 1 (Winter), 2012, pp. 153, 186, who also cites cautious support from J. Lusthaus, "Shifting Sands: Sea Level Rise, Maritime Boundaries and Inter-state Conflict", *Politics*, 30 (2020), pp. 113, 115–118. (Footnote in the Johannesburg Report.)

boundary exception".[61] From this, Lisztwan draws the conclusion that: "States cannot invoke coastline shift as grounds to terminate a maritime boundary".[62] She also cites the dicta in the Aegean Sea case, where the International Court of Justice (ICJ) said

> Whether it is a land frontier or a boundary line in the continental shelf that is in question, the process is essentially the same, and inevitably involves the same element of stability and permanence, and is subject to the rule excluding boundary agreements from fundamental change of circumstances.[63]

The 2016 Johannesburg Report summarizes the Sea Level Rise Committee's initial consideration of this question:

> As a matter of existing law, the Committee took the view that changes in coastlines arising through sea-level rise or any other reason did not affect, and had not to date affected, the validity of existing maritime boundary agreements. On the grounds of legal stability and certainty the Committee was skeptical as to whether this rule should be re-examined de lege ferenda or be subject to revision at this stage, since this could involve a revision of both the LOSC and the Vienna Convention on the Law of Treaties, Article 62(2).[64]

However, the Report noted that

> Re-examination of the above rules may, however, become necessary in connection with the extreme scenario of sea-level rise leading to the submergence of the entire territory of low-lying small island States or to their territory becoming uninhabitable. The Sea-level Rise Committee will revert to and address this issue in its final report (in 2018), in conjunction with the discussion of statehood.[65]

Nevertheless, the Committee's initial view was that, "for any treaty negotiated since the time that the issue of climate change and consequent sea-level rise has been in the public domain, the negotiators must be deemed to have been

61 Lisztwan, "Stability of Maritime Boundary Agreements", p. 189. (Footnote in the 2016 Report.) On this point see further I. Sinclair, *The Vienna Convention on the Law of Treaties* (2nd ed., Manchester, Manchester University Press, 1984), pp. 194–195.

62 *Ibid.* (Footnote in the Johannesburg Report.)

63 *Aegean Sea Continental Shelf* (Greece v. Turkey), [1978] ICJ Rep. 3, p. 85 (Dec. 19). (Footnote in the Johannesburg Report.)

64 Johannesburg Report, p. 17.

65 *Ibid.*

aware of it, hence sea-level rise could not in that event be regarded as a fundamental change of circumstances".

The Report then noted that

> The key question in this regard is to assess at which point in time it could be considered that sea-level rise, of such a magnitude as to impact objective criteria for negotiated boundaries, became a matter which was in the "public domain". The First World Climate Conference hosted by the World Meteorological Organization (WMO) was in 1979, while the IPCC was established by UNEP and WMO in 1988 – and the First Assessment Report by the IPCC was published two years later, in 1990.[66] As already mentioned, however, in 2014 the IPCC revised its predictions published in AR4 (2007), and posited sea-level rise of up to approximately one metre by 2100. IPCC estimates in AR5 are, already as of 2016, considered to be rather conservative.
>
> This time could arguably be as early as at the time of the WMO First World Climate Conference in 1979. However, it has been noted that "during the Third UN Conference on the Law of the Sea [which was concluded in 1982], there was no widespread recognition of the possible problems of sea-level rise, and negotiators did not anticipate that there would be a significant global regression of coastlines".[67] There are therefore no provisions in the 1982 LOS Convention itself for dealing with possible rises in sea level;[68] it can therefore hardly be expected that maritime boundary negotiators before that time could have been deemed to have been aware of the issue of sea-level rise as potentially relevant to the agreement they were negotiating.
>
> On the other hand, it could be argued that the latest estimates by the IPCC, as contained in AR5 of 2013/2014, are now based on significantly improved understanding of components that contribute to sea-level rise, and have for the first time introduced in the public domain a well-founded scientific explanation of all components of sea-level rise and its projections for the 21st century, of such a magnitude as to have the potential

66 Regarding findings on sea-level rise in that and subsequent IPCC assessment reports, see Part I of the 2016 report, section B.2.1. Note also General Assembly Resolution 44/206, Possible Adverse Impacts of Sea-Level Rise on Islands and Coastal Areas, Particularly Low-lying Coastal Areas, UN Doc A/44/206, adopted on Dec. 22, 1989, 85th plenary meeting; and General Assembly Resolution 43/53, Protection of Global Climate for Present and Future Generations of Mankind, UN Doc A/43/905, adopted on Dec. 6, 1988, 70th plenary meeting. (Adapted from footnote in the Johannesburg Report.)

67 Hayashi, "Sea Level Rise and the Law of the Sea – Future Options", p. 194, as well as the literature cited there. (Footnote in the Johannesburg Report.)

68 *Ibid.* (Footnote in the Johannesburg Report.)

> to affect the objective criteria on which the negotiation of maritime boundaries is based.
>
> There could also be various other points in time – before, between, or after those already discussed above – at which the knowledge or presumed awareness of the negotiators of the maritime boundaries about sea-level rise as the relevant circumstance might be regarded as being "foreseen by the parties" at the time of the conclusion of the agreement. Moreover, in those instances in which negotiators are clearly aware that there may be changes to the relevant baselines subsequent to the conclusion of a boundary treaty, the fact that they opted for a defined boundary instead of allowing for later changes indicates a preference for legal stability.

This Report concluded that "[t]he views, suggestions and opinions of the Committee members on this wide range of issues will be required in the further work of the Committee towards its final report in 2018".[69]

7 Consideration of Sea Level Rise at the UN

The 18th UN Informal Consultations on Oceans and the Law of Sea (UNICPOLOS or ICP 18) met May 15–19, 2017 with the theme "The effects of climate change on the oceans". In preparation for that meeting the Secretary-General's report described how the LOS Convention does not deal with those effects of sea level rise (SLR), as follows:

> 54. Neither UNCLOS nor customary international law address the impact of a total or partial loss of land territory that may result from SLR on maritime limits. UNCLOS specifies the maximum breadth of maritime zones and the sovereignty, sovereign rights and jurisdiction that coastal States can exercise therein. It also establishes an obligation for a coastal State to deposit with the Secretary-General of the United Nations charts or lists of geographical coordinates of its straight baselines, as well as outer limits and delimitation lines of its maritime zones. As a consequence of SLR, the land territory of coastal States may be dramatically diminished or, in extreme cases, disappear. Baselines which may have been fixed and deposited with the Secretary-General, and the outer limits of maritime zones or delimitation lines measured from there, may represent the configuration of the coastline prior to SLR. With the exception of article 7, paragraph 2, concerning unstable coastlines caused by deltas and

69 Johannesburg Report, p. 18.

other natural conditions, UNCLOS does not address variations in coastal geography.[70]

At ICP 18 there was a presentation on this subject by an attorney from Foley Hoag,[71] and concerns expressed briefly both before and after the presentation by Nauru on behalf of the PSIDS, Tonga and the Marshall Islands. Nauru stated on May 15:

> There are additional legal implications of the effects of climate change on the oceans. International law is presently silent on the question of how rising sea levels will impact maritime boundaries, including sovereignty. This is an especially crucial issue for small island developing states, which depend on the ocean jurisdictions for sustainable development, and should be addressed.[72]

Also on May 15, Tonga said:

> The adverse impacts of the climate on ocean ecosystems, such as "rising sea levels …", continue to affect and pose significant threats not only to Tonga's conservation and sustainable development efforts, but as with a number of other SIDS, its very existence as a nation state.[73]

The Marshall Islands stated on May 18:

> As emphasized in the Pacific regional statement by Nauru, there are important legal issues under the UN Law of the Sea regarding sea-level rise, other climate-driven impacts, and the continued definition of maritime boundaries. In this regard, the Republic of the Marshall Islands has recently submitted full boundaries as required by the UN Law of the Sea instrument. As a low-lying island nation, with no higher ground, these issues are of obvious relevance. It is important that state parties, and

70 Report of the Secretary-General on Oceans and Law of the Sea, A/72/70, March 6, 2017, para. 54 online: http://www.un.org/depts/los/consultative_process/documents/ICP18_report. No input from the ILA is noted in this report.

71 Christina Hioureas, "Effects of rising sea levels on maritime boundaries and deterritorialization", online: http://www.un.org/depts/los/consultative_process/documents/ICP-18_biosandabstracts.pdf. A summary of her remarks appears in *Earth Negotiations Bulletin* summary of ICP 18, vol. 25 no. 130 (May 22, 2017) at p. 9, online: http://iisd.cmail19.com/t/i-l-ukkkktt-qjjkladd-h/.

72 http://statements.unmeetings.org/media2/14684728/nauru-on-behalf-of-psids.pdf.

73 http://statements.unmeetings.org/media2/14684734/tonga.pdf.

> the international community as a whole, take into close account that our national boundaries, and the political identity therein, will remain assured.[74]

It was reported that Venezuela, supported by Tonga and Fiji, in their oral interventions suggested the state sovereignty and maritime boundaries issues raised in the presentation be added to the agenda of the Sixth Committee and referred to the International Law Commission.[75]

The report of the co-chairs of ICP 18 included the following summary of these interventions:[76]

> 25. Concern was expressed by many delegations about the possibility of the total or partial loss of land territory as a result of sea level rise, and its effect on maritime zones and boundaries, in particular for low-lying islands and coasts. It was also noted that the sea level was not rising uniformly, with some areas experiencing three times the global average. In addition to the loss of sovereign territory, the effects on populations living in coastal areas and the possible loss of hundreds of billions of dollars in infrastructure were highlighted. It was noted that in some countries, communities were already being earmarked for possible relocation as a result of sea level rise.
>
> 26. Some delegations called for the issue of sea level rise and its legal implications for small island developing States to be discussed by the Sixth Committee of the General Assembly or the International Law Commission. A delegation made specific reference to the need to consider the legal implications of sea level rise for maritime delimitation. However, concern was also expressed regarding those proposals, and the need to further discuss them was underlined.

The co-chairs summarized the relevant portions of the panel presentations and discussion as follows:

> 87. .... Christina Hioureas of Foley Hoag gave a presentation on the effects of rising sea levels on maritime boundaries and deterritorialization, including the legal consequences of shifting baselines, changes in the characterization of maritime features and disappearing States. ...

---

74 http://statements.unmeetings.org/media2/14684789/marshall-islands.pdf.

75 Earth Negotiations Bulletin summary of ICP 18, vol. 25 no. 130 (May 22, 2017) at p. 4, online: http://iisd.cmail19.com/t/i-l-ukkkktt-qjjkladd-h/. The summary also reported that the United States does not have a position regarding discussion of these issues at the Sixth Committee, but was open to discussions. *Id.* at p.5.

76 A/72/95, June 16, 2017.

108. A delegation queried how to enhance legal certainty and stability and address gaps in the current international legal framework in the light of the impacts of climate change on maritime boundaries. Ms. Hioureas recommended concluding maritime delimitation agreements. Another delegation noted that maritime delimitation negotiations were costly and time-consuming, while joint development agreements concluded with a neighbouring country could apply to all unresolved maritime issues, such as fisheries. Ms. Hioureas suggested that efforts should be taken to enter into or ensure that joint development agreements remained in existence despite changing circumstances owing to climate change. She noted, for example, that in some cases entitlements might no longer overlap as a result of shifting baselines. A delegation expressed the view that entering into delimitation agreements too soon might require having to reopen negotiations when additional changes took place owing to sea level rise and expressed a preference for joint development agreements. Differing views were expressed concerning the legal implications of arbitral awards relating to the law of the sea.

109. In response to a question concerning the consequences of sea level rise in the case of fixed coordinates included in maritime boundary treaties, Ms. Hioureas noted that, while a fundamental change of circumstances could not be invoked as a ground for terminating or withdrawing from a treaty establishing a boundary, possible grounds for doing so might exist, considering that the issue of geographic uncertainty had not been contemplated during the negotiations of the United Nations Convention on the Law of the Sea. She also noted the possibility of including relevant clauses in maritime boundary treaties to address changing circumstances and prevent disputes in the future. A delegation reiterated that a fundamental change of circumstances could not be invoked as a ground for terminating or withdrawing from a treaty establishing a boundary. The delegation also stressed the need to ensure stability in international law and expressed caution about overemphasizing the use of joint development zones as an alternative to maritime boundary delimitation, noting that they presupposed the legitimacy of claims. …

111. Noting the need for in-depth legal analysis of the impacts of sea level rise on the law of the sea, some delegations suggested that the topic could be included on the agenda of the Sixth Committee of the General Assembly or the International Law Commission. A delegation noted that the workload of the Sixth Committee was already heavy and that a future meeting of the Informal Consultative Process could focus on that issue.

8 Second Intersessional Meeting of the Sea Level Rise Committee

Chair of the Sea Level Rise Committee, Professor Vidas, arranged and chaired the second intersessional meeting of the Sea Level Rise Committee on the Adriatic island of Lopud, Croatia, September 14–16, 2017, where the 23 Committee members present (of the 36 members and alternates) considered proposals for inclusion in the committee's report for the August 19–24, 2018 Sydney biennial conference. At this meeting the members considered both the law of the sea and the human rights issues.

8.1 *Law of the Sea Issues*

Co-rapporteur Freestone led the discussion on proposals to freeze the baseline or just the outer limit of the maritime zones measured from the baseline, and separately on stability of maritime boundary agreements agreed by negotiation or by judicial/arbitral decision.

David Caron introduced a draft comparing the pros and cons of the possibilities for freezing, particularly on minimizing deviations from the law of the sea convention. A revised text will be circulated to the committee.

Differences of view were expressed on whether freezing of the baseline should apply only to the normal baseline (i.e., low water line along the coast), or whether it should also apply to straight baselines (article 7) and archipelagic straight baselines (article 47). In the latter case, should there be a different rule for archipelagic States constituted solely of low lying islands (such as Tuvulu) than to the Philippines and Indonesia.

The discussion on maritime boundaries addressed several issues. First, whether the exception in article 62(2) of the Vienna Convention on the Law of Treaties that fundamental change of circumstances applied to maritime boundary agreements. Second, any evidence that the effect of sea level rise had been considered by negotiators. Third, at what point in time could it considered that the negotiators took the possible effect of sea level rise into consideration. It was noted that ITLOS and Annex VII arbitral boundary delimitation cases are final and binding. Fourth, note was taken of several maritime boundary agreements where it was anticipated that modern surveys might require subsequent adjustment of the coordinates of the boundary. Under these circumstances language was suggested by which negotiators could anticipate movement of the low water line by sea level rise, and decide in advance that the coordinates would nevertheless be frozen. It was noted that these areas were mostly limited to coastlines that gradually changed gradient and not rocky coastlines such as the cliffs of Dover.

There was insufficient time to consider the issue of loss of statehood through total submersion of a low lying island State's entire land territory.

The members acknowledged such a discussion required additional expertise as resolution would ultimately be a political decision. The Chair suggested proper consideration of the issues would require a four-year extension of the committee's mandate, with which the members present agreed. In that regard, consideration is being given to a fourth intersessional meeting in Zagreb in 2019.

8.2 *Human Rights Issues*

A sub-committee under the leadership of Jane McAdam prepared with Bruce Burson (New Zealand) and Walter Kälin (Switzerland) for this intersessional meeting "Draft Principles on the protection of persons displaced in the context of sea-level rise". Members made a number of suggestions that will be considered and a revised text submitted for consideration at the third intersessional meeting.

9 Third Intersessional Meeting of the Sea Level Rise Committee

The Sea Level Rise Committee is planning to conduct a third intersessional meeting in conjunction with an international conference on ocean law issues concerning climate change and sea level rise, particularly on small island developing states (SIDS), to be held in Singapore March 13–14, 2018, to be convened by the Centre for International Law, National University of Singapore.

The draft report of the Sea Level Rise Committee to the Sydney conference will considered at that time. The final of the report is due to ILA Headquarters in London in June 2018.

10 Possible Future Role of the International Law Commission

At the June 2017 meeting of the International Law Commission several members, including one who is also a member of the ILA Sea Level Rise Committee, informally discussed the ILC undertaking a study of the issues. The matter has not yet been formally discussed by the ILC.

CHAPTER 19

Climate Change and the Arctic: Legal Regulations in Changing Times

*Elise Johansen**

Abstract

The effects of climate change are faster and more severe in the Arctic than in most of the rest of the world. The Arctic is warming at a rate almost twice the global average, leading, e.g., to changes in weather conditions and the thinning and receding of the sea ice, which again has severe effects on the marine environment and the Arctic biodiversity. However, changes in the Arctic also come with new opportunities for different types of activities, such as fishing farther north than ever before, the possibility to navigate through sea routes that are ice-free for longer periods of the year, prospects of hydrocarbon activities in areas previously covered by ice, and increased interest from the tourist industry in remote and pristine destinations, making Arctic seaborne tourism a new potential Arctic industry. The opportunity for economic growth in the region must be balanced against the stressors and environmental risks increased activity in the Arctic region present. This Article examines the regulatory framework and its capability to deal with new and emerging climate change-related challenges in the Arctic region.

1 Introduction

The Arctic is experiencing a profound transformation driven largely by the interacting forces of climate change and globalization. The rapid melting of the sea ice is affecting Arctic ecosystems and species directly as well as having global environmental implications that are causing global warming and changes in weather conditions. At the same time, the effects of climate change create new opportunities for human activities, such as shipping, fishing, mining, and the exploitation of oil and gas resources. Scientists forecast ice-free

* Associate Professor, Ph.D. in Law, K.G Jebsen Centre for the Law of the Sea, UIT The Arctic University of Norway. The author's PowerPoint presentation is available at http://www.virginia.edu/colp/pdf/yogya-johansen.pdf.

 | DOI:10.1163/9789004366619_024

summers in the Arctic as early as the late 2030s, which will open the region up to greater commercialization.[1]

The Arctic is warming faster than anywhere else on the planet, making the Arctic region more vulnerable and accessible at the same time. From being a relatively remote area with little or no human activities, the Arctic has undergone major development and changes during the last two decades. In 2004, the Arctic Climate Impact Assessment (ACIA) delivered its first comprehensive multi-disciplinary assessment of the impacts of climate change in the Arctic.[2] The latest report in May 2017 on the Snow, Water, Ice and Permafrost in the Arctic (SWIPA) contains some alarming findings, including that "it becomes increasingly clear that the Arctic as we know it is being replaced by a warmer, wetter, and more variable environment. This transformation has profound implications for people, resources, and ecosystems worldwide".[3]

This article will discuss and assess the current regulatory regime for the Arctic and examine how it is affected by climate change-related effects and challenges. This will be done by describing the current and developing regulatory regime for different activities, such as hydrocarbon activities, fisheries and shipping, and for different relevant issues, including the protection and preservation of the marine Arctic. By describing the current applicable legal regime and the ongoing climate change prompted legal processes, the objective is to assess the Arctic region's potential to respond to the climate change-related challenges by legal means and how the new uses of the Arctic along with the new challenges of the Artic region have affected how the Arctic States are cooperating among themselves and with non-Arctic players. The "Arctic States" are: Canada, the Kingdom of Denmark, Finland, Iceland, Norway, the Russian Federation, Sweden and the United States.[4] The term "Arctic coastal States" (also the "Arctic five") refers to the Arctic States that have maritime zones in the Arctic region. The five Arctic coastal States are: Canada, Denmark/Greenland, Norway, the Russian Federation and the United States.

The next section (2) of this Article presents a brief overview of the legal framework applicable to the Arctic region. It describes the main components of the existing regime and discusses the form and level of Arctic regional

1 The Arctic Council, *The Snow, Water, Ice and Permafrost in The Arctic – Summary for Policy makers*, 2017, p. 3, available at; http://www.amap.no/documents/doc/Snow-Water-Ice-and-Permafrost.-Summary-for-Policy-makers/1532.

2 The Arctic Council, *The Impacts of a Warming Arctic*, 2004. Available at http://www.amap.no/documents/doc/impacts-of-a-warming-arctic-2004/786.

3 The Arctic Council, *The Snow, Water, Ice and Permafrost in The Arctic – Summary for Policy-makers, op. cit.*

4 These are also the eight Member States of the Arctic Council.

cooperation, focusing mainly on the role of the Arctic Council. In the following sections, examples from different sectors of climate change-related challenges and how they are addressed by legal means are discussed. The section on hydrocarbon activities (3) discusses the issues of engaging in oil and gas-related activities in the Arctic and the challenges this entails both in regard to the environmental risks and in regard to which State will be entitled to engage in activities and where. The next section (4) deals with the situation of new and potential fishing activities taking place farther and farther north in the Arctic due to the ice melting, changes in fish migrating patterns and the constant demand for protein sources, by discussing the ongoing process of reaching an agreement for regulating fisheries in the Central Arctic Ocean. The section on navigation (5) looks at how ice melting in the Arctic is creating new possibilities for shipping routes and the extent to which the legal framework in place is capable of dealing with the challenges that increased traffic and routes farther north present. The section on protecting the marine environment (6) discusses how and the extent to which the Arctic States have taken steps to protect the Arctic marine environment from the challenges that new and increased activities in the region present, especially focusing on the use of area-based measures. The last section (7) contains some concluding remarks.

2 The Legal and Institutional Framework Applicable to the Arctic

The rapid changes that the region is undergoing have also challenged the legal framework governing the Arctic region. In 2008, the Arctic coastal States, through the Ilulissat Declaration decided not to work for the development of a treaty regime governing the Arctic, but to work within the framework provided for by the existing legal regime.[5] The five Arctic coastal States also expressed in the declaration that they have a stewardship role in protecting the Arctic Ocean and that they will take steps "in accordance with international law both nationally and in cooperation among the five states and other interested parties to ensure the protection and preservation of the fragile marine environment of the Arctic Ocean".[6]

Without a comprehensive governance regime, the characteristic of the Arctic as a region is complex. It is a system of fragmented international, regional

5 *Ilulissat Declaration*, Arctic Ocean Conference, Ilulissat, Greenland, 27–29 May 2008. Available at: https://www.regjeringen.no/globalassets/upload/ud/080525_arctic_ocean_conference-_outcome.pdf.

6 *Ibid.*

and bilateral, and domestic regulations, complemented by non-binding soft law mechanisms. The cornerstone of the legal framework applicable to the Arctic is the 1982 United Nations Convention on the Law of the Sea (UNCLOS)[7] and its two implementing agreements, the Part XI Deep Seabed Mining Agreement[8] and the Fish Stock Agreement.[9] The UNCLOS and its implementing agreements set out the jurisdictional framework for all marine activities and the main principles for regulating activities in the different maritime zones. However, as they are framework conventions, the substantive standards must to a large degree be found elsewhere. This has created the need for legally binding agreements on different levels: global, regional and domestic.

Complementing the global legal framework with regional and/or bilateral agreements comes with certain challenges. Due to the *Pacta tertiis* principle, regional regulation can never have the same global application and effect as an international treaty signed by a majority of the states in the world. This can make it difficult to ensure compliance. The process of negotiating an agreement on unregulated high seas fishing in the Central Arctic Ocean has also demonstrated that the parties to a regional agreement can be unwilling to commit themselves, if there are too little implications for the non-party States.[10] However, there are also clear benefits to taking regional approaches, as the rules and regulations can be designed to meet the challenges and needs of the specific area. Taking regional approaches also gives the parties a stronger role in the process simply by the fact that there will be fewer actors around the negotiating table, which creates a level playing field.

Through their vicinity, and hence their interest in the geographical area, the Arctic States play a vital role in the governance of Arctic areas. Particularly the Arctic coastal States have a prominent role, and corresponding obligations, in questions related to the governance of the Arctic due to the fact that large parts of the Arctic region are areas under national jurisdiction. The Arctic States cooperate on many different issues and in many different fora in which the Arctic Council serves as the leading intergovernmental forum

---

7 United Nations Convention on the Law of the Sea, opened for signature 1 1982; in force 16 November 1994, 1833 UNTS 397.

8 Agreement Relating to the Implementation of Part XI of the United Nations on the Law of the Sea of 10 December 1982, adopted 28 July 1994, in force 28 J UNTS 42.

9 Agreement for the Implementation of the Provisions of the United Nations on the Law of the Sea of 10 December 1982 relating to the Conservation and M of Straddling Fish Stocks and Highly Migratory Fish Stocks, adopted 4 Au force 11 December 2001, 2167 UNTS 3.

10 See the discussion below, Section 4.

for promoting cooperation, coordination and interaction among the Arctic States.[11] The Arctic Council was established in 1996 by the Ottawa Declaration and has, with the two primary objectives of promoting environmental protection and sustainable development, taken upon itself an important role as a policy maker collaborating with other organizations and bodies.[12]

The Arctic Council's mandate is quite broad in regard to both the geographical mandate and the substantial mandate. The Ottawa Declaration speaks of the "Arctic" and the "Arctic marine environment" without providing any geographical definition of these terms. As there are no generally accepted geographical definitions of the terms "Arctic" and "marine Arctic", the different components and outputs of the Arctic Council can have different geographical scopes.[13] The substantial mandate is described in the Ottawa Declaration as promoting "cooperation, coordination and interaction among the Arctic States", to "oversee and coordinate the programs established" referring to its working groups, and to "disseminate information, encourage education and promote interest in Arctic-related issues".[14] The Council does not have the capacity to adopt legally binding obligations, and its activities on marine issues have taken the form of mainly scientific monitoring and assessments. The Council has nevertheless shown itself important to the development of the rules and regulations applicable to the Arctic region by being a united force in international negotiations and due to the fact that legally binding instruments have been undertaken under the Council's auspices.[15]

---

11 The eight member states are: Canada, the Kingdom of Denmark, Finland, Iceland, Norway, the Russian Federation, Sweden and the United States. In addition to the eight member states, six organizations representing Arctic Indigenous peoples have status as Permanent Participants: the Aleut International Association, the Arctic Athabaskan Council, the Gwich'in Council International, the Inuit Circumpolar Council, the Russian Arctic Indigenous Peoples of the North, and the Saami Council. It is also possible to gain observer status for non-Arctic states; inter-governmental and inter-parliamentary organizations, global and regional; and non-governmental organizations.

12 Declaration on the establishment of the Arctic Council (the Ottawa Declaration). 19 September 1996. 35 ILM 1387 (1006), Ottawa, Canada.

13 E.J. Molenaar, "The Arctic the Arctic Council, and the Law of the Sea", in R.C. Beckman, T. Henriksen, K.D. Kraabel, E.J. Molenaar and J.A. Roach (eds.), *Governance of Arctic Shipping – Balancing Rights and Interests of Arctic States and user States* (Leiden/Boston: Brill Nijhoff, 2017), pp. 24–67.

14 The Ottawa Declaration, *op. cit.,* Article 1 a), b) and c).

15 Three legally binding agreements have been negotiated under the auspices of the Arctic Council: Agreement on Cooperation in Aeronautical and Maritime Search and Rescue in the Arctic, adopted 12 May 2011, entered into force 19 January 2013; Agreement on Cooperation on Marine Oil Pollution Preparedness and Response in the Arctic, adopted

3 Hydrocarbon Activities

The Arctic is estimated to hold the world's largest remaining untapped gas reserves and some of the world's largest undiscovered conventional oil resources.[16] This makes the Arctic an area rich in resources. One of the effects of climate change is that it will be possible to access previously inaccessible areas due to the decrease in sea ice. This again is expected to make the exploitation of oil and gas deposits economically feasible.[17] The desire to control these resources has led to what is called "the race for the Arctic".[18] The outcome of the question of who owns the North Pole, meaning the continental shelf under the Arctic high seas, relies on the legal regime governing the continental shelf.

The regime governing the continental shelf is described in Part VI of the UNCLOS. According to UNCLOS Article 77, the coastal States exercise sovereign rights to explore it and to exploit its natural resources. The outer limit of the continental shelf is at least 200 nautical miles from the coastal States' baselines. However, as the continental margin extends beyond 200 nautical miles, the coastal States can establish an extended continental shelf based on the procedures described in UNCLOS Article 76.[19] If a coastal State intends to establish the outer limit of the continental shelf beyond 200 nautical miles, or if the limits of the outer continental shelf constitute the delimitation line towards opposite States, the coastal State shall submit information on such limits to the Commission on the Limits of the Continental Shelf (CLCS).[20]

The legal status of the extended continental shelf in the Arctic region, including the sea bed under the Arctic high seas, is yet to be decided. So far, only Norway has submitted and received the recommendations from CLCS

15 May 2013, entered into force 25 March 2016; and Agreement on Enhancing International Arctic Scientific Cooperation, adopted 11 May 2017, not in force yet.

16 United States Geological Survey estimates that the Arctic contains approximately 13 percent of the world's undiscovered conventional oil resources and about 30 percent of its undiscovered conventional natural gas resources. United States Geological Survey, available at: http://geology.com/articles/arctic-oil-and-gas/.

17 A.G. Oude Elferink, "The Continental Shelf in the Polar Regions: Cold War or Black-Letter Law?" *Netherlands Yearbook of International Law,* Vol. 40, 2009, pp. 121–181.

18 *Ibid.* pp. 123–124.

19 Definition of Continental margin by Oude Elferink; "The continental margin comprises the submerged prolongation of the land mass of the coastal State, and consists of the seabed and subsoil of the shelf, the slope and the rise. It does not include the deep ocean floor with its oceanic ridges or the subsoil thereof". *In* A.G. Oude Elferink, "The Continental Shelf in the Polar Regions: Cold War or Black-Letter Law?" *Netherlands Yearbook of International Law,* Vol. 40, 2009, pp. 121–181.

20 UNCLOS Article 76(8).

(2009), and has hence established an outer continental shelf.[21] Russia submitted its renewed Arctic claims to the CLCS in August 2015.[22] The claim covers approximately 1.2 million square kilometers of territory, including the seabed under the North Pole, and is partly overlapping with the claim submitted by Denmark in December 2014.[23] In 2013, Canada filed a partial submission to the CLSC, which left out the Arctic part of Canada's continental shelf. Canada's then Foreign Affairs Minister John Baird stated at a news conference in relation to the partial submission that the country's scientists had been asked to do further work mapping the continental shelf, thereby making it clear that Canada has no intention of forfeiting the geographic North Pole. That was the first time a government cabinet minister has publicly declared that Canada wants to include the sea bed under the Arctic high seas in its submission.[24] Canada plans to submit its Arctic continental shelf claim in 2018, and it is expected to include the seabed under the North Pole and to overlap with both Russian and Danish submissions.

The United States is not a party to the UNCLOS, which raises the questions of the extent to which the United States is entitled to a continental shelf beyond 200 nm under customary law and what legal relationship exists between the entitlement to a continental shelf and the establishment of its outer limits.[25] The official American position is that UNCLOS Article 76, provision 1–7, reflects international customary law, and the United States plans to explore, based on these provisions, its right to establish a continental shelf beyond 200 nm.[26] Even though the answer to the question whether it is possible for the United States to establish an outer continental shelf is not straightforward, Magnùsson argues that there are compelling arguments that Article 76(1–7)

21 Statement by the Chairman of the Commission on the Limits of the Continental Shelf on the progress of work in the Commission, 20 April 2009, CLCS/62.

22 Partial Revised Submission of the Russian Federation to the Commission on the Limits of the Continental Shelf in respect of the Continental Shelf of the Russian Federation in the Arctic Ocean, Executive summary (3 August 2015) (Subm.no. 1b).

23 Partial Submission of the Government of the Kingdom of Denmark together with the Government of Greenland to the Commission on the Limits of the Continental Shelf, The Northern Continental Shelf of Greenland, Executive summary (15 December 2014) (subm. no. 76).

24 "Arctic claim will include North Pole, Baird pledges as Canada delays full seabed bid", *The Globe and Mail*, 9 December 2013. Available at: https://beta.theglobeandmail.com/news/politics/ottawa-delays-full-bid-for-claim-to-north-pole/article15824139/?ref=http://www.theglobeandmail.com&

25 See the discussions in B.M. Magnùsson, "Can the United States Establish the Outer Limits of Its Extended Continental Shelf Under International Law?" *Ocean Development & International Law*, Vol. 48, No. 1, 2017, pp. 1–16.

26 U.S. Department of State, "FAQ". Available at: www.continentalshelf.gov/faq/index.htm.

has become part of customary international law.[27] Perhaps the most convincing arguments in favor of the American viewpoint is that the International Court of Justice (ICJ) in the 2012 Nicaragua/Colombia Case declared that Article 76(1) has the status of customary international law[28] and that the United States and its neighbors have engaged in several maritime boundary delimitations that involve the extended continental shelf, including the so-called doughnut hole in the western Gulf of Mexico.[29] The latter area was made available for hydrocarbon development in 2001.

Based on the abovementioned claims, it is likely that the resources of the continental shelf under the central Arctic Ocean "belong" to one of the Arctic coastal States. Which State remains to be seen. The Arctic coastal States have thus far refrained from oil and gas-related activities on parts of the shelf where there are overlapping claims. In contrast, on the 200 nm part of the continental shelf, the Arctic coastal States have demonstrated very different attitudes towards hydrocarbon activities in the Arctic.

Both Russia and Norway have an offensive policy for Arctic hydrocarbon activities. During the last two years, Norway has been actively awarding licenses for oil and gas exploration and extraction on the northern part of their continental shelf. The delimitation agreement between Norway and Russia made it possible to start awarding licenses in the southeast part of the Barents Sea as well.[30] The Norwegian Ministry of Petroleum and Energy publicly proposed in June 2017 to include a total of 93 Barents Sea blocks in the forthcoming License Round.[31] These came in addition to the ten new licenses, consisting of 54 blocks, awarded in 2016 for the production of oil and gas in the Barents Sea.[32] According to the Barents Observer, "Never before has the country opened up for Arctic drilling at this magnitude".[33]

27 *Supra* note 26, p. 12.

28 *Nicaragua v. Colombia* (Judgment) [2012] I.C.J. Reports 624.

29 *Supra* note 26, p. 12.

30 Joint Statement on Maritime Delimitation and Cooperation in the Barents Sea and the Arctic Ocean, Oslo 27 April 2010, available at https://www.regjeringen.no/globalassets/upload/ud/vedlegg/folkerett/030427_english_4.pdf.

31 The 24 license round was the second time licenses was awarded in the newly opened area in the south east part of the Barents Sea. Information available online: https://www.regjeringen.no/en/aktuelt/announcement-of-the-24th-licensing-round/id2558309/.

32 The 23 license round consisted of 57 blocks. These where distributed with 34 blocks in the south-eastern Barents Sea (the formerly disputed area towards Russia), 20 blocks in other parts of the Barents Sea and 3 blocks in the Norwegian Sea. Information available online: https://www.regjeringen.no/en/aktuelt/new-major-opportunities-for-northern-norway/id2362200/.

33 A. Staalesen, "Going All In: Norway proposes massive opening of Arctic shelf", *The Barents Observer*, March 2017. Available at https://thebarentsobserver.com/en/industry-and-energy/2017/03/going-all-norway-opens-its-arctic-shelf-wide-open-oil-drilling.

Russia has implemented the same progressive policy for opening new fields for oil and gas activities. The Arctic shelf accounts for 87% of Russia's total offshore petroleum resources, and Russia has developed fields in both the Barents Sea (e.g., the Shtokman field and the Prirazlomnoye field), and in the Kara Sea.[34] The Kara Sea is more challenging to explore compared to the Barents Sea due to the ice coverage and harsh meteorological conditions.

Canada and the United States have taken a very different approach and declared their Arctic waters off limits for oil exploration.[35] President Obama announced in a joint statement that the United States is designating the vast majority of American waters in the Chukchi and Beaufort Seas as indefinitely off limits to offshore oil and gas leasing.[36] President Trudeau followed suit by saying that Canada will designate all Arctic Canadian waters as indefinitely off limits to future offshore Arctic oil and gas licensing.[37] Canada put in a small reservation saying that the declaration to make the Arctic indefinitely off limits to oil and gas activities will be "reviewed every five years through a climate and marine-science-based life-cycle assessment".[38] The Canadian and American standpoint towards Arctic hydrocarbon activities are hence totally different from the Norwegian and Russian oil and gas policies. However, since President Trump came to office, the official American attitude towards Arctic offshore oil drilling has changed. Trump started in April 2017 the process of reversing the Obama ban by signing an executive order intended to let energy companies drill in the previously protected Arctic areas.[39]

The decision of the CLCS regarding which state will be able to exercise sovereign rights for the purpose of exploring the continental shelf and exploiting its natural resources will be important in determining the role that the oil and gas industry will play in the further development and use of Arctic areas.

The risk of oil pollution from hydrocarbon related activities is dealt with at both the regional level through the Arctic Council and through bilateral agreements.[40] In 2013, the Arctic Council made the Agreement on Cooperation on Marine Oil Pollution Preparedness and Response in the Arctic (MOPPR), which

34 V.V. Ananev, V.E. Verzhbitsky, V.E. Vasilev, and I.A. Tanygin, "Russian Arctic Seas Petroleum Potential and Development Prospects", *Search and Discovery Article,* No. 10735, 2015.

35 United States-Canada Joint Arctic Leaders' Statement, December 20, 2016. Available at https://obamawhitehouse.archives.gov/the-press-office/2016/12/20/united-states-canada-joint-arctic-leaders-statement.

36 *Ibid.*

37 *Ibid.*

38 *Ibid.*

39 Executive order 13795, Implementing an America-First Offshore Energy Strategy, signed April 28, 2017.

40 M. Knol and P. Arbo, "Oil spill response in the Arctic: Norwegian experiences and future perspectives", *Marine Policy*, Vol. 50, Part A, 2014, pp. 171–177.

was the second legally binding agreement negotiated under the auspices of the Arctic Council.[41] The objective of the agreement is to strengthen cooperation, coordination and mutual assistance among the Parties on oil pollution preparedness and response.[42] It provides a framework for the Arctic States to work from in the event of an oil spill and establishes the methods and procedures for cooperation and quick response. The agreement mandates that each state "shall maintain a national system for responding promptly and efficiently to oil pollution incidents and to establish a system of authorities and contact points for prompt and effective response and notification".[43] Hence, it leaves the implementation to the national level and to the capabilities of the parties and the availability of relevant resources.[44] Since the agreement does not add any new obligations to existing regulations, it has been characterised as a "weak and incomplete response" to the risks associated with Arctic offshore oil drilling.[45]

In addition to the legally binding MOOPR agreement, the Arctic Council has initiated several projects and developed guidelines that aim to support the agreement and to continue the work of improving the safe operation of the oil and gas industry in Arctic areas. According to Article 21 of the MOPPR, the Parties shall develop and maintain Operational Guidelines "to assist in the implementation of" the agreement. The Operational Guidelines are included in Appendix IV of the MOPPR agreement and consist of recommended measures to facilitate an effective cooperative oil pollution incident response. Another important set of guidelines is the Arctic Offshore Oil and Gas Guidelines (AOOGG).[46] The AOOGG were written as a response to the concern regarding the potential impacts related to future increases in offshore petroleum activity in the Arctic. The AOOGG presents a set of recommended practices for the regulation of offshore oil and gas activities and is intended to encourage Arctic States and the industry to set the highest standards currently available when initiating and executing oil and gas related activities. This includes risk assessments, environmental impact assessment, procedures and plans for emergencies and responses.

41 Agreement on Cooperation on Marine Oil Pollution Preparedness and Response in the Arctic, May 15, 2013. Available at https://oaarchive.arctic-council.org/handle/11374/529.

42 *Supra* note 41, Article 1.

43 *Supra* note 41, Article 4.

44 *Supra* note 41, Article 15c.

45 C. Pelaudeix, "Governance of Arctic Offshore Oil & Gas Activities: Multilevel Governance & Legal Pluralism at Stake", *Arctic Yearbook*, 2015, p. 4.

46 The Arctic Offshore Oil and Gas Guidelines (AOOGG) was adopted in 1997 and later revised in 2002 and 2009. Available at: https://www.pame.is/index.php/document-library/arctic-offshore-oil-and-gas-documents.

Regional arrangements are also in place in regard to hydrocarbon activities. One example is the Norwegian-Russian oil spill response cooperation in the Barents Sea, established in 1994 by the Agreement on Cooperation to Combat Oil-Spills in the Barents Sea.[47] The Norwegian-Russian oil spill response cooperation has developed into an effective bilateral regime. The general framework for cooperation, the roles and functions of the cooperating parties, and the operative coordination procedures in an emergency situation have been established through a set of formal institutional procedures. These procedures have been followed up by the parties and practiced regularly through bilateral and multilateral training exercises.

A Joint Norwegian–Russian Contingency Plan for the Combatment of Oil Pollution in the Barents Sea (henceforth the Joint Contingency Plan) was also established in 1994 as an integral part of the Agreement on Cooperation to Combat Oil-Spills in the Barents Sea.[48] Later, a Memorandum on Maritime Safety was adopted by the parties in 2006. The Agreement on Cooperation to Combat Oil-Spills in the Barents Sea, the Joint Contingency Plan, and the Memorandum on Maritime Safety constitute the core of the substantive and operative components of the regime. The Norwegian-Russian oil spill response regime has developed into an effective regime and there is a strong belief that the regime has the ability to respond to incidents.[49]

4 Resource Management

The global components of international fisheries law and bodies apply to the marine Arctic, and the starting point is hence UNCLOS, which offers the jurisdictional framework that makes marine capture fisheries within a State's maritime zone part of its sovereignty and sovereign rights.[50] In areas beyond national jurisdiction, which in the Arctic consist of large parts of the Central Arctic Ocean, the starting point in the UNCLOS is the freedom of all states to

47 Agreement between the Government of the Kingdom of Norway and the Government of the Russian Federation on cooperation to combat oil-spills in the Barents Sea, signed in Moscow in April 1994. Available (in Norwegian): https://lovdata.no/dokument/TRAKTAT/traktat/1994-04-28-1?q=NorgeRussland.

48 Joint Norwegian-Russian Contingency Plan for the Combatment of Oil Pollution in the Barents Sea, an integral part of the 1994 agreement signed in Moscow in April 1994.

49 A.K. Sydnes and M. Sydnes, "Norwegian–Russian cooperation on oil-spill response in the Barents Sea", *Marine Policy*, Vol. 39, 2013, pp. 257–264.

50 UNCLOS, Article 2 (internal waters and territorial sea), Article 56(1)(a) (Exclusive Economic Zone), Article 77 (Continental shelf).

engage in fishing.[51] To implement the provisions of the UNCLOS, the UN Fish Stock Agreement was adopted in 1995.[52]

Without any key substantive standards of fisheries regulation, such as the total allowable catch (TAC), allocation procedures, or restrictions to avoid by-catch of non-target species, the Fish Stock Agreement and the UNCLOS rely upon States to carry out actual fisheries regulation, either individually or jointly, including through Regional Fishery Organisations (RFMOs).[53] The Fish Stock Agreement strengthens the RFMOs' role as the forum or organization for States to fulfil their legal duty to cooperate and to set the appropriate conservation and management measures to combat some of the limitations and challenges of the strict zonal approach in UNCLOS, and to facilitate a much needed way to regulate high seas fisheries.

There are a number of bilateral or regional arrangements under the jurisdiction of a number of RFMOs or RFMAs (arrangements as defined in Article 88 of the Fish Stocks Agreement) in place covering parts of the Arctic.[54] The North East Atlantic Fisheries Commission (NEAFC), which covers parts of the Central Arctic Ocean, and the Joint Norwegian-Russian Fisheries Commission, which covers the most important fish stocks of Norway and Russia in the Barents Sea and the Norwegian Sea, are two examples. However, most parts of the Central Arctic Ocean are not covered by any RFMO or RFMA. In addition to gaps in high seas coverage of RFMOs and RFMAs, the major challenges and shortcomings of the existing regime of Arctic fishing regulation are lack of data, inadequate domestic regulation, and gaps in Arctic Ocean coastal State fora and instruments.[55]

So far, most fisheries have been undertaken in the EEZ or the territorial sea of the Arctic coastal States. However, as ice melts, waterways into the Arctic Ocean will open, and it is predicted that many fish stocks will migrate north due to warmer temperatures. Thus, the possibility of unregulated commercial fishing in previously unreachable areas could become the reality in the near

51 UNCLOS, Article 87(1)(e).

52 Agreement for the Implementation of the Provisions of the United Nations Convention on the Law of the Sea of 10 December 1982 Relating to the Conservation and Management of Straddling Fish Stocks and Highly Migratory Fish Stocks, New York, adopted 4 August 1995, entered into force 11 December 2001, 2167 UNTS.

53 E.J. Molenaar, "Arctic Fisheries and International Law: Gaps and Options to Address Them", *Carbon & Climate Law Review*, Vol. 6, No. 1, *Climate Change, Arctic Change: Law and Policy* (2012), pp. 63–77.

54 For an overview, see E. Molenaar, "Arctic Fisheries Conservation and Management: Initial Steps of Reform of the International Legal Framework", 1 *Yearbook of Polar Law*, 2009, pp. 427–463.

55 *Ibid.*, at p. 70.

future. This has also served as the impetus for addressing the above-mentioned gaps in the existing regime of international fisheries fora and instruments relating to the Arctic Ocean.

After several years of discussions and searching for a suitable mechanism, on 16 July 2015, the Arctic coastal States signed in Oslo the Declaration Concerning the Prevention of Unregulated High Seas Fishing in the Central Arctic Ocean (the Oslo Declaration).[56] The declaration acknowledged that climate change is putting pressure on Arctic ecosystems, which have so far been relatively unexposed to human activities. The Arctic coastal States commit themselves to conduct only commercial fishing in the high seas portion of the Arctic Ocean "pursuant to one or more regional or sub regional fisheries management organizations or arrangements that are or may be established to manage such fishing in accordance with recognized international standards". This is not to be read as a commitment to ban or freeze commercial fishing, but to point out that fishing can occur only pursuant to the rules and regulations of an RFMO or RFMA.[57] This again means that the NEAFC, which is specifically referred to in the declaration, is to be considered an existing RFMO under the Oslo Declaration regime, and that no fishing can take place in the Arctic high seas as long as no RFMO/RFMA covering the area exists. Furthermore, the Oslo Declaration commits the Arctic coastal States to establish a joint scientific research program to promote ecosystem understandings.

After the adoption of the Oslo Declaration, the so-called Arctic five were criticised for not including other states in a negotiation regarding areas that are beyond national jurisdiction. The Arctic five then expanded Central Arctic Ocean precautionary fisheries discussions to include China, Japan, South Korea, Iceland and the EU. The first meeting of the so called "Arctic five plus five" occurred in Washington in 2015, and the parties have since worked on negotiating a text.[58] There was hope that a final agreement could be reached in the Reykjavik meeting in March 2017, but the parties were not able to adopt a final agreement. A new meeting is scheduled to occur in November 2017 in

56 Declaration Concerning the Prevention of Unregulated High Seas Fishing in the Central Arctic Ocean, signed July 2015, Oslo. Available at https://www.regjeringen.no/globalassets/departementene/ud/vedlegg/folkerett/declaration-on-arctic-fisheries-16-july-2015.pdf.

57 E.J. Molenaar, "The Oslo Declaration on High Seas Fishing in the Central Arctic Ocean", *Arctic Yearbook,* 2015, pp. 1–5.

58 E.J. Molenaar, "The December 2015 Washington Meeting on High Seas Fishing in the Central Arctic Ocean" post of 5 February 2016 on The JCLOS Blog. Available at https://site.uit.no/jclos/2016/02/05/the-december-2015-washington-meeting-on-high-seas-fishing-in-the-central-arctic-ocean/.

Washington, DC, and again there is hope for finalizing the process through a signed agreement. Looking at the process, the negotiations have moved from the Oslo Declaration's prohibition of all fisheries unless regulated through an RFMO or an RFMA to an agreement that will seemingly be open to some degree of commercial exploitation.

According to observers who have been following the negotiation process, disagreement about the role that the Arctic coastal States shall play in allowing new fisheries is what prevents the parties from entering into a final agreement. For eight years, the parties have been trying to arrive at an international instrument that will regulate high seas fisheries in the Central Arctic Ocean. Although no agreement is on the table, the parties seem to be closer than ever.

5 Shipping

The melting of the sea ice in the Arctic comes with an expectation of new and increased opportunities for shipping. Transit through the Northern Sea Route[59] and the Northwest Passage[60] has increased due to longer ice free periods. In the future, if the melting rates continue, navigating through the "North Pole" in the Central Arctic Ocean might be a real alternative for some periods of the year. The trans-Arctic sea routes offer shorter alternatives going from east to west than using the Panama and the Suez Canals, and hence represent an opportunity for operational savings. The increased traffic in Arctic areas is also a result of the expansion of other activities and sectors, such as the oil and gas activities in the Barents Sea, fishing activities, and also an increased interest in Arctic seaborne tourism.

Protecting the marine environment from the risks posed by navigation in the Arctic has been strengthened by the adoption of the International Code for Ships Operating in Polar Waters (Polar Code).[61] With the adoption of the Polar

59 The Northern Sea Route encompasses all routes across the Russian Arctic coastal seas from Kara Gate (at the southern tip of Novaya Zemlya) to the Bering Strait.

60 The Northwest Passage is the marine routes between the Atlantic and Pacific Oceans along the northern coast of North America that span the straits and sounds of the Canadian Arctic Archipelago.

61 The International Code for Ships Operating in Polar Waters (Polar Code), entered into force 1 January 2017, The Polar Code and SOLAS amendments were adopted during the 94th session of IMO's Maritime Safety Committee (MSC), in November 2014; the environmental provisions and MARPOL amendments were adopted during the 68th session of the Marine Environment Protection Committee (MEPC) in May 2015.

Code, rules and regulations applicable to Arctic shipping are in place.[62] However, several hazards were not addressed in the Polar Code, such as banning heavy fuel oil and navigation in areas that are vulnerable due to their ecological significance or in the migratory areas of marine mammals.[63] Furthermore, the Polar Code does not apply to all types of ships.[64] Hence, regulatory gaps still exist.

Increases in Arctic shipping will introduce direct near-surface emissions of pollution, including Black Carbon. Direct emissions in the Arctic have severe warming effects. Compared to Co2, the warming effect of Black Carbon in the Arctic is at least double, as it absorbs incoming and snow-reflected radiation.[65] In recent years, the International Maritime Organization (IMO) has introduced international regulations to reduce emissions of nitrogen oxides and the sulphur content of fuel, but so far there is no ban on the use of heavy fuel oil in the Arctic.[66] At the Marine Environment Protection Committee meeting (MEPC70) in October 2016, the IMO took steps toward phasing out the use

62 In addition to the Polar Code, various IMO agreements on safety of navigation and prevention of marine pollution clearly apply to the Arctic Ocean, such as the International Convention for the Safety of Life at Sea (SOLAS), the International Convention for the Prevention of Pollution from Ships (MARPOL), and the London Convention on Ocean Dumping.

63 T. Henriksen, "Protecting polar environment: coherency in regulating Arctic shipping", in R. Rayfuse (ed.), *Research Handbook on International; Marine Environmental Law* (Cheltenham/Northampton: Edward Elgar Publishing, 2015), pp. 363–384, at p. 371.

64 The Code does not apply to ships less than 500GT, fishing vessels or those entitled to sovereign immunity.

65 IMO study Report, *Investigation of appropriate control measures (abatement technologies) to reduce Black Carbon emissions from international shipping*, Prepared by Litehaus in co-operation with D.A. Lack, J. Thuesen and R. Elliot, 2012. Available at http://www.imo.org/en/OurWork/Environment/PollutionPrevention/AirPollution/Documents/Air%20pollution/Report%20IMO%20Black%20Carbon%20Final%20Report%2020%20November%202012.pdf.

66 The IMO finalized in 2009 a package of specific technical and operational reduction measures for the control of GHG emissions from ships. The MEPC started in March 2010 the consideration making the technical and operational measures mandatory for all ships irrespective of flag and ownership. This work was completed in July 2011. The adopted measures add to MARPOL Annex VI a new Chapter 4 entitled "Regulations on energy efficiency for ships", making mandatory the Energy Efficiency Design Index (EEDI) for new ships and the Ship Energy Efficiency Plan (SEEMP) for all ships. The regulations entered into force through the tacit acceptance procedure on 1 January 2013 and apply to all ships over 400 gross tonnage and above. Read more on the IMO webpage at: http://www.imo.org/en/OurWork/Environment/PollutionPrevention/AirPollution/Pages/GHG-Emissions.aspx

of heavy fuel oil by deciding to cut ship sulphur content in shipping fuels to five percent by 2020.[67] This is a step towards reducing harmful Co2 and Black Carbon emissions in the Arctic and limiting the risk of a heavy fuel oil-spill, but it does not amount to the ban of heavy fuel oil that exists throughout Antarctica. However, the IMO Marine Environmental Protection Committee agreed during a meeting in London in July 2017 to a proposal submitted by seven IMO members to start the rule-making process.[68] The issue will be discussed at the Committee's next meeting in April 2018, and, according to the most ambitious statements, a ban on heavy fuel oil in the Arctic, following the Antarctic model, can be adopted in 2020 and go into effect 18 months thereafter.[69]

On many occasions, the EU has challenged the IMO progress, or lack thereof, and pushed for EU regulations where there has been an absence of adequate international measures. On 16 March 2017, the European Parliament (EP) adopted a resolution on an integrated European Union (EU) policy for the Arctic.[70] This resolution looks at many issues related to that region, one of which is the use of heavy fuel oil in maritime transport.[71] The resolution "calls on the Commission and the Member States to take all necessary measures to facilitate actively the ban on the use and carriage of HFO as ship fuel in vessels navigating the Arctic seas".[72] It has been suggested that such a ban can be enforced by the use of port state control measures. This raises several legal questions regarding the scope of the EU and its member states' jurisdiction.[73] However, the EP resolution represents only the first step in a long regulatory process, and the parties have not yet agreed to a final and binding document.

67 "MEPC 70: IMO Moves Towards Arctic Phase Out of World's Dirtiest Fuel". Press release from the Environmental Investigation Agency (EIA). Available at: https://eia-global.org/press-releases/mepc-70-imo-phase-out.

68 International Maritime Organization's (IMO) Marine Environmental Protection Committee (MEPC 71) in London, July 2017.

69 It was Canada that submitted the proposal with Finland, Germany, Iceland, the Netherlands, Norway and the United States as co-sponsors. The proposal gained support from all the Arctic States and also from France, Spain, Estonia, Czech Republic, Poland and Singapore.

70 European Parliament resolution of 16 March 2017 on an integrated European Union policy for the Arctic, 16 March 2017 (2016/2228(INI)).

71 N.F. Coelho, *Remarks on the European Parliament's proposal to ban Heavy Fuel Oils in the Arctic*, blogpost in *The JCLOS Blog*, posted 28 April 2017. Available at: http://site.uit.no/jclos/2017/04/28/remarks-on-the-european-parliaments-proposal-to-ban-heavy-fuel-oils-in-the-arctic/#more-439.

72 European Parliament resolution of 16 March 2017 on an integrated European Union policy for the Arctic, 16 March 2017 (2016/2228(INI)).

73 See the discussions in N.F. Coelho, *op. cit.*

When talking about regulating shipping in the Arctic, mention must be made of Article 234 of the UNCLOS, which is a provision devoted to "ice-covered areas".[74] Both Canada and Russia have justified unilateral action on the basis of Article 234 and adopted measures mandatory for ships sailing in Canadian and Russian Arctic waters, such as the Northern Canada Vessel Traffic Services (NORDREG) and the Russian rules of navigation in the water area of the Northern sea route.[75] In general, Article 234 offers a legal basis for regulations for the prevention, reduction, and control of marine pollution from vessels in ice covered areas. Looking back at the UNCLOS negotiations, it is clear that this Article was a result of some Arctic States' desire to have a legal basis for stricter environmental protection rules in the Arctic area.[76] The need for such rules is more pressing today than ever.

6 Protection of the Arctic Marine Environment

UNCLOS provides in Article 192 a general obligation to protect and preserve the marine environment. Among the major challenges facing the UNCLOS are keeping up with pressing challenges of today and intercepting new and very necessary approaches to ocean management, such as climate change related challenges that call for new approaches that can provide holistic and integrated management with measures that are flexible and can be implemented in dynamic ways. The Convention on Biological Diversity (CBD) complements the UNCLOS's provisions regarding the governance of marine resources.[77] The CBD

74 The wordings of article 234 is perceived as ambiguous and its content and the meaning of its different terms have been analyzed in legal literature: See e.g., D. McRae, "The negotiation of Article 234", in F. Griffiths (ed.), *Politics of the Northwest Passage* (McGill-Queen's University Press, 1987), pp. 98–114; K. Bartenstein, "The 'Arctic Exception' in the Law of the Sea Convention: A Contribution to Safer Navigation in the Northwest Passage", 42 *Ocean Development & International Law,* 2001, pp. 22–52; E.J. Molenaar, "Options for Regional Regulation of Merchant Shipping outside IMO, with Particular Reference to the Arctic Region", 45 *Ocean Development & International Law,* 2014, pp. 272–298.

75 See J.J. Solski, "Russia", and D.R. Rothwell, "Canada and the United States", in: R.C. Beckman, T. Henriksen, K.D. Kraabel, E.J. Molenaar and J.A. Roach (eds.), *Governance of Arctic Shipping – Balancing Rights and Interests of Arctic States and user States* (Leiden/Boston: Brill Nijhoff, 2017), pp. 173–216 and 217–244.

76 On the negotiations relating to Art. 234 see Nordquist, Myron, Rosenne, Shabtai, and Yankov, Alexander (eds.), *United Nations Convention on the Law of the Sea – A Commentary,* vol. IV (Dordrecht: Martinus Nijhoff Publishers, 1991), p. 302.

77 The United Nations Convention on Biological Diversity. The Convention was opened for signature at the Earth Summit in Rio de Janeiro on 5 June 1992 and entered into force on 29 December 1993.

addresses the protection of marine biodiversity, hereby ecosystems, habitats and populations of species and directs the parties to adopt a precautionary approach. The CBD goes even further than the UNCLOS in recognizing the need to protect the marine environment from a wide range of human activities and to conserve marine biodiversity.[78] This includes mechanisms for the use of area-based management tools such as marine protected areas (MPAs). All of the Arctic States, with the exception of the United States, are parties to the CBD. The geographical scope for the CBD is mainly in areas within national jurisdiction and is thus comprised of the internal waters, the territorial sea, the continental shelf and the EEZ of the Arctic States. For areas beyond national jurisdiction, such as the Central Arctic Ocean, the obligations in the CBD are less clear.[79]

The use of marine protected areas (MPAs) as a tool for protecting the Arctic marine environment has been on the agenda since the regional cooperation between the Arctic States started in 1991.[80] In 2009, the Arctic Council adopted the Arctic Marine Shipping Assessment (AMSA), which focused on the risks posed by Arctic marine shipping.[81] The AMSA concluded that the threats to the Arctic marine ecosystems that are posed by Arctic shipping effectively can be "mitigated through careful planning and effective regulation in areas of high risk".[82] The outcome of this assessment was a list of recommendations that was developed to provide a guide for future actions by the Arctic Council, where both recommendations II C and II D highlighted the use of area-based measures as the proper tool for managing the threats and challenges to the Arctic marine environment. As a response to recommendation II C, the Arctic Council adopted in 2015 a framework for a Pan-Arctic network of marine protected areas.[83] The objective of this framework is to set out a common vision for international cooperation in MPA network development and management. It aims to inform the development of MPAs and networks of MPAs that are located within the national jurisdiction of the Arctic States. The framework

78 I.U. Jakobsen, *Marine Protected Areas in International Law An Arctic perspective* (Leiden/Boston: Brill Nijhoff, 2016), p. 95.

79 *Ibid.*, p. 96.

80 L. Wenzel, N. Gilbert, L. Goldsworthy, C. Tesar, M. Mcconell and M. Okter, "Polar opposites? Marine conservation tools and experiences in the changing Arctic and Antarctic", *Aquatic Conservation- marine and Freshwater Ecosystems*, Vol. 26, No. 52 (2016), 61–84.

81 Arctic Council, *The Arctic Marine Shipping Assessment 2009 Report* (AMSA Report 2009), p. 5. Available at https://oaarchive.arctic-council.org/handle/11374/54.

82 *Ibid.*, p. 152.

83 The Arctic Council, *Framework for a Pan-Arctic Network of Marine Protected Areas*, April 2015. Available at: http://hdl.handle.net/11374/417.

offers guidance and is not legally binding, which is only natural, since it deals with issues under national jurisdiction. Hence, the framework relies on the Arctic States to pursue MPA development based on their own authorities, priorities, and timelines.

For areas beyond national jurisdiction, the Arctic Council working group PAME (Protection of the Arctic Marine Environment) has explored the need for and the possibility of using area based management tools in the Arctic. Focusing on recommendation II D of the AMSA report, PAME has assessed the possibility and possible effect of using area-based measures to protect the Arctic high seas marine areas from the risks posed by shipping. A report from Det Norske Veritas (DNV), ordered by PAME, reviewed potential available IMO measures suited to protect the vulnerable areas.[84] The measures that were assessed were MARPOL Special Areas and three options for protecting the high seas of the Arctic by designating an area as a Particularly Sensitive Sea Area (PSSA). As for Special Areas under MARPOL, the report concluded that no significant gap exists between the current levels of protection offered through the designation of Special Areas in the Arctic high seas and the normal MARPOL requirements for ships operating in the Arctic high seas, especially in light of the new requirements stemming from the Polar Code.[85] In regard to the three options for protecting the high seas area by designating an area a PSSA with associated protective measures, option number one is described as potentially very effective. This option would imply designating the entire Arctic high seas area a PSSA with associated protective measures, such as vessel traffic service (VTS) with ships reporting systems (SRS) to monitor traffic and areas to be avoided (ATBA), established and enforced in a dynamic fashion reflecting the movement of the ice edge.[86] The other two options offered more moderate versions. The report questioned whether option number one is possible due to the restrictions that it would impose on the freedom of navigation and that such restrictions would make it difficult to gain political acceptance.[87] Designating a large part of the Central Arctic Ocean a PSSA with associated protective measures, such as ATBA, raises several legal questions. Do the Arctic high seas areas meet the criteria for being designated a PSSA? May a PSSA be established on the high seas? Would a PSSA covering most of the Arctic high seas violate the

84 Specially Designated Marine Areas in The Arctic High Seas, Report No./DNV Reg No.: 2013-1442/17JTMM1D-26, Rev 2. (Hereinafter DNV report). Available at https://www.cbd.int/doc/meetings/mar/ebsaws-2014-01/other/ebsaws-2014-01-submission-finland-en.pdf.

85 *Ibid.*, p. 53.

86 *Ibid.*, pp. 55–56.

87 *Ibid.*

freedom of navigation? Is it possible to adopt dynamic ATBAs?[88] The Arctic Council has recognized that pursuing the designation of a high seas PSSA raises several legal questions, and it is safe to say that the suggestion of designating the entire Arctic high seas as a PSSA was not embraced by the Arctic States.

The work regarding protection of the marine environment in the high seas areas of the Arctic is still pending.[89] The Arctic Council established a task force on Arctic marine cooperation in 2015. Its objective was "to assess future needs for a regional seas program or other mechanism, as appropriate for increased cooperation in Arctic marine areas".[90] After the Fairbank meetings in April 2017, a report describing the next steps was made.[91] The report stressed the need to take an ecosystem-based approach and pointed out the need for holistic, integrated Arctic regional cooperation to complement sectoral mechanisms, as the issues affecting the Arctic marine environment today are addressed on a sectoral basis.[92] The report furthermore stated that "the Arctic States, including the Permanent Participants and Arctic inhabitants, are uniquely placed to serve as stewards for Arctic marine ecosystems as a whole, including through taking an integrated approach that cuts across sectors and jurisdictional boundaries, and applying our unique knowledge of Arctic marine ecosystems, including traditional knowledge".[93] From the statements in this report, it seems clear that the Arctic Council is assessing the possibility of establishing a regional sea programme for the Arctic that will encompass all sectors and all jurisdictions, including areas beyond national jurisdiction.

A new task force has been established with a mandate to build upon its work to date, to negotiate the terms of reference for a new subsidiary body and to identify complementary enhancements to existing Arctic Council mechanisms.[94] Prior to the Fairbanks meeting, it was debated whether the Council should begin negotiations on a cooperation mechanism for Arctic marine areas and to "make recommendations on the nature and scope of any such

---

88 For an analysis and discussion of these questions see: I.U. Jakobsen and E. Johansen, "Efforts of the Arctic Council to Protect Sensitive Arctic High Sea Areas from the Impact of Shipping", *MarIus*, No. 471, 2016.

89 Arctic Council, Status on Implementation of the AMSA 2009 Report Recommendations, April 2015.

90 Iqaluit Declaration, 2015, paragraph 43.

91 Arctic Council, *Report to Ministers of the Task Force on Arctic Marine Cooperation*, 2017. Available at https://oaarchive.arctic-council.org/bitstream/handle/11374/1923/2017-04-30-Edocs-4079-v3-TFAMC-report-to-ministers-with-cover-and-colophon.pdf?sequence=1&isAllowed=y.

92 *Ibid.*, p. 3.

93 *Ibid.*

94 Senior Arctic Officials' Report to Ministers, Iqaluit, Canada, 24 April 2015, at pp. 77–78.

mechanisms".[95] The response stressed that the needs identified in the Fairbanks report "could best be addressed in a manner consistent with our shared value of complementing, rather than duplicating, the Arctic Council's existing marine cooperation by situating any new marine cooperation mechanisms within the Arctic Council".[96] This can be interpreted as a willingness to look into new ways of cooperating and strengthening the Arctic Council's marine stewardship of the entire Arctic region.

7 Concluding Remarks

The Arctic region is clearly facing some new and emerging challenges due to climate change. Perhaps the most difficult task is how to mitigate the severe effects of climate change in the Arctic and in the rest of the world. The Arctic States cannot alone reverse the effects affecting the Arctic or even slow them down; it is a global issue and concern, and the reduction of greenhouse gas emissions must happen based on both national and global actions in line with the Paris Agreement.[97] All of the Arctic States are parties to the Paris Agreement and can through their nationally determined contribution contribute to the reach of the Paris Agreement's objective of limiting global warming to "well below 2°C and to pursue efforts to limit the temperature increase to 1.5°C above pre-industrial levels".[98] However, President Trump announced on 1 June 2017 that the United Sates will withdraw from the Paris Agreement.[99] Trump announced that he wanted to negotiate a better deal for the United States, and that America remains committed to robust efforts to protect the environment.[100] The leaders of France, Germany, and Italy responded immediately to Trump's statement saying that the Paris Agreement is irreversible and could not be renegotiated.[101] Withdrawing from the Paris Agreement is, however, a four-year long process. Any Party to the Agreement can withdraw, but only

95 Report to Ministers of the Task Force on Arctic Marine Cooperation, *op. cit.*, p. 8.

96 *Ibid.*

97 The United Nations, The Paris Agreement, Accepted Report of the Conference of the Parties on its twenty-first session, held in Paris from 30 November to 13 December 2015, Decision 1/CP.21 (in force 4 November 2016), Decision 1/CP.21, Adoption of the Paris Agreement (UN Doc. FCCC/CP/2015/xx).

98 Paris Agreement, Article 2(1)(a).

99 Statements and Releases from the White House Press Office, *President Trump Puts American Jobs First,* 1 June 2017.

100 M. Shear, "Trump Will Withdraw U.S. From Paris Climate Agreement", *New York Times*, 1 June 2017. Available at https://www.nytimes.com/2017/06/01/climate/trump-paris-climate-agreement.html.

101 *Ibid.*

after three years from the date on which the Agreement entered into force for a Party.[102] Furthermore, it will take one year from the notification of withdrawal until the withdrawal takes effect.[103] For the United States, the Paris Agreement entered into force on 4 November 2016, meaning that the earliest a withdrawal can be effective is 4 November 2020.

The Paris Agreement has faced heavy criticism for leaving the determination of the size of the emissions reductions to the States to decide, for not being ambitious enough when setting the overall temperature targets, and for lacking enforcement mechanisms.[104] With limited accountability for States that fail to fulfil their obligations, the Paris Agreement relies on other compliance mechanisms, such as transparency, accountability, and precision.[105] Whether these compliance mechanisms will accomplish the objectives of the Paris Agreement can perhaps not be answered until 2030, when the first review of the nationally determined contributions and their effect happens.

The Paris Agreement recognises a balance between mitigation and adaptation to address climate change. As this Article has demonstrated, the Arctic States are addressing climate change challenges by adapting to the new situation with an incentive to be precautionary in the way new activities are regulated. The Arctic States are able to engage in discussions and processes on many issues concerning Arctic matters and through many forms of cooperation, in which the Arctic Council serves as the leading forum for high level cooperation. Although cooperation among the Arctic States in general is viewed as good and well-functioning, it is not totally without friction. The disagreement between Norway and several other parties to the Spitsbergen Treaty on the treaty's geographical scope of application serves as one example. Details from a Russian Defense Ministry report show that Moscow is not happy with Norway's attempt to establish absolute national jurisdiction over Svalbard and its shelf.[106] Another example is the disagreement about the legal status of the Northwest Passage and waters within the Northern Sea Route.[107]

---

102 Paris Agreement, Article 28(1).

103 Paris Agreement, Article 28(2).

104 A.M. Slaughter, "The Paris Approach to Global Governance", *Project- Syndicate blog*, posted 28 December 2015. Available at https://www.project-syndicate.org/commentary/paris-agreement-model-for-global-governance-by-anne-marie-slaughter-2015-12.

105 See the discussion in D. Bodansky and S.D. O'Connor (2016) "The Legal Character of the Paris Agreement", *Review of European, Comparative, and International Environmental Law*, Vol. 25, No. 2, 2016, pp. 142–150. DOI: 10.1111/reel.12154.

106 T. Nilsen, "Kommersant: Russia lists Norway's Svalbard policy as potential risk of war", *The Barents Observer*, posted Oktober 2017. Available at: https://thebarentsobserver.com/en/security/2017/10/kommersant-russia-lists-norways-svalbard-policy-potential-risk-war.

107 E.J. Molenaar, 2017, *op. cit.*

Despites these disagreements, the Arctic States are cooperating in many areas and on several issues in regard to policy making, legal regulation, implementation and enforcement, which makes the Arctic region an interesting example of a well-functioning region in relation to cooperation among States. This well-functioning and high level cooperation might be explained by the fact that the Arctic States share the same core objectives, namely the well-being of the Arctic region.

CHAPTER 20

An Overview of Ocean Acidification: Relationships

Stephen A. Macko and Christina M. Fantasia***

Abstract

The current concentration of atmospheric carbon dioxide (CO_2) surpasses 400 ppm, resulting in increasingly higher levels of this gas being absorbed into the ocean and reacting with ocean water. This process is changing the acidity of the global ocean, with increases already observed in higher latitudes and more changes expected in the future. Historically, ocean pH has averaged near 8.1, with historical changes occurring gradually. However, over the past few decades, the rate of change has increased dramatically, lowering the pH by about 0.1 (a 30% increase in acidity) and affecting the buffering capability of the ocean. Many studies have found negative physiological effects on a wide range of calcareous organisms, from phytoplankton to shellfish to predators. Ecologically, potential trophic cascades have been observed in laboratory experiments. These studies usually utilize the lower pH predictions derived from the International Panel on Climate Change (IPCC) evaluations of the continued rise in the atmospheric carbon dioxide content. Owing to the well-understood physico-chemical processes of ocean acidification, future pH predictions are especially robust, relying chiefly on the amount of CO_2 entering the atmosphere and the rate of oceanic absorption by the ocean.

1 Introduction

Carbon dioxide (CO_2) levels in the atmosphere have now seasonally surpassed 400 ppm (parts per million), a direct consequence of the increased levels of carbon dioxide emissions by human activities, including the burning of fossil fuels. Directly relating to these increases is the increased level of dissolved carbon dioxide in the ocean, which results from marine waters scavenging the gas from the atmosphere (Caldeira & Wickett, 2003). Over the past 100 years, the oceans have absorbed approximately one third of the carbon dioxide emitted by anthropogenic sources (Le Quere *et al.*, 2012). This absorption of the

* Professor, Department of Environmental Sciences, University of Virginia.
** Doctoral student, Department of Environmental Sciences, University of Virginia.

 | DOI:10.1163/9789004366619_025

gaseous carbon dioxide can have great and predictable responses in the ocean water chemistry, affecting carbonate ion concentrations, calcite and aragonite mineral saturation levels, and eventually influencing the pH of the ocean water. The inorganic carbon cycle is particularly relevant when discussing ocean acidification (OA) for it includes the many forms of dissolved CO_2. When CO_2 dissolves, it reacts with water to form ions from the dissolved carbon dioxide: carbonic acid (H_2CO_3), bicarbonate (HCO_3-), and carbonate ($CO_3{}^{2}$-; Figure 20.1). Each of these species' relative abundance is dependent upon other factors such as seawater temperature and alkalinity (Tyrrell, 2008).

Although the natural absorption of CO_2 by the world's oceans has helped mitigate the atmospheric climatic effects of increased CO_2, it is believed that increased levels in the ocean have caused a decrease in pH of approximately 0.1 units (Doney, 2006), or a 30% increase in acidity due to the logarithmic nature of this scale (Figure 20.2). Such a decrease is expected to have negative consequences for many species, as outlined below, due to the "skeleton" of

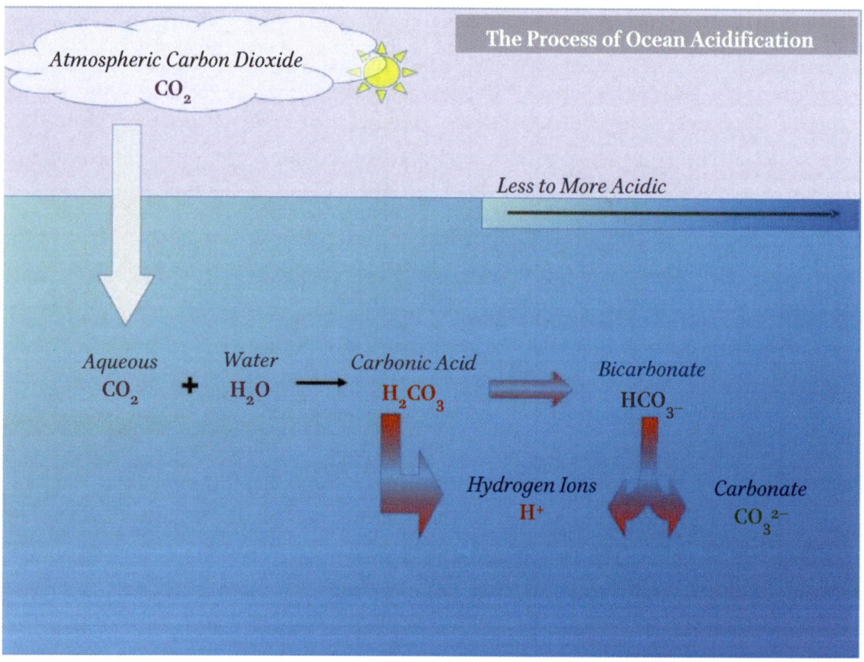

FIGURE 20.1 The Process of Ocean Acidification. As carbon dioxide diffuses into ocean water, it reacts with water to form carbonic acid. Carbonic acid dissociates into hydrogen ions, bicarbonate, and carbonate. The increasing hydrogen concentrations increases the acidity of the ocean. Greater acidity causes calcium carbonate, a key ingredient in shell, to become less stable and dissolve. SOURCE: C. FANTASIA

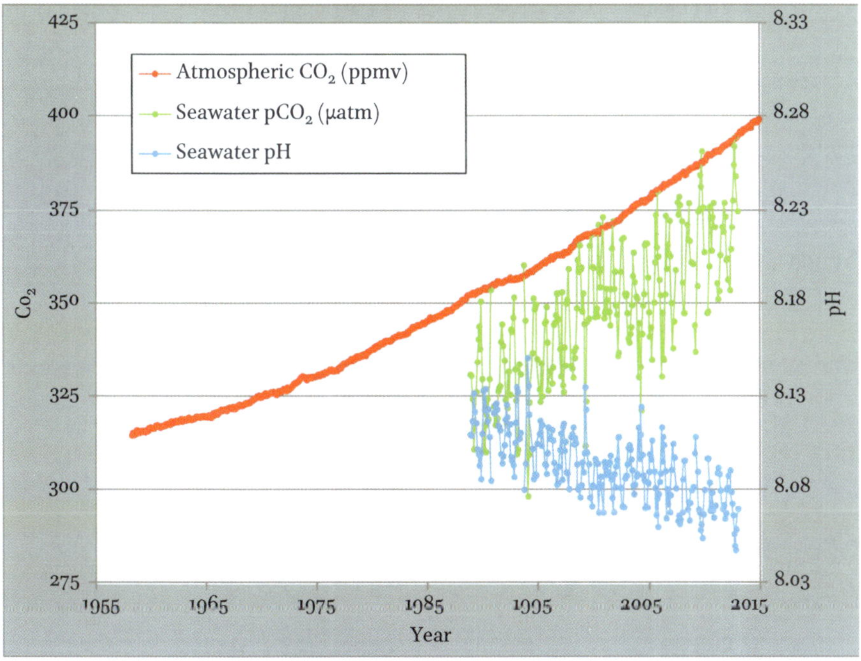

FIGURE 20.2 Increasing CO_2 concentrations at the Mauna Loa Observatory. This graph shows the correlation between rising levels of carbon dioxide (CO_2) in the atmosphere at Mauna Loa with rising CO_2 levels in the nearby ocean at Station ALOHA. As more CO_2 acumulates in the ocean, the pH of the ocean decreases.

SOURCE: FROM THE NOAA ESRL GLOBAL GREENHOUSE GAS REFERENCE NETWORK

these organisms being composed of calcite and aragonite, two mineral forms of calcium carbonate. Currently these skeletons are stable in surface waters because the carbonate ion is present in supersaturated concentrations. However, as ocean pH falls further, so will the concentration of this ion. When carbonate becomes undersaturated, structures made of calcium carbonate are vulnerable to dissolution (Feely *et al.*, 2004). In the current ocean system, supersaturation occurs in the surface waters, with undersaturation developing in the deeper waters. The boundary line between the two is named the aragonite saturation horizon and is anticipated to creep nearer the surface as time progresses (Feely *et al.*, 2016).

An important aspect of this "other carbon dioxide problem" is that, unlike models of climatic warming which are based on complex mathematics of many forcings and feedbacks, heightened acidity in the ocean is fairly predictable. The mechanisms for increasing acidity are well-established, physical chemical processes: increasing carbon dioxide in the atmosphere will increase the amount dissolved in the ocean. The pH of the ocean is dependent on that amount of dissolved CO_2. The unknowns are merely the levels that atmospheric CO_2 will reach and the rate at which the surface ocean attains equilibrium with that level. As such, with continued fossil fuel usage, carbon dioxide will continue to increase in our atmosphere and the pH of the ocean will continue to decline.

2 A Brief History of Atmospheric Concentrations of Carbon Dioxide and Its Effect on Ocean Acidity

Historically, the climate has fluctuated in concordance with atmospheric carbon dioxide concentrations, as understood via ice records containing air pockets of ~400,000 years old (Honisch & Hemming, 2005). One such ice core, the Vostok ice core, has been analyzed over the past 50 years to determine carbon dioxide concentrations in ancient history. Several studies have been published on the data within the ice; one such study found that in the past 420,000 years, no CO_2 concentrations have surpassed the current 400 ppm of CO_2 (Petit *et al.*, 1999). Due to well-understood physico-chemical processes, a relationship between ocean pH and atmospheric carbon is established, allowing for the use of known principles to translate atmospheric CO_2 into pH values (Honisch *et al.*, 2012). To evaluate older carbon dioxide concentrations and thus pH levels, marine organismal proxies are commonly utilized, with analyses of foraminifera shells widely accepted. Oftentimes, stable isotopes of carbon in the shells may be used to communicate the ancient pH and atmospheric carbon levels. A recent advancement utilizes boron isotopes incorporated into foraminifera tests (Sanyal *et al.*, 2000; Honisch *et al.*, 2012), but due to this chronometer's newness, some researchers suggest more parameterization is required to better constrain the information obtained (Pagani *et al.*, 2005). Others have argued that when careful data collection, management and analysis procedures are maintained, this tool is quite accurate (Honisch *et al.*, 2007). Within the scientific literature, it appears that researchers are slowly beginning to accept this new technique; one study compared their results from boron isotope procedures to the widely accepted standard of the Vostok ice core. They found that both methods covaried strongly and yielded similar atmospheric CO_2 concentrations, suggesting that when the fossils are properly cleaned and collected from well-preserved sediment cores, using only one species and all of the same size, this method is equal with ice core data (Honisch & Hemming, 2005). As such, our understanding of the atmosphere in older time periods may be

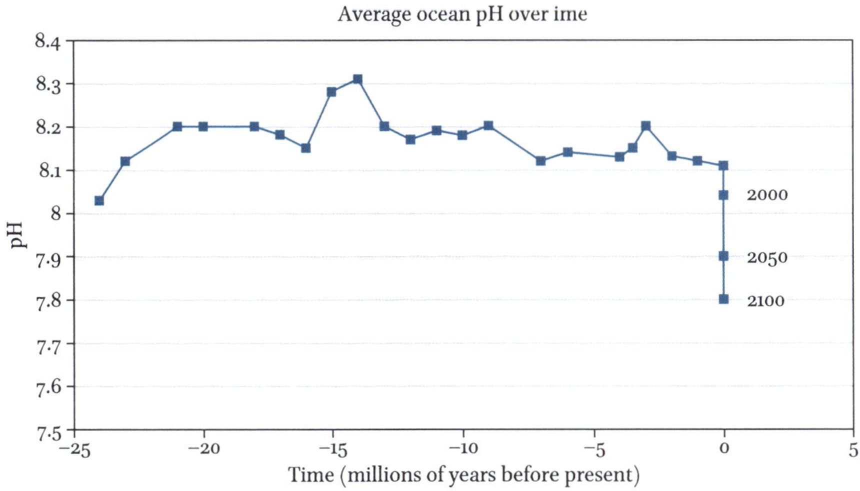

FIGURE 20.3 Historic changes in ocean pH. Oceanic pH values have remained relatively constant until the most recent times, with dramatic declines predicted as atmospheric carbon dioxide increases.
SOURCE: C. FANTASIA

expanded (Honisch *et al.*, 2012). When an epibenthic species of foraminifera incorporating high amounts of boron into their tests were used to determine whether there is any difference between interglacial and glacial pH values, the authors discovered pH and carbonate concentrations to be similar and displayed the expected interglacial periods (Honisch *et al.*, 2008). Other studies have suggested variation of 0.3 pH units (Sanyal *et al.*, 1995, 1997), but most authors suggest a difference between 0.07 and 0.1 pH units over millions of years (Figure 20.3). All in all, this time of earth's history contained bottom waters near 8.0 pH, quite close to today's values (Honisch *et al.*, 2008). Another study looked at a profile of middle Miocene to late Pleistocene epochs and found the pH of oceans to be similar to modern oceans (Palmer *et al.*, 1998), although during the Cenozoic era atmospheric CO_2 concentrations underwent wider and more erratic swings (Pearson & Palmer, 2000).

Ocean pH has been pretty stable with increasing atmospheric CO_2 stability starting ~24 million years ago (Pearson & Palmer, 2000). Even so, interdecadal variations of pH have been observed on reefs in the Pacific Ocean, where they co-vary closely with the Interdecadal Pacific Oscillation (IPO) that changes every ~50 years. Using a 300-year-old coral, they looked at boron isotope composition and suggest that these natural pH variations may help mitigate the effects of ocean acidification on this coral reef in the Southwest Pacific (Pelejero *et al.*, 2005).

Although it is known that ocean pH varies with atmospheric CO_2, there are some that suggest modern variation may be orders of magnitude faster than historic changes (Honisch *et al.*, 2012). Over the past 300 million years, carbon dioxide never topped 7500 ppm, with those changes in concentration occurring over millions of years in a gradual fashion owing to tectonic and biological influences. Using three models, a generic ocean circulation model, a geochemical model, and a four-box model, one study shows the carbonate/pH system is best suited to buffering gradual changes in CO_2, with abrupt changes in CO_2 concentrations resulting in swings of ocean pH. Between the three models, the following factors were included: changes in temperature, sedimentation, and weathering; carbonate/silicate terrestrial mineral weathering, production of these minerals in shallow water, production/oxidation of biogenic carbonate minerals, air-sea gas exchange, and transport via advection, mixing, and biological processes (Caldeira & Wickett, 2003).

3 A Brief Summary of Biological Impacts

Within the past decade, single species studies studying the effect of ocean acidification have filled peer-reviewed journals, with a wide variety of taxonomies represented. Represented here will be a cursory synopsis of affected organisms based on broad taxonomic standing.

3.1 *Phytoplankton*

Although high CO_2 has shown to be primarily beneficial for photosynthetic species (Burkhardt *et al.*, 2001; Rost *et al.*, 2003; Trimborn *et al.*, 2008), certain phytoplankton have reacted negatively to increased acidity, such as *Trichodesmium*, a genus of cyanobacteria that is prevalent throughout tropical and subtropical oceans. In laboratory studies, low pH environments of 7.8 showed that growth and N_2-fixation significantly decreased as a result of decreased efficiency of the enzyme nitrogenase, which increased in concentration with decreased pH. Growth is highly dependent on limiting nutrients like iron, and may have synergistic limiting effects on growth. In the center of large amounts of growth of cells, called blooms, the carbon dioxide may become the limiting factor and mitigate negative impacts of CO_2-derived acidity. In associated field studies, nitrogen fixation was again significantly reduced, especially in iron-poor regions. This suggests the combination of iron-poor and CO_2-rich regions may interact to reduce N_2-fixation more than in the presence of one variable. Since *Trichodesmium* contributes up to 50% of new nitrogen to marine systems, this decline in nitrogen may seriously affect organisms dependent upon this new nitrogen (Hong *et al.*, 2017).

Another type of phytoplankton, a coccolithophore *Emiliania huxleyi*, is also likely negatively impacted by changes in acidity, as has been shown in laboratory experiments (Meyer & Riebesell, 2015). A recent mesocosm experiment within a Norwegian fjord supports this laboratory observation. As the most significant producer of calcium carbonate in pelagic zones, their population reduction could impact oceanic albedo, CO_2 uptake, and carbon flux. In this mesocosm experiment, not only did population size decline, but the survivors were unable to form blooms. These blooms, or exponential colony growth, currently occur anytime there is plentiful resources, but these findings show that in the future this may not happen. This diminishment of blooms meant a reduction of particle sinking velocities by up to 30% and organic matter sinking to the sediment was reduced by 25% (Riebesell *et al.*, 2017). Such losses could significantly impact benthic organisms that are highly dependent on detritus filtering from above and thus contain trophic repercussions.

In the Southern Ocean off the coasts of Antarctica, researchers have initiated studies on diatoms, a siliceous phytoplankton, by evaluating the combination of ocean acidification and changes in irradiance. One such study looked at constant and dynamic light regimes combined with pCO_2 of 390 and 1000 ppm in order to determine changes in growth, elemental composition, primary production, and photophysiology. It was found that under dynamic, or changing, light with increased pCO_2 (lower pH), diatoms (*Chaetoceros debilis*) experienced reduced growth and strongly reduced primary productivity. Additionally, energy transfer efficiency was substantially reduced, which affected the ability to transfer light energy into biomass (Hoppe *et al.*, 2015). In another study, three important diatom species (*Chaetoceros debilis*, *Fragilariopsis kerguelensis* and *Phaeocystis antarctica*) were tested for their response to the combination of low or high CO_2 and low or high irradiance. Despite increased CO_2, neither growth or particulate organic carbon (POC) increased under either light scenario for any species; in fact POC content declined in one (*F. kerguelensis*) and growth declined in another (*C. debilis*). Additionally, both of these species experienced decreased photochemical efficiency. As a result of these environmental changes, both species employed diverging photoacclimation strategies, with *C. debilis* increasing Chlorophyll a (Chl *a*) and fucoxanthin and *F. kerguelensis* decreasing Chl *a* and reducing rates of electron transport. The third species, *P. antarctica*, showed more tolerance for increased acidity combined with irradiance changes (Trimborn *et al.*, 2017).

3.2 *Zooplankton*

One intensively studied zooplankton well known for its negative response to ocean acidification are pteropods. In particular, these organisms already show shell dissolution in natural environments along the California coast, with severe dissolution expected to triple by 2050. As an important prey group for birds, fishes, and whales, their mortality due to increased CO_2 will significantly influence trophic systems in high latitude regions (Bednaršek *et al.*, 2014). Since this system is naturally high in CO_2 due to various environmental factors such as upwelling and coastal circulation patterns, it was difficult to determine how much of these impacts are directly related to anthropogenic sources until recently. New data suggest that average surface carbon derived from anthropogenic sources is nearly twice the natural amount of carbon present, which has caused the natural aragonite saturation horizon to move up the water column by 30–50 m since the pre-industrial period. This has resulted in pteropod shell dissolution increasing between 19–26% since that time (Feely *et al.*, 2016).

Although prior studies have shown potentially neutral outcomes for calanoid copepods in the year 2100 at a pCO_2 of 1000 ppm (Weydmann *et al.*, 2012; McConville *et al.*, 2013), and others show lethal and sub-lethal effects only occurring at extreme pCO_2 levels that are beyond any climate scenarios (Watanabe *et al.*, 2006; Pascal *et al.*, 2010), a recent study suggests negative impacts on the juvenile stage (nauplii), a stage that had been previously ignored. They found mortality tripled under predicted CO_2 levels by the year 2100 (1000 ppm). Other stages, including eggs and adult stages, did not share this trend until pCO_2 reached 3000 ppm. Since copepods are one of the primary means by which biomass moves up the food chain, meaning they are a primary food source, their sensitivity to OA will likely strongly impact trophic systems as well as biogeochemical cycles (Cripps *et al.*, 2015).

Although brine shrimp are not indigenous to marine systems, this small crustacean is often used as a model for determining stress responses on the zooplankton community as a whole. There are several species of brine shrimp, with *Artemia salina* being the most commonly used species for scientists and as feed in aquaculture. Although no significant changes in morphology were observed, hatching rates were significantly reduced, survival decreased, and growth slowed during later stages in comparison to control groups. Additionally, key enzymes were impacted by acidification stress, showing changes in activity in tests groups at 7.6 and 7.8 pH (Zheng *et al.*, 2015). Currently, the IPCC predicts a pH of 7.8 for average ocean values in 2100 (IPCC, 2007); 7.6 pH is a more localized pH prediction. In a different model species, *Artemia franciscana*, hatching rates were also reduced, most significantly at pH 7.0. In their highest pH test group of 7.6, the hatch rate was reduced compared with the control, but less drastically (Salma *et al.*, 2012).

3.3 *Shellfish*

Certain populations of shellfish have shown to be susceptible to increased acidity in laboratory conditions (Gazeau *et al.*, 2013), while other populations have been more resistant, suggesting a species by species approach may be necessary for determining how much of this group will experience detrimental effects. In sea urchins, early developmental stages are the most at risk, with deformed physiology and delayed development occurring under IPCC climate scenarios for 2100, which projects ocean acidity will reach a lower limit of 7.5 pH (Zhan *et al.*, 2016). Oysters are a primary study group that show high susceptibility to changes in acidity. One model bivalve, *Crassostrea virginica,* highlights these sensitivities. When exposed to pH 7.5, mortality significantly increased, with reduced shell mass in comparison to the control as well as reduced soft tissue mass (Beniash *et al.*, 2010). Many oyster species reside within estuaries and coastal regions, which fluctuate widely in daily pH values as the currents and tides of the system changes. In another study with *Crassostrea virginica,* a long-term dataset of pH values for the Chesapeake Bay shows that pH is already commonly lower than the average ocean pH prediction of the IPCC for 2100 (IPCC, 2007). Their own laboratory data support the other findings, namely that after a change in pH of ~0.5 units, oysters are negatively impacted. However, they did suggest that increased salinity and warmer water temperatures helped offset the decrease in biocalcification (Waldbusser *et al.*, 2011). An earlier study highlights the sensitivity of early life stages for shellfish in general after evaluating the response of clam, oyster, and scallop species to predicted 21st century pH values (Talmage & Gobler, 2009). However, bivalves are not the only early life stage shellfish to experience negative impacts; at least one species of juvenile squid (*Doryteuthis pealeii*) had delayed development, reduced size, and abnormal statolith shape. This bone-like structure is important for balance and motion detection (Kaplan et al., 2013).

3.4 *Fishes*

As the most commercially important groups, a wide proportion of fish families have been studied in laboratory settings. One such species is the little skate (*Leucoraja erinacea*). This skate is especially sensitive to temperature changes, but when researchers included increasing acidity, these impacts were significantly exacerbated on the organisms from a Gulf of Maine population. However, less of an impact was observed on the organisms collected from George's Bank slightly farther south in North America. These findings show that even within the same species, different populations may show divergent effects of increased CO_2 (Santo, 2015).

In the liver and gill cells of cod, the effects of elevated CO_2 (hypercapnia) resulted in decreased metabolism in liver cells, while respiration was depressed in gill cells. Additionally, ion regulation became increasingly expensive in liver cells, but became less expensive for gill cells (Stapp *et al.*, 2015), showing that varied responses exist within the cells of organisms. This may enhance, mitigate, or cancel out any organism's responses, but at the current time this is uncharted territory.

A population of Atlantic silversides (*Menidia menidia*) showed what may be better news. Over the breeding season, it was found this wild population exhibited seasonally changing responses to ocean acidity. Their native breeding grounds naturally become more acidic (pH 7.67) over the summer season, with young adapting to this change naturally. Young that were from the early offspring fared far worse in lower pH conditions compared with young from later batches. It was concluded there may be maternal provisioning or epigenetic transgenerational plasticity (TGP) occurring within the genome. If it is the latter, this is the first time it has been observed in a wild population, although it has been observed in laboratory settings (Murray *et al.*, 2014). TGP describes changes in the DNA that allows for organisms to adapt more quickly than previously thought. However, few other species have shown such plasticity in acidic conditions, with far more displaying detrimental effects. Even so, scientists are finding some neutral effects, as with some tropical prey fishes and predators with their oxygen intake. Under hypercapnic conditions, there were few signs of oxygen deprivation, at rest or at maximum capacity (Couturier *et al.*, 2013).

Using natural daily variability of pH conditions on a coral reef, reef fishes currently experience short daily bursts of high CO_2, higher than what has been determined to be critical CO_2 levels in laboratory settings (~600 ppm). In current conditions of 400 ppm, the length of time exposed to this level of CO_2 is short enough to show little effect on the fishes in this shallow reef system. However, both the exposure time and the magnitude of CO_2 will increase in the future, requiring further studies that look at pH values even lower than the average ocean pH predicted for the year 2100 (Shaw *et al.*, 2013).

One group of fishes that has been severely underrepresented in acidification studies are the scavengers. As the recyclers of the oceans, their reaction to increased acidity is an important component to maintaining the health of oceanic systems. Even less is known regarding abyssal creatures' responses, but at least one study has conducted an *in situ* experiment to determine whether deep-sea scavengers are affected by ocean acidification. The two scavengers studied (an eelpout and octopus species) showed no response to increased acidity, although the prey fishes all died, regardless of treatment (Barry & Drazen, 2007). Thus, at least two species of scavengers are potentially safe from increased CO_2, although follow-up studies are needed to confirm these results. In general, fishes are able to compensate in the short-term to decreased pH, but after chronic exposure, significant effects are seen across a variety of physiological systems. Over time, it is possible for them to adapt, but more studies are needed to assess this assumption (Heuer & Grosell, 2014).

One of the most iconic organisms that people associate with the ocean is the shark. Even this cartilaginous fish may not be entirely immune to the impacts of acidification, with studies showing reduced odor tracking abilities (Dixson *et al.*, 2015), reduced growth, and impaired hunting behavior (Pistevos *et al.*, 2015). A third study suggested some effects on neurophysiology, though no negative impacts were observed on growth, metabolic rate, aerobic scope, or skin morphology (Green & Jutfelt, 2014).

3.5 *Marine Mammals*

Other predators of the sea, such as marine mammals, are also expected to feel the impacts of a warming ocean and increasing acidity. Although they may not experience direct effects to their physiology or behavior, they will experience changes in prey distribution. This is especially significant for species with limited habitat like the vaquita (*Phocoena sinus*), or species that live at high latitudes. Sea ice is important habitat for narwhales, beluga whales, bowhead whales, and polar bears; changing sea temperatures are expected to further decrease its extent. Additionally, many baleen whale species migrate to the high latitudes to feed on plankton, of which many important species show high sensitivity to increased acidity (Simmonds & Isaac, 2007). As a quarter of the world's cetaceans are currently considered endangered, prey loss or change in habitat places further stress on an already stressed community (Simmonds & Eliott, 2009). One study suggested some harbor porpoises may be starving as a result in prey population declines (sand eels), stating that these populations have been reduced as a direct result of climate forcing (Macleod *et al.*, 2007). However, it is difficult to statistically say that porpoise starvation is a direct result of declining prey populations or climate forcing (Simmonds & Eliott, 2009).

3.6 *Summary*

Meta-analysis compiled on 228 studies on ocean acidification from various taxa determined how biological responses may affect the ocean as a whole. Looking at older meta-analyses, they discuss how responses are highly variable between organisms and trophic levels, with differing responses between early and later life stages and geography. When they pooled at taxa together, OA significantly affected growth, development, survival, abundance, and calcification. Upon separating them, corals, coccolithophores, and mollusks show the greatest average loss in calcification. When they included elevated temperature, the above variables showed dampened values, but were not statistically significant. They did show that some fleshy algae and diatoms may marginally benefit from OA, though overall systems as a whole are negatively impacted (Kroeker *et al.*, 2013).

4 The Ecological Effects from Ocean Acidification

Within the past ten years, more studies have focused on the effects of acidification throughout an ecosystem. Some have prioritized certain systems, such as coral reefs systems, while others evaluate cascading effects through trophic levels. These studies differ from biological studies in that physiological responses of individuals play a reduced role in determining how a system will respond as a whole to decreased pH (Ferrari *et al.*, 2015). Additionally, many of the studies incorporate warmer waters in order to simulate as much as possible natural environmental conditions of future oceans. In keeping with this goal of natural environmental conditions, mesocosm studies are the most common, where organisms are placed within an existing natural structure, such as a reef or fjord, and CO_2 is pumped into their netted cages.

4.1 *Microbial*

Due to the mixed responses of primary producers to ocean acidification, heterotrophic bacteria have the potential to be heavily influenced by changes in their feed availability. In environments already reduced in nutrients, phytoplankton are the sole source for organic carbon (Hoikkala *et al.*, 2009). As such, increasing inorganic carbon does not mean increasing organic carbon for secondary consumers (Thingstad *et al.*, 2008), but may mean significant alterations to current biogeochemical cycles. A recent study has evaluated nutrient limited mesocosms in the Baltic Sea in an attempt to understand this underrepresented group in OA literature. Their data suggest that bacteria are indirectly impacted by acidification and may be altering the composition of the bacterial community, with further statistical analysis showing there will very likely be changes in functional groups of bacteria and phytoplankton. They conclude that the future ocean may have increasing autotroph biomass during low nutrient periods, with decreased heterotrophic composition and biomass, ultimately showing a decoupling between phytoplankton and microbial dynamics. The paramount result of these changes would be impacts on the global carbon pump (Hornick *et al.*, 2017), although more research is required to quantify these impacts.

A review of papers highlighting the interaction of bacteria and OA offers some clues to potential effects on microbially-mediated processes. They found that bacterial primary production shows mixed responses to increased acidity, suggesting species-specific responses. In those species that respond well to OA, increasing eutrophication and anoxia may result in the surface oceans. Some N_2-fixers gained energy from increased carbon availability and as a result were able to increase nitrogen fixation (Das & Mangwani, 2015), although this is not true for all nitrogen fixers (Hong *et al.*, 2017). In other studies, trace gas emissions were shown to significantly decrease. One such trace gas is dimethyl sulfide (DMS), an important atmospheric gas that assists in the formation of clouds by acting as a platform for water vapor condensation. This gas, a by-product of microbial metabolism, is directly sourced from marine bacteria and may decrease from indirect impacts of OA. The study also highlighted the changes in microbial diversity and competition as a direct result of pH changes, with pathogenic microorganisms significantly increasing. This also relates to biofilms, which are an important community structure that support microorganisms and eukaryotes. As they change in diversity, reverberations would be felt throughout the trophic levels (Das & Mangwani, 2015).

Quorum sensing (QS) is a method many species of bacteria utilize in order to communicate with host species such as corals, sponges, algae, and animals (Dobretsov *et al.*, 2009). In low pH environments, it appears that QS processes are affected, shifting the dominant bacterial species to sub-dominant positions (Krause *et al.*, 2012). This may result in higher order species composition also changing, as in the case of algae *Ulva* species, which respond to QS by increased spore settlement (Das & Mangwani, 2015).

4.2 *Phytoplankton*

In oligotrophic environments such as the Mediterranean Sea, the presence of limiting nutrients such as iron are more important to phytoplankton growth than increasing carbon dioxide amounts. However, when these limiting nutrients are freely available, increased carbon dioxide may influence phytoplankton populations. In a series of mesocosm studies in oligotrophic and mesotrophic environments, only limited benefits were observed from increased CO_2 during this short-term experiment. However, including vertical mixing and weather conditions over long-term studies may change these results (Gazeau *et al.*, 2017). In a long-term mesocosm experiment within the North Sea off of Sweden, total primary production did not increase, nor did photoacclimation, although phytoplankton biomass increased under OA conditions. As a whole, their results suggest the phytoplankton community is affected by OA during certain seasons, which may affect overall ecosystem functions in the long-term (Eberlein *et al.*, 2017). Another study highlighting iron-limited regions in the Weddell Sea in Antarctica found that iron richness with increased CO_2 resulted in significant taxonomic shifts. When iron was limited and CO_2 abundant, species shifts occurred, despite no change in primary production. As such, this shows strong differences in species composition with CO_2 levels changing. The species that are more abundant in high CO_2 conditions are heavier species, which translates to faster sinking rates into the deep sea and may alter the current biological pump rates in certain areas of the Southern Ocean (Hoppe *et al.*, 2013). Such modifications could reverberate through the wider ecosystem, as the Southern Ocean is an important feeding ground for a wide variety of organisms. With faster sinking rates for the primary food source, food availability may be shortened.

Many open ocean regions in the world experience iron-depleted conditions. Other studies have found that decreasing pH may ultimately reduce the bioavailability of iron in these same spots. Laboratory experiments with acidified water and diatoms show decreased iron uptake as is predicted according to iron chemical reactions. A secondary set of experiments in the Atlantic determined that diatoms and coccolithophores show decreased iron uptake in natural environments enriched with CO_2. Since iron is an important mineral in photosynthesis, further reduction in its concentrations will increase iron stress in phytoplankton populations throughout oligotrophic regions (Shi *et al.*, 2010). A later study also focusing on bioavailability conducted a mesocosm experiment in the northeast subarctic Pacific, a region of high nutrients and low chlorophyll concentrations. In acidified conditions, Chl *a* was significantly reduced and particulate organic carbon was reduced. Certain species of phytoplankton were reduced compared to other species (Melancon *et al.*, 2016), supporting earlier findings regarding species composition changes (Hoppe *et al.*, 2013). The reduction of Chl *a* would be a direct result of decreased iron availability due to its importance in photosynthetic processes. This set of researchers used iron rich dust in order to simulate natural iron deposition (Melancon *et al.*, 2016).

Overall, phytoplankton have exhibited varied responses to decreasing pH, making it difficult to determine a holistic response. In a meta-analysis publication, observations of diverse taxonomic groups were assembled to determine responses of phytoplankton as a whole. Their diverse range of responses were then incorporated into a global marine ecosystem model in order to determine how current phytoplankton populations may be impacted in future oceans.

Their model showed that OA is more potent than warming water or low nutrients (Dutkiewicz *et al.*, 2015).

4.3 *Zooplankton*

Far fewer studies have explored the ecological impact on zooplankton communities, but one recent report has begun to bring light into this void. A mesocosm study over 113 days off the coast of Sweden using end-of-the-century CO_2 levels found that young copepods are the most sensitive organisms, with little to no effect on adults. They found community structure remained relatively stable, but the sensitivity of young copepods may affect biomass transfer to higher trophic levels in future oceans (Alguero-Muniz *et al.*, 2017).

4.4 *Fishes*

In a predator-prey relationship study, one juvenile prey species increased its maximum oxygen intake under acidified conditions, with one other juvenile prey species and its predator showing neutral effects. As such, near-future oceans do not seem to have negative metabolic effects on any of these species. Additionally, it is easy to observe there is varied aerobic responses between species, with one juvenile prey even displaying augmented aerobic scope (Couturier *et al.*, 2013). This suggests that some fishes will not experience direct physiological aerobic stress. Another study looked at CO_2 vents in the northern and southern hemispheres and corroborated other studies' reduced predator avoidance, but they also found that in regions with shelter-rich habitats, this altered behavior does not truly impact their survival. Overall, altered abundances of predators and prey were observed, in addition to habitat availability. These changes could have wide reaching consequences throughout ecosystems (Nagelkerken *et al.*, 2015).

Beyond predator-prey interactions, at least one study has shown that increasing CO_2 has significant impacts on sound absorption within ocean waters. Sound absorption in water is dependent upon dissolved oxygen levels, sulfur and nitrogen levels, pH, and temperature. One of the two buffering systems involved in regulating pH, the boric acid system, is more effective at absorbing low-frequency noises in higher pH waters. This means that in future oceans with a conservative pH decrease of 0.15 units would result in 20% reduction in sound absorption. Adding increased temperature of 3° C results in further loss between 5–10% (Hester *et al.*, 2008). Such changes could significantly alter underwater communication between a variety of organisms. It is already known that cetacean communication is disrupted from underwater ship communication (McQuinn *et al.*, 2011); with amplified noises in future oceans, this may be further accentuated.

4.5 *Coral Reef Systems*

A fair amount of research has been dedicated to the effect of OA on coral reefs. Since these systems are the most productive regions in otherwise oligotrophic waters, the potential loss of the foundation of this system, calcareous corals, would be detrimental to the entire region. Ecologically, these reefs are nurseries for a wide variety of important fishes. Beyond fishery importance, most reefs surround island nations in tropical waters whose primary revenue is highly dependent upon tourism (Macko *et al.*, 2017). Without these reefs, several island nations would lose tourist revenue as well as an important local food source. One recent study used an ecological-economic model to determine commercial fishery losses in reef systems, with a high emission, worst case scenario resulting in $40–69 billion lost globally (Speers *et al.*, 2016). Within OA studies, a wide variety of results have been observed, with some species showing sensitivity and others displaying greater resistance (Macko *et al.*, 2017). Using two coral species in one taxa and two calcified algae in another, one mesocosm study looked at their responses in Hawaii, French Polynesia, and Japan, where environmental conditions differ. These four species are common to reefs throughout the Pacific. Their results show the two coral species and one algae species were insensitive to OA, with the second algae species only sensitive in certain regions. Although warming temperature was not included, it appears that at least some common species will survive in future oceans and although OA may reduce biodiversity on reef systems, certain species may be resilient (Comeau *et al.*, 2014). Even so, another mesocosm study in southern Australia hypothesized that corals in general are currently experiencing decreased calcification rates in comparison with pre-industrial rates. In order to determine this, they used sodium hydroxide (NaOH) to increase total alkalinity within a secluded lagoon and evaluated the response of corals in the lagoons. Over 15 days, NaOH was introduced to the system in order to restore total alkalinity to known pre-industrial values. Net community calcification increased as a result, suggesting they are currently experiencing reduced calcification rates and that certain isolated lagoons and bays may be able to undergo remediation via NaOH (Albright *et al.*, 2016). Beyond reef calcifiers, fishes in reefs may be sensitive to synergistic effects of temperature and elevated carbon dioxide. With both stressors present, some fish species showed much higher predation rates (30–70%) than present rates. Although one species consumed more oxygen under high temperature/CO_2, they did not experience greater predation, which is an unexpected finding. As such, predicting the effects of temperature and CO_2 on the community and ecological level is not as simple as extrapolating from prior physiological experiments (Ferrari *et al.*, 2015), accentuating the need for more mesocosm studies that incorporate heightened temperature and elevated CO_2.

5 Trusting Climate Models

Comprehensively discussing climate models and their accuracy is beyond the scope of this paper, but an introduction to this topic is important for any conversation about the future state of the oceans. This final section is meant to introduce how models are developed, their parameterization, and how models are verified. Some models were introduced over twenty years ago and scientists may now assess their accuracy, which will also be briefly explored.

5.1 *How Models Are Developed and Parameterized*

A good portion of the complexity within models is due to the systems they are attempting to explain and then predict. An accurate assessment of forest systems, for example, requires knowledge of forest ecology, biogeochemical cycling, hydrology, radiative forcing, and a host of other biotic and abiotic factors. The same broad expanse of specialized knowledge is required for ocean models and includes information on ocean currents, atmosphere-ocean interactions, land-ocean interactions, biogeochemical cycling, geology, and more.

Starting near the middle of the 20th century, researchers began to build models to understand how aspects of a system affect other parts of the same or differing system. In ocean studies, some of the earliest models assumed a simple ocean without any circulatory dynamics. In 1987, at least one model began incorporating simple circulation, with atmospheric cloud layers, boundary layers, and solar radiation and its seasonal variation. They also used mixed layer that accounts for heat distribution and movement in the upper 50 meters of the ocean. Sea ice formation and melting was also accounted, in addition to its associated effects on albedo and therefore solar heat absorption and loss (Wilson & Mitchell, 1987). Since then, models have only become more complex with the ability of computers to perform more advanced calculations in shortened time frames. The mathematics involved in each of the variables within a model are based on decades of field observation combined with theory that have then been tested against natural systems. Those formulas that have consistently tracked with natural observations are trusted as fairly good mathematical representations of natural events. Some equations are based on tried and true Newtonian physics formulas, with others based on other, trusted, formulations. In the thirty years since 1987, each variable within models has been refined and verified via the scientific community at large, with continued refinement arriving in many publications, increasing our understanding of climate sensitivity at its upper and lower bounds (Knutti & Hegerl, 2008).

One extremely important parameter for climate models involves the pathways carbon travels through the ecosystem. The global carbon cycle is a familiar concept for anyone having taken an introductory biology course, but most are not familiar with the methods determining those pathways and how the mass of carbon moving through systems are evaluated. Since carbon dioxide is a component of this cycle, accurately quantifying where carbon goes after entering the ocean is valuable information for creating an accurate climate model. However, even quantifying the global carbon cycle is a monumental task that requires detailed knowledge and quantifiable data in many sub-disciplines of environmental science, thus demanding a community of experts sharing information and critiques. Fortunately, such a community exists, and has established a baseline for analyzing global movement of carbon. Its model of carbon cycling accounts for fossil fuel combustion, cement production, emissions associated with goods and services, deforestation, fire, and vegetation photosynthesis and respiration. To predict future changes in carbon amounts, community researchers also determined the rates of change in land use, ocean sink, terrestrial sink, and atmospheric changes. Their study shows that carbon from human usage has increased steadily and at times abruptly between 1959 and 2011, with their model suggesting no abatement. Additionally, their work highlighted the importance of oceans as a sink for carbon dioxide (Le Quere *et al.*, 2012).

This work on the global carbon budget has significantly improved current understanding of the carbon cycle and thus less parameterization is required in this region. However, other large-scale processes are unresolved and require usage of simplified physical models in order to include these processes within climate models. Each of these are able to be mathematically "tuned" in order to best fit past observations, since the mechanisms of these processes are not well understood (IPCC, 2007).

5.2 *Coupling Atmosphere and Ocean Models*

Evaluating the future ocean's well-being is dependent on not one, but two models – atmospheric and oceanic. Coupling the two models requires the same collaborative effort as with global climate models and many researchers have done so. Coupled models have been developed at least since the 1970s with moderate success at replicating current system dynamics (Bryan *et al.*, 1975) and have continued to undergo refinement as dynamic oceans are incorporated with sea ice (Wilson & Mitchell, 1987), and newer atmospheric developments are incorporated (Tardif *et al.*, 2014). More recently, researchers have been able to better quantify the effect of glacial processes on the rate of climate change (Ganopolski & Rahmstorf, 2001), with studies seemingly agreeing that more feedback mechanisms are activated (Cox *et al.*, 2000) and that the overall result is accelerating changes on a global scale (Ganopolski & Rahmstorf, 2001; Cox *et al.*, 2000).

5.3 *Model Verification and Prediction Accuracy*

At this point in the scientific discussion, a wide plethora of studies have published their own model's predictions, refined parameters, evaluated local, regional, and global scales, and determined feedback mechanisms. Many of these suggest wide ranging negative consequences, accelerated pace, and describe the culpability of humans using isotopic methodology for proof. However, a model's results must match reality for anyone to believe and react to this information; in short, these results must be verified. This is well understood and thus every respected published model has been verified against observational data pertaining to the correct scale and system (Bryan *et al.*, 1975; McNeil & Matear, 2008; Knutti & Sedlacek, 2013).

Even when a model is describing well-defined systems, with proper parameterization, and has been verified through more than one relevant dataset, a model may still be wrong. Nature still has the capability to follow undefined oscillations in atmospheric cycles, affecting a wide range of physical processes. With such complexity contained within global climate models, concern over their accuracy is legitimate and must be addressed. Some studies have collated several model predictions to determine the overall predictability of climate models, determining that although projected CO_2 level pathways are widespread, this is a consequence of the climate system and cannot be reduced (Roe & Baker, 2007). Additionally, accuracy improves when models such as the CMIP5 (Coupled Models Intercomparison Project 5) are applied at or above the regional and decadal scale, due to averaging out of errors (Sakaguchi *et al.*, 2012). At this point in climate studies, some older models have made predictions that may now be compared to current climate. The authors took 18 climate models, split the dataset into 15-year sections starting in the 1950s, and determined how well they predicted El Nino (ENSO) oscillations in the Pacific Ocean. In general, they did very well (Risbey *et al.*, 2014). Climate models have been doing extremely well predicting global surface temperature for over 40 years, and in fact there are some studies that suggest IPCC estimates are underestimating the effect of climate changes on the ocean, particularly in regards to sea level rise (Roe & Baker, 2007). Looking at just three simple climate models from climate scientist James Hansen shows that his predictions in the 1980s track well with observed average global temperatures, with observations reasonably correlating with his moderate CO_2 scenario (Mann & Gaudet, 2017). Beyond these, the IPCC AR5 contains an entire chapter discussing the validity of climate models, evaluating how well they have performed and are performing, and conclude that they are reasonably trustworthy (Flato *et al.*, 2013).

6 Summary and Conclusion

The prospect of an acidifying ocean is one that civilization must directly confront to maintain current norms in economy, commercial fisheries, tourism, and ecological diversity. The very vastness of the ocean means any changes to its body will reach out and touch a wide variety of systems upon which humans and animals depend. From microbes to marine mammals, coral reefs to high latitude feeding grounds, and coastal cities to rural communities relying on ecotourism, many important aspects are likely to be affected. Based on the extensive work within climate modeling, we can accept certain high and low CO_2 predictions, allowing us to prepare our future generations for the likelihood of an altered environment and how that will affect our species. As we deal with a problem that may impact the trajectory of our known world, we cannot not be hesitant in action, for it is better to enact rational policies now than to wait until we pass the tipping point.

References

Albright, R., L. Caldeira, J. Hosfelt, L. Kwiatkowski, J.K. Maclaren, B.M. Mason, Y. Nebuchina, A. Ninokawa, J. Pongratz, K.L. Ricke, T. Rivlin, K. Schneider, M. Sesboue, K. Shamberger, J. Silverman, K. Wolfe, K. Zhu, & K. Caldeira. 2016. Reversal of ocean acidification enhances net coral reef calcification. *Nature*, 531:362–364, doi:10.1038/nature17155.

Alguero-Muniz, M., S. Alvarez-Fernandez, P. Thor, L.T. Bach, M. Esposito, H.G. Horn, U. Ecker, J.A.F. Langer, J. Taucher, A.M. Malzahn, U. Riebesell, & M. Boersma. 2017. Ocean acidification effects on mesozooplankton community development: Results from a long-term mesocosm experiment. *PLoS ONE*, 12:e0175851, https://doi.org/10.1371/journal.pone.0175851.

Barry, J., & J. Drazen. 2007. Response of deep-sea scavengers to ocean acidification and the odor from a dead grenadier. *Marine Ecology Progress Series*, 350:193–207.

Bednaršek N., R.A. Feely, J.C.P. Reum, B. Peterson, J. Menkel, S.R. Alin, & B. Hales. 2014. *Limacina helicina* shell dissolution as an indicator of declining habitat suitability owing to ocean acidification in the California Current Ecosystem. *Proceedings of the Royal Society B*, 281:20140123, http://dx.doi.org/10.1098/rspb.2014.0123.

Beniash, E., A. Ivanina, N. Lieb, I. Kurochkin, & I. Sokolova. 2010. Elevated level of carbon dioxide affects metabolism and shell formation in oysters *Crassostrea virginica* (Gmelin). *Marine Ecology Progress Series*, 419:95–108.

Bryan, K., S. Manabe, & R.C. Pacanowski. 1975. A global ocean-atmosphere climate model. Part II. The oceanic circulation. *Journal of Physical Oceanography*, 5:30–46.

Burkhardt, S., G. Amoroso, U. Riebesell, & D. Sultemeyer. 2001. CO2 and CO3- uptake in marine diatoms acclimated to different CO2 concentrations. *Limnology and Oceanography*, 46:1378–1391.

Caldeira, K., & M.E. Wickett 2003. Anthropogenic carbon and ocean pH. *Nature*, 425:365.

Comeau, S., R.C. Carpenter, Y. Nojiri, H.M. Putnam, K. Sakai, & P.J. Edmunds. 2014. Pacific-wide contrast highlights resistance of reef calcifiers to ocean acidification. *Proceedings of the Royal Society B*, 281:20141339. http://dx.doi.org/10.1098/rspb.2014.1339.

Couturier, C.S., J.A.W. Stecyk, J.L. Rummer, P.L. Munday, & G.E. Nilsson. 2013. Species-specific effects of near-future CO2 on the respiratory performance of two tropical prey fish and their predator. *Comparative Biochemistry and Physiology, Part A*, 166:482–489.

Cox, P.M., R.A. Betts, C.D. Jones, S.A. Spall, & I.J. Totterdell. 2000. Acceleration of global warming due to carbon-cycle feedbacks in a coupled climate model. *Nature*, 408:184–187.

Cripps, G., P. Lindeque, & K.J. Flynn. 2015. Have we been underestimating the effects of ocean acidification in zooplankton? *Global Change Biology*, 20:3377–3385. doi: 10.1111/gcb.12582.

Das, S., & N. Mangwani. 2015. Ocean acidification and marine microorganisms: responses and consequences. *Oceanologia*, 57:349–361.

Dixson, D.L., A.R. Jennings, J. Atema, & P.L. Munday. 2015. Odor tracking in sharks is reduced under future ocean acidification conditions. *Global Change Biology*, 21:1454–1462. doi: 10.1111/gcb.12678.

Dobretsov, S., M. Teplitski, & V. Paul. 2009. Mini-review: quorum sensing in the marine environment and its relationship to bio-fouling. *Biofouling* 25:413–427.

Doney, S.C. 2006. The dangers of ocean acidification. *Scientific American*, 294:58–65.

Dutkiewicz, S., J.J. Morris, M.J. Follows, J. Scott, O. Levitan, S.T. Dyhrman, & I. Berman-Frank. 2015. Impact of ocean acidification on the structure of future phytoplankton communities. *Nature Climate Change*, 5:1002–1006. DOI: 10.1038/NCLIMATE2722.

Eberlein, T., S. Wohlrab, B. Rost, U. John, L.T. Bach, U. Riebesell, D.B. Van de Waal. 2017. Effects of ocean acidification on primary production in a coastal North Sea phytoplankton community. *PLoS ONE*, 12:e0172594, doi:10.1371/journal.pone.0172594.

Feely, R.A., C.L. Sabine, K. Lee, W.L. Berelson, J. Kleypas, V.J. Fabry, & F.J. Millero. 2004. Impact of anthropogenic CO_2 on the $CaCO_3$ system in the oceans, *Science*, 305:362–366.

Feely, R.A., S.R. Alin, B. Carter, N. Bednarsek, B. Hales, F. Chan, T.M. Hill, B. Gaylord, E. Sanford, R.H. Byrne, C.L. Sabine, D. Greeley, & L. Juranek. 2016. Chemical and biological impacts of ocean acidification along the west coast of North America. *Estuarine, Coastal and Shelf Science*, 183:260–270.

Ferrari, M.C.O., P.L. Munday, J.L. Rummer, M.I. McCormick, K. Corkill, S.-A. Watson, B.J.M. Allan, M.G. Meekan, & D.P. Chivers. 2015. Interactive effects of ocean acidification and rising sea temperatures alter predation rate and predator selectivity in reef fish communities. *Global Change Biology*, 21:1848–1855, doi: 10.1111/gcb.12818.

Flato, G., J. Marotzke, B. Abiodun, P. Braconnot, S.C. Chou, W. Collins, P. Cox, F. Driouech, S. Emori, V. Eyring, C. Forest, P. Gleckler, E. Guilyardi, C. Jakob, V. Kattsov, C. Reason and M. Rummukainen, 2013: Evaluation of Climate Models. In: Climate Change 2013: The Physical Science Basis. Contribution of Working Group I to the Fifth Assessment Report of the Intergovernmental Panel on Climate Change [T.F. Stocker, D. Qin, G.-K. Plattner, M. Tignor, S.K. Allen, J. Boschung, A. Nauels, Y. Xia, V. Bex and P.M. Midgley (eds.)]. Cambridge University Press, Cambridge, United Kingdom and New York, NY, USA.

Ganopolski, A., & S. Rahmstorf. 2001. Rapid changes of glacial climate simulated in a coupled climate model. *Nature*, 409:153–158.

Gazeau, F., L.M. Parker, S. Comeau, J.-P. Gattuso, W.A. O'Connor, S. Martin, H.-O. Portner, & P.M. Ross. 2013. Impacts of ocean acidification on marine shelled molluscs. *Marine Biology*, 160:2207–2245.

Gazeau, F., A. Sallon, P. Pitta, A. Tsiola, L. Maugendre, M. Giani, M. Celussi, M.L. Pedrotti, S. Marro, & C. Guieu. 2017. Limited impact of ocean acidification on phytoplankton community structure and carbon export in an oligotrophic environment: Results from two short-term mesocosm studies in the Mediterranean Sea. *Estuarine, Coastal, and Shelf Science*, 186:72–88.

Green, L., & F. Jutfelt. 2014. Elevated carbon dioxide alters the plasma composition and behaviour of a shark. *Biology Letters*, 10:20140538. http://dx.doi.org/10.1098/rsbl.2014.0538.

Hester, K.C., E.T. Peltzer, W.J. Kirkwood, & P.G. Brewer. 2008. Unanticipated consequences of ocean acidification: A noisier ocean at lower pH. *Geophysical Research Letters*, 35:L19601, doi:10.1029/2008GL034913.

Heuer, R.M., & M. Grosell. 2014. Physiological impacts of carbon dioxide and ocean acidification on fish. *American Journal of Physiology-Regulatory, Integrative and Comparative Physiology*, 307, doi:10.1152/ajpregu.00064.2014.

Hoikkala, L., H. Aarnos, & R. Lignell. 2009. Changes in nutrient and carbon availability and temperature as factors controlling bacterial growth in the northern Baltic Sea., *Estuaries and Coasts*, 32:720–733. doi:10.1007/s12237-009-9154-z.

Hong, H., R. Shen, F. Zhang, Z. Wen, S. Chang, W. Lin, S.A. Kranz, Y.-W. Luo, S.-J. Kao, F.M.M. Morel, & D. Shi. 2017. The complex effects of ocean acidification on the prominent N2-fixing cyanobacterium *Trichodesmium*. *Science,* 356:527–531.

Honisch, B., & N.G. Hemming. 2005. Surface ocean pH response to variations in pCO2 through two full glacial cycles. *Earth and Planetary Science Letters*, 236:305–314.

Honisch, B., N.G. Hemming, & B. Loose. 2007. Comment on "A critical evaluation of the boron isotope-pH proxy: the accuracy of ancient ocean pH estimates" by M. Pagani, D. Lemarchand, A. Spivack, & J. Gaillerdet. *Geochimica et Cosmochimica Acta,* 71:1636–1641.

Honisch, B., T. Bickert, & N.G. Hemming. 2008. Modern and Pleistocene boron isotope composition of the benthic foraminifer *Cibicidoides wuellerstorfi. Earth and Planetary Science Letters*, 272:309–318.

Honisch, B., A. Ridgwell, D.N. Schmidt, E. Thomas, S.J. Gibbs, A. Sluijs, R. Zeebe, L. Kump, R.C. Martindale, S.E. Greene, W. Kiessling, J. Ries, J.C. Zachos, D.L. Royer, S. Barker, T.M. Marchitto Jr., R. Moyer, C. Pelejero, P. Ziveri, G.L. Foster, & B. Williams. 2012. The geological record of ocean acidification. *Science*, 335:1058–1063. DOI: 10.1126/science.1208277.

Hoppe, C.J.M., C.S. Hassler, C.D. Payne, P.D. Tortell, B. Rost, & S. Trimborn. 2013. Iron limitation modulates ocean acidification effects on Southern Ocean phytoplankton communities. *PLoS ONE*, 8:e79890. doi:10.1371/journal.pone.0079890.

Hoppe, C.J.M., L.-M. Holtz, S. Trimborn, & B. Rost. 2015. Ocean acidification decreases the light-use efficiency in an Antarctic diatom under dynamic but not constant light. *New Phytologist*, 207:159–171. doi: 10.1111/nph.13334.

Hornick, T., L.T. Bach, K.J. Crawfurd, K. Spilling, E.P. Achterberg, J.N. Woodhouse, K.G. Schulz, C.P.D. Brussaard, U. Riebesell, & H.-P. Grossart. 2017. Ocean acidification impacts bacteria-phytoplankton coupling at low-nutrient conditions. *Biogeosciences*, 14:1–15. doi:10.5194/bg-14-1-2017.

IPCC, 2007: Climate Change 2007: The Physical Science Basis. Contribution of Working Group I to the Fourth Assessment Report of the Intergovernmental Panel on Climate Change [Solomon, S., D. Qin, M. Manning, Z. Chen, M. Marquis, K.B. Averyt, M. Tignor and H.L. Miller (eds.)]. Cambridge University Press, Cambridge, United Kingdom and New York, NY.

Kaplan, M.B., T.A. Mooney, D.C. McCorkle, & A.L. Cohen. 2013. Adverse effects of ocean acidification on early development of squid (*Doryteuthis pealeii*). *PLoS ONE*, 8:e63714. doi:10.1371/journal.pone.0063714.

Knutti, R., & G.C. Hegerl. 2008. The equilibrium sensitivity of the Earth's temperature to radiation changes. *Nature Geoscience*, 1:735–743, doi:10.1038/ngeo337.

Knutti, R., & J. Sedlacek. 2013. Robustness and uncertainties in the new CMIP5 climate model predictions. *Nature Climate Change*, 3:369–373. DOI: 10.1038/NCLIMATE1716.

Krause, E., A. Wichels, L. Giménez, M. Lunau, M.B. Schilhabel, & G. Gerdts. 2012. Small changes in pH have direct effects on marine bacterial community composition: a microcosm approach. *PLoS One*, 7:e47035, http://dx.doi.org/10.1371/journal.pone.0047035.

Kroeker, K.J., R.L. Kordas, R. Crim, I.E. Hendriks, L. Ramajos, G.S. Singh, C.M. Duartes, & J.-P. Gattuso. 2013. Impacts of ocean acidification on marine organisms: quantifying sensitivities and interaction with warming. *Global Change Biology*, 19:1884–1896, doi: 10.1111/gcb.12179.

Le Quere, C., R.J. Andres, T. Boden, T. Conway, R.A. Houghton, J.I. House, G. Marland, G.P. Peters, G. van der Werf, A. Ahlstrom, R.M. Andrew, L. Bopp, J.G. Canadell, P. Ciais, S.C. Doney, C. Enright, P. Friedlingstein, C. Huntingford, A.K. Jain, C. Jourdain, E. Kato, R.F. Keeling, K. Klein Goldewijk, S. Levis, P. Levy, M. Lomas, B. Poulter, M.R. Raupach, J. Schwinger, S. Sitch, B.D. Stocker, N. Viovy, S. Zaehle, & N. Zeng. 2012. The global carbon budget, 1959–2011. *Earth System Science Data Discussions*, 5:1107–1157. doi:10.5194/essdd-5-1107-2012.

Macko, S.A., C. Fantasia, & G. Xue. 2017. Potential global economic impacts of ocean acidification. *Center for Oceans Law & Policy: International Marine Economy*. Eds. M.H. Nordquist, J.N. Moore, & R. Long. 20:79–97. doi: 10.1163/9789004323445_003.

Macleod, C.D., M.B. Santos, R.J. Reid, B.E. Scott & G.J. Pierce. 2007. Linking sand eel consumption and the likelihood of starvation in harbour porpoises in the Scottish North Sea: could climate change mean more starving porpoises? *Biological Letters*, 3:185–188.

Mann, M., & B. Gaudet. "Validating Climate Models". Department of Meteorology and Atmospheric Science. Penn State College of Earth and Mineral Sciences. 2017. Web. 27 Sept. 2017.

McConville, K., C. Halsband, E.S. Fileman, P.J. Somerfield, H.S. Findlay, & J.I. Spicer. 2013. Effects of elevated CO2 on the reproduction of two calanoid copepods. *Marine Pollution Bulletin*, 73:428–434.

McNeil, B.I., & R.J. Matear. 2008. Southern Ocean acidification: A tipping point at 450-ppm atmospheric CO2. *PNAS*, 105:18860–18864. http://www.pnas.orgcgidoi10.1073pnas.0806318105.

McQuinn, I.N., V. Lesage, D. Carrier, G. Larrivee, Y. Samson, S. Chartrand, R. Michaud, & J. Theriault. 2011. A threatened beluga (*Delphinapterus leucas*) population in the traffic lane: Vessel-generated noise characteristics of the Saguenay-St. Lawrence Marine Park, Canada. *Journal of the Acoustical Society of America*, 130:3661–3673.

Melancon, J., M. Levasseur, M. Lizotte, M. Scarratt, J.-E. Tremblay, P. Tortell, G.-P. Yang, G.-Y. Shi, H. Gao, D. Semaniuk, M. Robert, M. Arychuk, K. Johnson, N. Sutherland, M. Davelaar, N. Nemcek, A. Pena, & W. Richardson. 2016. Impact of ocean acidification on phytoplankton assemblage, growth, and DMS production following Fe-dust additions in the NE Pacific high-nutrient, low-chlorophyll waters. *Biogeosciences,* 13:1677–1692.

Meyer, J., & U. Riebesell. 2015. Responses of coccolithophores to ocean acidification: a meta-analysis. *Biogeosciences,* 12:1671–1682.

Murray, C.S., A. Malvezzi, C.J. Gobler, & H. Baumann. 2014. Offspring sensitivity to ocean acidification changes seasonally in a coastal marine fish. *Marine Ecology Progress Series,* 504:1–11.

Nagelkerken, I., B.D. Russell, B.M. Gillanders, & S.D. Connell. 2015. Ocean acidification alters fish populations indirectly through habitat modification. *Nature Climate Change,* 6:89–93, DOI:10.1038/NCLIMATE2757.

Pagani, M., D. Lemarchand, A. Spivack, & J. Gaillardet. 2005. A critical evaluation of the boron isotope-pH proxy: the accuracy of ancient ocean pH estimates. *Geochimica et Cosmochimica Acta,* 69:953–961.

Palmer, M.R., P.N. Pearson, S.J. Cobb. 1998. Reconstructing past ocean pH-depth profiles. *Science,* 282:1468–1471.

Pascal, P.-y., J.W. Fleeger, F. Galvez, K.R. Carman. 2010. The toxicological interaction between ocean acidity and metals in coastal meiobenthic copepods. *Marine Pollution Bulletin,* 60:2201–2208.

Pearson, P.N., & M.R. Palmer. 2000. Atmospheric carbon dioxide concentrations over the past 60 million years. *Nature,* 406:695–699.

Pelejero, C., E. Calvo, M.T. McCulloch, J.F. Marshall, M.K. Gagan, J.M. Lough, & B.N. Opdyke. 2005. Preindustrial to modern interdecadal variability in coral reef pH. *Science,* 309:2204–2207.

Petit, J.R., J. Jouzel, D. Raynaud, N.I. Barkov, J.-M. Barnola, I. Basile, M. Benders, J. Chappellaz, M. Davis, G. Delaygue, M. Delmotte, V.M. Kotlyakov, M. Legrand, V.Y. Lipenkov, C. Lorius, L. Pepin, C. Ritz, E. Saltzman, & M. Stievenard. 1999. Climate and atmospheric history of the past 420,000 years from the Vostok ice core, Antarctica. *Nature,* 399:429–436.

Pistevos, J.C.A., I. Nagelkerken, T. Rossi, M. Olmos, & S.D. Connell. 2015. Ocean acidification and global warming impair shark hunting behaviour and growth. *Nature-Scientific Reports,* 5:16293. DOI: 10.1038/srep16293.

Riebesell, U., L.T. Bach, R.G.J. Bellerby, J.R.B. Monsalve, T. Boxhammer, J. Czerny, A. Larsen, A. Ludwig, & K.G. Schulz. 2017. Competitive fitness of a predominant pelagic calcifier impaired by ocean acidification. *Nature Geoscience,* 10:19–24.

Risbey, J.S., S. Lewandowsky, C. Langlais, D.P. Monselesan, T.J. O'Kane, & N. Oreskes. 2014. Well-estimated global surface warming in climate projections selected for ENSO phase. *Nature Climate Change,* 4:835–840. DOI: 10.1038/NCLIMATE2310.

Roe, G.H., & M.B. Baker. 2007. Why is climate sensitivity so unpredictable? *Science,* 318:629–632. DOI: 10.1126/science.1144735.

Rost, B., U. Riebesell, S. Burkhardt, & D. Sultemeyer. 2003. Carbon acquisition of bloom-forming marine phytoplankton. *Limnology and Oceanography*, 48:55–67.

Sakaguchi, K., X. Zeng, & M.A. Brunke. 2012. The hindcast skill of the CMIP ensembles for the surface air temperature trend. *Journal of Geophysical Research*, 117:D16113. doi:10.1029/2012JD017765.

Salma, U., H. Md G. Uddowla Y. Yeo Lee, & H.-W. Kim. 2012. Effects of pH change by CO2 Induction and Salinity on the hatching rate of *Artemia franciscana*. 2012. *Fisheries and Aquatic Sciences*, 15:177–181.

Santo, V.D. 2015. Ocean acidification exacerbates the impacts of global warming on embryonic little skate, *Leucoraja erinacea* (Mitchill). *Journal of Experimental Marine Biology and Ecology*, 463:72–78.

Sanyal, A., N.G. Hemming, G.N. Hanson, and W.S. Broecker. 1995. Evidence for a higher pH in the glacial ocean from boron isotopes in foraminifera. *Nature*, 373:234–236.

Sanyal, A., N.G. Hemming, W.S. Broecker, & G.N. Hanson. 1997. Changes in pH in the eastern equatorial Pacific across stage 5–6 boundary based on boron isotopes in foraminifera. *Global Biogeochemical Cycles*, 11:125–133, DOI:10.1029/97GB00223.

Sanyal, A., M. Nugent, R.J. Reeder, & J. Bijma. 2000. Seawater pH control on the boron isotopic composition of calcite: Evidence from inorganic calcite precipitation experiments. *Geochimica et Cosmochimica Acta*, 64:1551–1555.

Shaw, E.C., P.L. Munday, & B.I. McNeil. 2013. The role of CO_2 variability and exposure time for biological impacts of ocean acidification. *Geophysical Research Letters*, 40:4685–4688.

Shi, D., Y. Xu, B.M. Hopkinson, & F.M.M. Morel. 2010. Effect of ocean acidification on iron availability to marine phytoplankton. *Science*, 327:676–679. DOI:10.1126/science.1183517.

Simmonds, M.P., & S.J. Isaac. 2007. The impacts of climate change on marine mammals: early signs of significant problems. *Oryx*, 40:19–26. doi:10.1017/S0030605307001524.

Simmonds, M.P., & W.J. Eliott. 2009. Climate change and cetaceans: concerns and recent developments. *Journal of the Marine Biological Association of the United Kingdom*, 89:203–210.

Speers, A.E., E.Y. Besedin, J.E. Palardy, & C. Moore. 2016. Impacts of climate change and ocean acidification on coral reef fisheries: An integrated ecological-economic model. *Ecological Economics*, 128:33–43.

Stapp, L.S., C.M. Kreiss, H.O. Portner, & G. Lennig. 2015. Differential impacts of elevated CO2 and acidosis on the energy budget of gill and liver cells from Atlantic cod, *Gadus morhua*. *Comparative Biochemistry and Physiology, Part A*, 187:160–167.

Talmage, S.C., & C.J. Gobler. 2009. The effects of elevated carbon dioxide concentrations on the metamorphosis, size, and survival of larval hard clams (*Mercenaria*

mercenaria), bay scallops (*Argopecten irradians*), and Eastern oysters (*Crassostrea virginica*). *Limnology and Oceanography*, 54:2072–2080.

Tardif, R., G.J. Hakim, & C. Snyder. 2014. Coupled atmosphere-ocean data assimilation experiments with a low-order climate model. *Climate Dynamics*, 43:1631–1643. DOI 10.1007/s00382-013-1989-0.

Thingstad, T.F., R.G.J. Bellerby, G. Bratbak, K.Y. Borsheim, J.K. Egge, M. Heldal, A. Larsen, C. Neill, J. Nejstgaard, S. Norland, R.-A. Sandaa, E.F. Skjoldal, T. Tanaka, R. Thyrhaug, & B. Töpper. 2008. Counterintuitive carbon-to-nutrient coupling in an Arctic pelagic ecosystem, *Nature Letters*, 455:387–391. doi:10.1038/nature07235.

Trimborn, S., N. Lundholm, S. Thoms, K.U. Richter, B. Krock, P.J. Hansen, & B. Rost. 2008. Inorganic carbon acquisition in potentially toxic and non-toxic diatoms; the effect of pH-induced changes in seawater carbonate chemistry. *Physiologia Plantarum*, 133:92–105.

Trimborn, S., S. Thoms, T. Brenneis, J.P. Heiden, S. Beszteri, & K. Bischof. 2017. Two Southern Ocean diatoms are more sensitive to ocean acidification and changes in irradiance than the prymnesiophyte *Phaeocystis antarctica*. *Physiologia Plantarum*, 160:155–170.

Tyrrell, T. 2008. Impacts of changing ocean chemistry in a high-CO_2 world. *Mineralogical Magazine*, 72:329–362.

Waldbusser, G.G., E.P. Voigt, H. Bergschneider, M.A. Green, & R.I.E. Newell. 2011. Biocalcification in the Eastern oyster (*Crassostrea virginica*) in relation to long-term trends in Chesapeake Bay pH. *Estuaries and Coasts*, 34:221–231.

Watanabe, Y., A. Yamaguchi, H. Ishidai, T. Harimoto, S. Suzuki, Y. Sekido, T. Ikeda, Y. Shirayama, M.M. Takahashi, T. Ohsumi, & J. Ishizaka. 2006. Lethality of increasing CO_2 levels on deep-sea copepods in the western North Pacific. *Journal of Oceanography*, 62:185–196.

Weydmann, A., J.E. Soreide, S. Kwasniewski, & S. Widdicombe. 2012. Influence of CO_2-induced acidification on the reproduction of a key Arctic copepod *Calanus glacialis*. *Journal of Experimental Marine Biology and Ecology*, 428:39–42.

Wilson, C.A., & J.F.B. Mitchell. 1987. A doubled CO_2 climate sensitivity experiment with a global climate model including a simple ocean. *Journal of Geophysical Research*, 92:13,315–313,343.

Zhan, Y., W. Hu, W. Zhang, M. Liu, L. Duan, X. Huang, Y. Chang, & C. Li. 2016. The impact of CO_2-driven ocean acidification on early development and calcification in the sea urchin *Strongylocentrotus intermedius*. *Marine Pollution Bulletin*, 112:291–302.

Zheng, C.-q., J. Jeswin, K.-l. Shen, M. Lablche, K.-j. Wang, & H.-p. Liu. 2015. Detrimental effect of CO_2-driven seawater acidification on a crustacean brine shrimp, *Artemia sinica*. *Fish & Shellfish Immunology*, 43:181–190.

Index